Risk and Energy Infrastructure

Cross-Border Dimensions

Consulting Editor **Thomas J Dimitroff**

Consulting editor
Thomas J Dimitroff

Publisher
Sian O'Neill

Editors
Carolyn Boyle
Veronique Musson

Marketing manager
Alan Mowat

Production
John Meikle, Russell Anderson

Publishing directors
Guy Davis, Tony Harriss, Mark Lamb

Risk and Energy Infrastructure: Cross-Border Dimensions
is published by
Globe Law and Business
Globe Business Publishing Ltd
New Hibernia House
Winchester Walk
London SE1 9AG
United Kingdom
Tel +44 20 7234 0606
Fax +44 20 7234 0808
Web www.globelawandbusiness.com

Printed by CPI Antony Rowe

ISBN 978-1-905783-48-9

Risk and Energy Infrastructure: Cross-Border Dimensions
© 2011 Globe Business Publishing Ltd

All rights reserved. No part of this publication may be reproduced in any material form (including photocopying, storing in any medium by electronic means or transmitting) without the written permission of the copyright owner, except in accordance with the provisions of the Copyright, Designs and Patents Act 1988 or under terms of a licence issued by the Copyright Licensing Agency Ltd, 6-10 Kirby Street, London EC1N 8TS, United Kingdom (www.cla.co.uk, email: licence@cla.co.uk). Applications for the copyright owner's written permission to reproduce any part of this publication should be addressed to the publisher.

DISCLAIMER
This publication is intended as a general guide only. The information and opinions which it contains are not intended to be a comprehensive study, nor to provide legal advice, and should not be treated as a substitute for legal advice concerning particular situations. Legal advice should always be sought before taking any action based on the information provided. The publishers bear no responsibility for any errors or omissions contained herein.

Table of contents

Executive summary ———— 5

Introduction ———————— 13
   Thomas J Dimitroff
   Infrastructure Development
   Partnership LLP

Part 1: Macro-considerations

Macro-risks after the panic ——— 17
of 2008
   Kevin Gardiner
   Barclays Wealth

Political risk in large projects —— 33
   Arve Thorvik
   Thorvik International Consulting

Part 2: Cross-border dimensions

Multilateral and bilateral ———— 51
investment agreements
   Tom Cummins
   Ronnie King
   Ashurst LLP

Intergovernmental agreements — 73
   Katie Baehl
   R Coleson Bruce
   George F Goolsby
   Baker Botts LLP

Trans-boundary energy ———— 91
projects and maritime
transport risk
   Glen Plant
   Legal consultant

Cross-border regulatory ———— 129
risk: the EU exemption regime
   Leigh Hancher
   Allen & Overy LLP

Part 3: Host state perspectives

Host government agreements: — 145
the investor perspective
   Charles Lindsay
   Allen & Overy LLP

Tax risks ——————————— 159
   Stuart F Schaffer
   Baker Botts LLP

A host state perspective ———— 173
on risk
   Judith H Kim
   Geoffrey Picton-Turbervill
   Ashurst LLP

Part 4: The investor and other stakeholder perspectives

Joint venture risks and ———— 183
responsibilities
   Thomas J Dimitroff
   Infrastructure Development
   Partnership LLP

Project finance and _____ 201
risk mitigation
    William E Browning
    Infrastructure Development
    Partnership LLP
    Alexandre Chavarot
    William J Clinton Foundation

Technical and operational _____ 221
risk
    Deborah L Grubbe
    Operations and Safety Solutions, LLC

Holistic project management _____ 229
and risk
    William E Browning
    Yashar Latifov
    Infrastructure Development
    Partnership LLP

A strategic approach to _____ 239
environmental and social
impact assessments
    David Blatchford
    Martin Lednor
    Infrastructure Development
    Partnership LLP

Security risk management _____ 257
    Antony FS Ling
    LPD Strategic Risk Ltd

Risk, project-affected _____ 273
communities and their land
    Henry Thompson
    Oxania Ltd

About the authors _____ 295

Executive summary

This book is divided into four key sections. Section 1 addresses two macro-risks that face project planners implementing large-scale energy infrastructure projects: macroeconomic and political risks. The next section examines a variety of horizontal (state-to-state) dimensions of risk encountered in the development of cross-border energy infrastructure projects. Section 3 proceeds to address the vertical (state-to-investor) dimensions of risk from the perspectives of investors and host states. The last section looks at risks posed to, and by, energy infrastructure from the perspectives of the investors and other stakeholders. The following is a chapter-by-chapter summary highlighting key points made by this book's authors.

Section 1: Macrorisks
The first chapter, by Kevin Gardiner, examines the range of macroeconomic and financial risks that planners of energy infrastructure projects should consider. Gardiner provides a 50-year summary of economic cycles that identify a variety of risk factors. These include:
- a sudden setback in aggregate demand and output;
- fluctuations in economic activity and prices;
- volatility; and
- other important macro-variables such as interest rates and exchange rates, and the prices of major factors of production such as the price of steel.

Gardiner suggests that other macroeconomic risks posed to planners are contextual in nature – for instance, domestic and cross-border legal and regulatory risk, monetary and fiscal policy fluctuations, and changes in results by amending the calculus for determining the internal rate of return or net present value. Other types of contextual risk can include a change in demand for a commodity due to a shift in consumer preference – for example, the viability of nuclear energy projects post-Fukushima has been undercut by safety concerns. The range and scale of macro-shocks surveyed by Gardiner demonstrate that there are always new 'unknown unknowns' and the best advice to planners is to stress-test their calculations extensively.

Arve Thorvik examines the topic of political risk within the context of large-scale energy infrastructure projects in the second chapter. Analytical tools are offered to help developers to understand and mitigate types of political risk presented in the development and implementation of projects. After mapping out a typology describing different political systems, ranging from those that are dysfunctional to transitional and Western democracies, Thorvik identifies a catalogue of root causes of political risk. Thorvik provides a variety of recommendations for preventing and reducing political risks before reaching two counter-intuitive conclusions: that political risks do not necessarily originate with politicians, but rather with the project itself, and also that a Western democracy may pose political risks that are at least as challenging as those posed by so-called 'less developed' political systems.

Section 2: Cross-border dimensions
The third chapter, by Tom Cummins and Ronnie King, examines risks that arise when infrastructure provides routes for energy transit between states. Unlike energy infrastructure developed within the context of a single state, transboundary infrastructure may involve states that have no direct interest in the function of the infrastructure and really have only a transit fee to gain as a benefit. Cummins and King observe that there are now over 2,600 bilateral investment treaties and several key multilateral treaties in force which provide protections against governmental interference with investments. Through an analytical survey of numerous arbitral awards, the co-authors review key provisions contained in most bilateral investment treaties, ranging from, among other things, the definitions of 'investor', 'investment' and 'expropriation' to the critical dispute resolution provisions. Three important multilateral investment treaties are referenced (the North American Free Trade Agreement, the Association of Southeast Asian Nations and the Energy Charter Treaty), before the authors proceed to detail the unique provisions of the Energy Charter Treaty in respect of investment in energy. The Energy Charter Treaty Model Intergovernmental and Host Government Agreements are referenced. The authors conclude by observing that jurisprudence associated with bilateral and multilateral investment treaties is evolving, as is the confidence of many emerging economies that may now be less willing to respect the protections offered.

The fourth chapter, by Katie Baehl, R Coleson Bruce and George F Goolsby, examines the use of tailored project-specific intergovernmental agreements to support large-scale strategic infrastructure. The chapter examines intergovernmental agreements (Vienna Convention treaties) as tools to mitigate risks posed to projects by allocating state-to-state risks to the party best placed to manage it – that is, the state. The authors stress that such agreements are not limited to energy infrastructure and may be used effectively to abate risks to all types of cross-border infrastructure project. Project risks are viewed from the perspective of the host states, including producing states, transit states and consuming states. Private sector stakeholders are fundamentally concerned to secure project rights that are certain and predictable over time. Intergovernmental agreements differ from bilateral investment treaties by providing efficiencies in respect of applicable home country taxes and uniformity in the application of fiscal, legal, technical and regulatory standards, and by supplanting domestic laws unsuitable for cross-border projects. The chapter concludes by examining ways in which intergovernmental agreements may be deployed alongside of host government agreements. Two case studies are presented:
- the Baku-Tbilisi-Ceyhan pipeline, where the intergovernmental agreement

provided for compatible host government agreements across Azerbaijan, Georgia and Turkey; and
- the West African Gas Pipeline, in which a single omnibus host government agreement was entered into between the states of Benin, Ghana, Nigeria and Togo and the project sponsors.

Glen Plant explores, in the fifth chapter, the risks posed to large-scale energy infrastructure projects that have maritime segments involving the carriage of crude oil, liquefied natural gas or refined products by tankers. Plant comprehensively examines the key risks posed to maritime energy cargoes, including:
- war risk, terrorism, piracy, armed robbery and hijacking;
- safety, environmental protection law and blockage risk; and
- direct action protest risk.

While many of the risks appear to be of high impact but low probability, Plant demonstrates through numerous well-researched and well-documented examples that these risks need to be factored into the assessments of project planners. Plant cites as examples two international chokepoints (the Straits of Hormuz and the Straits of Malacca) through which nearly 80% of all seaborne traded oil transits, leaving oil tankers vulnerable to many of the aforementioned risks.

Leigh Hancher analyses regulatory risks to terrestrial cross-border energy infrastructure in developed markets that have comprehensive competition regulation. This chapter therefore stands in marked contrast to other chapters that examine regions where there may be little more in place than applicable bilateral investment treaties. In developed markets, such as the European Union, competition policies are given a high priority and counterbalance the policy objectives associated with attracting investment. The EU electricity and gas exemption regime is designed to discourage bundled ownership between transmission owners, operators and shippers, and their long-term capacity reservations. Accordingly, developers of new energy infrastructure need to demonstrate why failing to secure an exemption to the foregoing principles would pose a risk that the investment would not be viable. Hancher expertly surveys current EU jurisprudence governing the electricity and gas exemption regime, and concludes that the new regime remains partial and unpredictable.

Section 3: Host states' perspectives

Charles Lindsay examines the range of interrelated investor risks from the investment perspective within the context of particular host government jurisdictions in which large-scale transboundary projects are developed. For Lindsay, the investment perspective includes the position of equity investors, their project financiers and shippers, or customers, utilising the infrastructure and thus underpinning the economics of the project. In particular, Lindsay focuses on the use of the host government agreement as a tool for reducing investment-related risks. In addition to discussing the ultimate practical value of host government agreements, this seventh chapter surveys various techniques that may be deployed to enhance the likelihood that such agreements' rights and obligations will ultimately be enforceable. Lindsay goes on to examine the utilisation of these agreements' provisions to mitigate a full range of specific host government risks, including political risk, government permitting risks, land rights, transit rights, foreign labour and materials, technical standards, changes in law taxation and expropriation.

The economics of any energy infrastructure investment are critically dependent upon the tax treatment afforded by the host government in which it is located and the home government where the investor is based and/or incorporated. The viability of cross-border infrastructure projects hinges upon tax stability and, frequently, tax concessions from host states. The host state's willingness to provide these concessions may depend upon its appetite for the venture. In the eighth chapter, Stuart Schaffer shares his extensive experience of, and practical insights into, tax issues and considerations that investors are likely to confront when seeking to minimise the tax treatment of energy infrastructure projects. For Schaffer, the pre-development stage is critical and a careful survey of all potential fiscal impacts upon a project should be undertaken, with all effects thoroughly identified and understood upfront. He goes on to examine the variety of 'host government take' and offers insights into likely negotiating positions, strategies and tactics that tax negotiators may wish to consider to minimise the risk of fiscal leakage. He offers approaches to the preparation of tax provisions in host government agreements, including ways to minimise the potential tax implications of non-tax provisions, stabilising the project's tax treatment over time and harmonising tax treatment in cross-border projects.

Energy infrastructure investments are often large scale and always long term, and therefore require a fair balance of interests in order to be sustainable. However clever the negotiators may be when the initial deal is struck, unfair advantages invariably become evident over time and will lead to the unravelling of the deal. In the ninth chapter, Judith Kim and Geoffrey Picton-Turbervill view the range of risks confronted by the host state offtaker in evaluating the benefits of a potential energy infrastructure investment. Kim and Picton-Turbervill suggest that the reader consider a hypothetical country and potential gas transit project, and examine the range of objectives and core conditions that the host state would likely want fulfilled. These include many of the same commercial concerns that a downstream offtaker would consider, including delivery schedule, gas quality and pricing. Whether the host state is an offtaker or merely a transit country, all states have additional concerns, including protecting the environment, health, safety and physical security. Kim and Picton-Turbervill neatly illustrate the policy tensions that host states face in balancing the need for investment and the interest in securing supplies against the sometimes countervailing interests of the investor. As host states in emerging markets become more experienced, they may become less willing to accept residual risks – that is, risks not fully within their control. The chapter concludes with the observation that where host states seek a more commercially balanced approach to investment, the result is more likely to prove sustainable over time.

Section 4: Investors' and other stakeholders' perspectives
In the 10th chapter, this book's consulting editor examines the internal risks that

joint venture participants pose to themselves when they join forces. While joint ventures offer the benefits of minimising cost, spreading risk and pooling expertise, they can also create obstacles that impede delivery on schedule, on standard and on budget. Proper management of internal risks is a core issue of corporate governance that should lie entirely within the control of the joint venture. Unfortunately, many joint ventures fail to achieve their business objective and many more remain dysfunctional. The chapter surveys a wide variety of factors that present internal risks to joint ventures, including corporate identity, ownership (western independent, listed and state-owned), business culture and corporate values. The chapter goes on to argue that a failure to align behaviour, standards, values and management systems at the formation stage of a joint venture may sow the seeds for its failure at later stages. The chapter closes by examining current trends and observes that requirements for higher standards in corporate behaviour are often more rigorous in US, UK and Western European contexts. This stands in marked contrast to applicable requirements in markets with the fastest growth in hydrocarbon production and largest reserves. Lastly, the ability of Western privately held and publicly traded oil companies to compete effectively and/or join forces with state-owned companies will be shown to depend upon policy and enforcement choices made by governments going forward.

The lack of adequate finance constitutes a key risk to the implementation of energy infrastructure. Financing can also be used as an effective risk mitigation tool. In the 11th chapter, William E Browning and Alexandre Chavarot discuss the use of project financing in energy infrastructure developments. After defining the core attributes of project finance, Browning and Chavarot examine two main reasons that project sponsors choose to project finance: to access capital (especially useful for weaker credit sponsors that participate with creditworthy sponsors) and to ensure proper mitigation through allocation of key project risks to the party most capable to manage the risks. The chapter expertly reviews the key pre and post-completion risks that financing institutions focus upon. It also explores the discrete roles that various participants in project financing play and the sources, purpose, products and constraints that sponsors should consider in sourcing debt in project financing. The chapter concludes by examining key criteria and processes involved in successful project financing as evidenced in three large-scale recent projects: Baku-Tbilisi-Ceyhan, Nord Stream (Phase 1) and Blue Stream. By posing a series of questions to the finance manager or adviser to each of the three projects under discussion, a series of truly fascinating case studies emerges from an already compelling chapter.

The 12th chapter, by Deborah L Grubbe, examines the nature of technical and operational risk by exploring the nature of economic value ascribed to these risks in the context of an M&A due diligence process. Ironically, Grubbe observes that these risks are referred to as 'non-financial risks' and she discusses them in the context of three broad categories. The first involves valuing people and includes an examination of the workforce, its leadership and culture. The second involves plant (ie, the facility and related hard assets), with subtopics including operations, supply chain and procurement processes. The third involves processes or the soft asset base. Grubbe offers useful tables with practical sample questions that will provide a highly effective generic aide to participants on M&A teams. This will enable a pragmatic assessment of this all-important dimension of risk in the context of any energy infrastructure project. Grubbe's tables will also be useful as reference points for mitigating technical and operational risk in the implementation of new energy infrastructure developments.

William E Browning and Yashar Latifov explore in the 13th chapter the core components of successful project management. The authors counsel that it is only through a deeply embedded project management system, strong leadership and a strong governance system that the successful technical delivery of a project is assured. The chapter goes on to survey the organisational management of risk, where all risks are comprehensively identified and coordinated activities required to manage those risks are accepted and agreed by the management team. The structure and goals of a successful project management system are also characterised by the authors as holistic planning. As major energy infrastructure projects evolve, the authors argue, the specific management system deployed must be sufficiently robust to evolve throughout the project's various stages. The authors further observe that the maximum opportunity to influence a project is at the project design phase. As expenditures increase, the ability to retrofit design becomes more difficult and costly. It is therefore imperative that all project assumptions be rigorously understood and challenged prior to the final investment decision. Browning and Latifov conclude by underscoring the value of iterative challenges to the project management process across all stages of a project's timeline.

Traditional approaches to environmental and social impact assessments are often a one-step process geared primarily towards fulfilling financing or regulatory requirements. This is in marked contrast to the more fundamental need to manage environmental and social performance throughout the lifecycle of a project. Accordingly, traditional environmental and social impact assessments are of questionable value. In the 14th chapter, David Blatchford and Martin Lednor propose a strategic approach to environmental and social impact assessments that cater all-inclusively to stakeholder interests and constitute a core component of a project's success. The authors argue that this strategic approach requires sponsors to view the assessment as an ongoing iterative risk management tool evolving alongside the project. From this perspective, risk factors gain or lose importance according to whether the project is in the pre-development, construction or decommissioning phase. Blatchford and Lednor propose that strategic assessments offer enhanced environmental and social performance, greater stakeholder acceptance and reduced project risk at a reduced cost to the project sponsors.

In Chapter 15, Anthony FS Ling counsels business to give priority to the causes of security failure and observes that all risks discussed in this book can lead to security problems if left unmitigated. In order to grasp the range of potential causes, Ling urges businesses to adopt a method that alleviates the risk of security failure by prioritising the recognition of potential causes as early as possible – preferably pre-entry; these causes are then logged onto a risk matrix. Categories within the prescribed risk matrix include regional assessments, country security risk assessment scenarios, reputational assessments, environmental, health and social impact assessments, and project security risk assessments. Dimensionality is enhanced in his discussion of each category by drawing upon a wealth of professional experience in Asia, South America, Africa, the former Soviet Union and the Middle East. Ling's chapter further offers a methodology to source and organise the data points into low, medium and high risks for each assessment category before turning to a discussion of risk mitigation. In particular, Ling suggests that for any risk graded medium or above, a risk assessment plan be recommended. With respect to transboundary projects, Ling observes that the causes of risk are likely to belong to any of three overlapping areas (social, political and project risks), and concludes by providing specific pragmatic recommendations on risk mitigation within each area.

Henry Thompson examines the interrelated dynamics between energy infrastructure projects and communities where he places a specific emphasis on land utilised for projects in the book's closing chapter. This 16th chapter is divided into two parts, with the first examining the risks that arise when there is a breakdown in communication between projects and communities, and the second examining the risks and issues that arise in connection with land tenure and land rights. Thompson elegantly enables the reader to share the differing viewpoints of business and communities by offering a case study as focal point for his analysis. His chapter's many insights are penetrating and explain with precision sources of misunderstanding, such as: "local communities are primarily agrarian, stable, traditional in outlook, and ostensibly conservative and …the land is quite literally the foundation of their life …[while] extractive companies are not so much connected to the land as to the resources that lie beneath it." In the second part of the chapter, Thompson skilfully analyses the types of ownership and usage of land – especially rural land – and observes that, in general, indigenous populations incur significant losses to livelihoods, status and stability as a direct result of major investments in oil and gas. He goes on to observe that indigenous people also constitute a significant risk. "The poorer, less well educated, less well resourced and legally empowered people are, the more likely they are to incur loss and create conflict." Thompson concludes by pointing out that the fundamental onus lies squarely on the company to address the multiplicity of issues and risks in an inclusive and appropriate manner. "Free prior informed consent is rooted in international human rights law and constitutes a core principle of corporate responsibility."

Introduction

Thomas J Dimitroff
Infrastructure Development Partnership LLP

Projecting global requirements for future primary energy is notoriously difficult and must ultimately rely upon some form of scenario analysis. The International Energy Agency has estimated that global requirements will increase by 1.6% per annum or 45% in total between 2007 and 2030. In addition to the replacement of ageing infrastructure, the development, financing, construction and operation of the future energy infrastructure required to deliver this energy to end users will place acute emphasis upon mobilising large-scale public and private sector investment. The agency further projects that total investment in energy supply infrastructure between 2007 and 2030 will require approximately $26.3 trillion.[1] Whatever the actual scale of demand growth in energy, the investment required to satisfy it will rest upon a staggering array of complex international trade, macroeconomic, political, technical, environmental and social challenges, presenting areas of risk that warrant careful analysis by policy makers and private sector investors.

The dislocations in global supply and demand for oil and gas are becoming increasingly marked. Remaining lower-cost hydrocarbon reserves are now concentrated in the Middle East, North Africa and the former Soviet Union, while the main consuming markets are located in North America, Europe and – increasingly – Asia-Pacific (China and India). In the aftermath of the 2008 financial crisis, economic growth in emerging markets has driven demand for oil higher and thus has created higher commodity prices. The upward pressure on oil prices has also been pushed by unprecedented instability across the Middle East and North Africa. In addition, that pressure – exacerbated by the Fukushima tragedy in Japan – has renewed the cloud over the utilisation of nuclear energy as a potential substitution for hydrocarbons.

The supply/demand dislocation has driven disparate and competing policy responses. On the one hand, the distance from conventional producing fields to markets is increasing as governments and international oil and gas companies are forced to seek out new reserves further afield. This requires the construction and operation of additional infrastructure such as oil and gas pipelines and related facilities, including liquefied natural gas facilities; this, in turn, raises further concerns, prompting additional policy considerations.

Diversification away from cheaper oil in the Middle East, North Africa and the former Soviet Union also forces governments and international oil companies to

---

1   www.iea.org/speech/2009/Tanaka/4th_OPEC_Seminar_speech.pdf.

develop and produce reserves in new, more politically fluid and hostile environments such as Iraq or in environmentally sensitive areas such as the Arctic, the ultra-deep water Gulf of Mexico or Brazil. While reserves diversification may in one sense further secure energy supplies and aid international oil companies to maintain their reserves' replacement ratios, the technological complexity and longer lead times also escalate costs.

By contrast, the exploitation of reserves closer to home reduces the distance from production source to market, ensures security of supply and alleviates strains on balance of payments associated in part with financing high-cost imported oil. On the other hand, the development and production of heavy oil sands in Canada coupled with shale gas in the United States and the European Union raise concerns relating to the environment, potential social impacts and the risk of climate change; each case requires, as a consequence, the cost-intensive deployment of technology.

The pragmatic reality is that the projected rise in primary energy demand appears to be a juggernaut that will require vast amounts of global investment to maintain pace. Governments will need to secure energy supplies for their populations and to do so they will need to work with investors. Decisions will need to be taken to minimise risks and lead times posed by legal, regulatory and fiscal requirements. Governments will also have equally compelling obligations to ensure that the interests of their populations are protected by ensuring fair terms for enabling investment while minimising potential detrimental impacts on health, safety, environmental, security and human rights of their population.

Policy makers, industrial players and financiers will need a clear understanding of the nature and range of potential risks involved. In particular, investors will need to quantify the cost associated with assuming risks that cannot be controlled (eg, political risk) and the cost associated with managing risks that can be controlled (eg, health, safety and environmental risks). The costs and lead times associated with mitigating and managing these risks will in part determine the ultimate return on investment. Where the risks (and associated costs) are so high, or the lead times so long, that acceptable returns are not possible, investors will simply not commit the required investment.

In reconciling these competing policy considerations, additional questions arise in connection with whether these risks are best managed bilaterally between states and investors. A common assumption prevails: as investors move from more mature members of the Organisation for Economic Cooperation and Development (OECD) to emerging markets, risks to investment increase. This enhanced risk is largely associated with perceptions about relative deficiencies in the application of the rule of law and the overall strength of the underlying economy.

The recent financial crisis has challenged this assumption as investors look to emerging markets for growth and greater returns on capital as more mature OECD countries appear to stagnate. The root cause of the financial crisis is the direct result of unregulated risk taking on Wall Street and other financial centres among OECD countries. Coincidental with the financial crisis, two of the most spectacular and tragic industrial accidents in the last decade (Texas City and Macando) occurred in the United States. Both were in part blamed on inadequate regulatory oversight by agencies otherwise charged with protecting health, safety and environment. Together these events have wreaked havoc on regional macroeconomic stability, and have also resulted in loss of life, injury and significant environmental effects upon the ecosystem of the Gulf of Mexico and the many people who depend upon it for their livelihood. With regulators either sleeping at the wheel or having inadequate regulatory oversight at their disposal, the idea that respect for the rule of law is greater among OECD members than in emerging markets may now have less of a deterrent effect on investment in developing economies.

Civil society is increasingly demanding that the interests of people, communities and the environment be taken more directly into account in establishing the legal architecture that defines how energy infrastructure projects are developed, financed, constructed and operated. The traditional bilateral arrangement – host state and investor, where rights to develop projects are granted to investors and the forward implementation of the project is defined entirely within the framework of that bilateral relationship – appears increasingly inadequate to protect these broader interests. The role of government as the exclusive protector of the interests of people, communities and the environment is prompting unprecedented scrutiny by all stakeholders. Direct consultations between affected communities and investors are being demanded. Norms are being put into place by governments, international organisations and, internally, companies to identify and manage the risks associated with failing to take into account the wider impacts of these vitally important projects.

This book brings together experts in the field of identifying various risks posed to, and by, complex energy infrastructure projects, and explores the mitigation of those risks. It is hoped that several objectives are thereby fulfilled. By asking each expert to offer a perspective on the risks lying within their particular area of expertise, and collecting these perspectives together into a single volume, a broad-spectrum analysis of the variety of risks facing the development, financing, construction and operation of large-scale energy infrastructure projects is presented. It is also hoped that this book may form a reference point for professionals in one discipline who wish to gain an understanding of the risks viewed through the lens of another professional.

It is surprising to note the absence of a holistic view of the risks posed to, and by, these large-scale projects among core players involved in their implementation: international oil and gas companies, governments and civil society. This book is intended to provide a comprehensive viewpoint on associated risks. It may also contribute to an informed dialogue among industry, policy makers and civil society to develop an enhanced understanding of the risks and pragmatic challenges associated with delivering key energy infrastructure projects.

Within international oil companies, the absence of a 360° vision may be due to the way that many such companies are organised and mobilised to deliver projects. Typically, the personnel deployed at the front end of a project (ie, in the feasibility and business development phase) are different from the personnel deployed to execute the project (ie, the construction phase) and the personnel brought in to operate the infrastructure. Moreover, the implementing company (the operator) will often procure vast numbers of third-party contractors to undertake the front-end engineering, design and construction of the project. Operator assurance over the execution of the contracted works and services may often fall between the cracks separating the operator's procurement department, its on-ground construction managers and assurance personnel. Lastly, given the 20 to 30-year economic lifespan of these projects, the operating personnel may change many times over the lifespan of the infrastructure developed. The utility of a holistic view of each phase of such a project will, it is hoped, be of undisputed value to the management at any given stage of development or operation.

Within governments, there is rarely an understanding of the core commercial drivers that require alignment in the correct sequence before a project may be realised. Time and again, governments fail to understand that when private sector finance is involved, politics and commerciality are both necessary conditions, but rarely is either sufficient on its own to deliver a large-scale energy infrastructure project. Without a clear line of sight to a commercial return on investment, private sector money will not be invested.

Within international civil society organisations, there is often an abundance of goodwill that is frequently undermined by a failure to understand the pragmatic dimensions of risk involved in the implementation of energy infrastructure projects. In those instances where the expertise is present, civil society organisations are often tempted to campaign against high-profile projects rather than focusing on the more mundane task of policing the day-to-day implementation of more conventional projects or commitments made by projects after they have been constructed and long after they disappear from the headlines.

Acknowledgements

I would like to thank each of the chapter authors who contributed their time and expertise to deliver this publication. I would also like to thank the following people for their reviews, edits, suggestions and support: Frederick Farryl Goodwin, Dr Helga Graham, Georgina Birkmyre Fraser, Prof Dr Ferid Muhic, Sir Mark Allen, Andrew Gowers, Bojan Shimbov, Martin Short, John Sherman, Charles Lindsay, William E Browning, Patricia Dimitroff, Susan Maples, Dr Glen Plant and Sian O'Neill. Finally, I would like to thank my wife Madeleine and our children John, Elizabeth and Catherine, as well as my parents, Lambro and Patricia, for their patience and support in completing this project.

Macro-risks after the panic of 2008

Kevin Gardiner
Barclays Wealth

The future is profoundly unknowable and risk analysis, no matter how diligent, cannot alter this inconvenient fact. From time to time, however, we persuade ourselves that perhaps it may not be – most recently, during the decade or so of remarkable economic stability that ended in 2007. But just as nemesis follows hubris, the 'Great Recession' duly followed hard on the heels of the 'Great Moderation': the crisis that broke out in 2008 was a brutal reminder both that economic and market volatility has not been tamed, and that the events triggering it cannot easily be identified in advance. 'Unknown unknowns' and 'black swans' (rare events with important repercussions) lie in wait for forecasters rash enough to reason by induction. Even if planners were lucky enough to live in a world in which all possible events had been identified in advance – a world in which the future were merely probabilistically risky, as opposed to genuinely uncertain – the difficulties of gauging the scale and likelihood of those events would still be daunting. (A fuller discussion of the distinction between statistical risk and uncertainty can be found in Peter L Bernstein (1996) and Nassim Nicholas Taleb (2007).)

The discussion in this chapter (of the range and extent of macroeconomic and financial risk facing the planners of large-scale transnational energy supply projects) was compiled in late 2010, with this re-learned lesson still fresh in analysts' minds. The checklist offered below is intended only to help organise the discussion and to foster a particular way of thinking about risk. While it carries no pretence to completeness or precision, even heuristically, it offers some practical advice nonetheless.

1. **Steering by the rear-view mirror**
In attempting to plan for the future, we start by looking at the past: we assume that history is known with some certainty and that it is relevant to what might happen in the future. In fact, outside the physical developments that can be traced with some confidence using the tools delivered by the natural sciences and the simple chronology of major events, our knowledge of the past is less soundly based than it often suits us to think. The identification of much causality in history is unavoidably subjective, for one thing. In the realm of economics and finance in particular, the further back we look, the less convincing are attempts at quantification. The older the data, the less credible it is – some heroic academic feats of estimation and collation notwithstanding (eg, Angus Maddison (1998), which offers quantitative analysis spanning a period of more than 1,000 years, or Sidney Homer (1963), whose

narrower focus allows the author to present data from the classical world and earlier). Ambiguity arises because much data is simply not available (the concept of 'gross domestic product' (GDP), for example, dates from the work of Simon Kuznets in the 1930s only and was calculated first for the United States in the early 1940s). When data is available, we cannot be sure that the things being measured are comparable. In what sense, for example, are the price indices used to calculate inflation rates in the medieval period comparable with today's consumer prices? More topically, what would an accountant today make of a set of corporate accounts that pre-dated the Great Depression and modern accounting standards? Natural scientists can often fill in gaps in their historical data series. For example, the temperature was not measured many thousands of years ago and, as with GDP, likely did not even exist as a concept (there were no humans around to experience it). Nonetheless, scientists are able to estimate its variation from today's level with some confidence using evidence from polar ice and fossils. Economic activities and rates of exchange, by contrast, leave few physical traces.

We use the past as a guide to gauge future risk because outside our imagination, it is all that we have. Thus, a brief review of some relatively recent history has to be the starting point for considering future risk. We know that neither the scale nor the nature of that future risk is necessarily bounded by this historical experience. Yet some approximate awareness of what has gone before might at least foster the open-mindedness and nimbleness needed to gauge and respond to the macro risks facing planned projects today.

2. **The road to 2010**

The scale of social, economic and financial change seen within a recent single lifetime has been almost unfathomable. It was largely unpredicted.

The cataclysm of World War II illustrated just how quickly the fabric of civilised society could unravel. The post-war decades contained their share of trauma too, starting even as Europe and Japan were rebuilding. The outbreak of a third world war was narrowly avoided in Korea, Cuba and, arguably, Vietnam; the industrial slaughter of World War I was revived on the Iran/Iraq border; there were examples of genocide in Bosnia, Cambodia and Rwanda; tens of millions died in the mass famine that accompanied China's Great Leap Forward; the Cold War and the doctrine of 'mutual assured destruction' cast a daunting shadow over the lives of many more. Localised conflicts and civil wars have been plentiful, and in some cases (as in the Arab/Israeli struggle) have been long-lasting and are ongoing. More recently, organised terrorism and failing nation-states, and the Western response to them, have fostered conflict in Afghanistan and Iraq. This list ignores altogether the numerous natural disasters that have punctuated the period.

Despite these many and varied crises, the world is less geopolitically volatile now than it was half a century ago. The tensions of the Cold War subsided after much high-profile negotiation (including a famous walk in the woods taken by a US and a Russian negotiator in Geneva in 1982) in the face of strong public hostility to nuclear arms and the eventual collapse, from within, of the Soviet bloc and the Berlin Wall. Widespread conflict between developed states has not erupted – partly, no doubt, because of the work of international bodies such as the United Nations, but partly also because the credibility and attractiveness of collectivism have been so visibly and widely undermined by its material failure. We may not have seen the end of history, as envisaged by Francis Fukuyama (1993), but we have seen the possible causes of conflict tempered at least. It may seem foolhardy or crass even to attempt a quantitative illustration of this trend, but agencies that monitor armed conflict have been able to trace a marked decline in battlefield deaths over the 1946-2007 period at least.

It is not necessary to subscribe fully to the thesis advanced by Fukuyama to accept the idea that governments are becoming less autocratic (though in itself this is no guarantee of stability). Fukuyama notes a longer-term tendency towards democratic forms of government: from 1775, when he suggests that there were no governments that would count as democracies by today's standards, the number of democracies reached 25 in 1919, fell to 13 in 1940, but subsequently rose to 61 as of 1990.

This does not mean, of course, that things will stay as they are. The period immediately ahead of World War I was marked by a widespread belief that conflict among advanced industrial nations was all but impossible (as Liaquat Ahamed recently reminded us (2009)). Nor does it mean that the destination arrived at was inevitable. It is easy to imagine how the outcome could have been very different had the protagonists at one of several global flashpoints acted differently. Sanity mostly prevailed, but it might not have.

The course of economic and financial developments in the past 60 years or so has also been punctuated by crises. As in the geopolitical arena, the primary trend has nonetheless been a favourable one. Here, too, it is clearly premature to conclude that things will necessarily stay as they are.

The immediate aftermath of World War II was characterised by the challenges of economic rebuilding and its financing, and those posed by mass demobilisation. On the latter front, it is largely forgotten that the first time that the new Keynesian doctrine of fiscal demand management was put into practice was not in an attempt to cut unemployment, but instead to try to suppress the potential inflation that might be unleashed as the suppressed consumer demand of the war years reverted to more normal levels of spending. The financial infrastructure designed at the 1944 negotiations in Bretton Woods fostered both reconstruction and financial stability, while avoiding the punitive reparations that had done so much to undermine the peace after 1918. Gradually, rationing ended (as relatively late as 1954 in the United Kingdom), international transactions revived, currency convertibility was resumed and extended, and tariffs and other trade barriers were slowly but steadily dismantled. In Europe, the long journey towards ever-closer union formally began with the Treaty of Rome in 1957, but the first moves in that direction had been made a decade earlier and the journey continues today. Fixed exchange rates – and links to gold – were eventually jettisoned in the 1970s. Capital markets became steadily more open, credit controls were phased out – altogether in some cases – and financial instruments became more complex and contingent, culminating in the widespread deregulation that characterised the millennium.

As with the geopolitical climate, the improved economic and financial climate seen in the past half century or so has been interrupted by numerous setbacks and crises. The general increase in living standards has been punctuated by business cycles and secular trends. Individual countries and regions have grown and inflated at markedly different rates; there have been many instances of industrial unrest and financial emergency – with the latest episode of the latter counting arguably as the most pressing since the bank failures of the 1930s.

Starting with the most obvious risk to project planners – namely, the possibility of a sudden setback in aggregate demand and output – there have been many individual instances of recession across the global economy since 1945. For the larger economies, these local business cycles can be tracked with some confidence. In the case of the United States, for example, the National Bureau of Economic Research identifies 11 distinct cycles since 1945. (Coincidentally, this matches the number of years when the level of real GDP registered an annual decline.) The bureau actually use a more careful definition of 'recession' ("significant declines in economic activity") that allows it to date the start and finish of recessions to a particular month, using a wider range of activity variables (including unemployment and personal incomes): the average length of the cycles that the bureau identifies is between 66 and 73 months, according to whether the dating is on a peak-to-peak or trough-to-trough basis. The declines in real annual GDP look modest – 2% is a large figure in this context. However, there is a distinction to be drawn between aggregate output (what national accounts statisticians term a 'net' or 'value-added' concept) and the number of gross transactions taking place: the latter tends to be much more volatile than the former, and it is the latter that can matter most for individual project revenues. The United States, of course, is also a large, diverse economy, and the movements even in GDP observed in smaller countries can be much more marked.

Generally, the business cycles observed in still more aggregated measures of economic activity, such as regional or global GDP indices, are more muted, reflecting the natural diversification imparted by a larger sample. An important qualification of particular relevance to cross-border projects relates to the much more synchronised nature of business cycles in neighbouring countries: geographical proximity is still a key determinant of trade and a country experiencing a severe downturn will usually cast an economic shadow across its borders. (There can be exceptions – for example, North Korea – but these are often non-market-based economies that would likely require a unique approach to project appraisal even in the absence of economic weakness.)

Inflation and interest rates also exhibited cyclical fluctuations alongside those in economic activity in the United States and elsewhere. The past is not an infallible guide to the future, but it would be reckless to assume that the business cycle has been tamed, particularly after 2008-2009: project planners thus need to take into account the fact that the level of demand, input costs, pricing power and capital charges are all likely to be subject to considerable cyclical variation.

Moreover, the relatively short-term gyrations of the business cycle can coincide with evolving longer-term or secular trends. The net result is that projections of the likely level of even economy-wide cash flow (let alone specific project revenues) several years or more into the future carry considerable margins of error. In the case of Japan, the level of nominal GDP in late 2010 was 7% lower than that in mid-1997, after a steady decline in the general price level had more than offset a very low rate of real economic growth. In the United States, the 10-year moving average of real GDP growth in late 2010 was at its lowest level in half a century: had it stayed at the level recorded over the 10 years to 2000, GDP in 2010 would have been almost one-fifth higher. (The recent low trend rate of US growth may itself be leading to further forecast risk: since 2008, many commentators have been talking of a 'new normal' in which economic growth will be low for some time to come, but it is a moot point as to whether they are doing anything other than extrapolating forward the trend rate of economic growth seen in the previous decade. Forecasts are often driven not by a fresh appraisal of the future, but by the experience of the recent past.)

Viewed in an even longer-term context, the cyclical and secular trends of the past half century seem, if anything, to be relatively modest in scale – assuming, of course, that the older data can be trusted. For the United States, estimates of GDP and consumer prices extending back to before World War II and into the 19th century show several more dramatic and long-lasting downturns than anything seen since. The impact of the brush with financial disaster in 2008, however, served notice that we cannot take even this apparent longer-term decline in volatility for granted: the drop in output in 2009 was the biggest since 1946 and at the low point many commentators were indeed suggesting that a 1930s-style episode was at hand. Similarly, the apparently modest scale of the decline in consumer prices alongside that seen in the 1930s and 1920s is not wholly reassuring: had the fall in output developed into something more dramatic, so too might deflation have gathered more momentum.

US GDP growth, 1850-2009 (%)

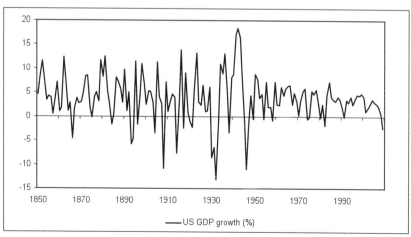

Source: Barclays Wealth Investment Strategy Group, BEA

US consumer price inflation, 1850-2009 (%)

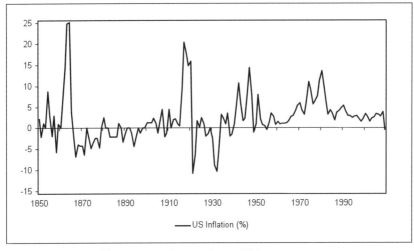

Source: Barclays Wealth Investment Strategy Group, BEA

In addition to, but often distinct from, fluctuations in economic activity and prices, planners need to take into account the possibility of independent (or exogenous) volatility in other important macro variables such as interest and exchange rates, and the relative prices of major factors of production such as labour, real estate and commodities (including, of course, energy). Such changes can be driven by, for example, trade imbalances, industrial relations and the operation of cartels (most visibly perhaps in the case of the Organisation of the Petroleum-Exporting Countries and the price of crude oil). Governments and central banks often intervene to try to reduce the volatility of interest and exchange rates in particular. However, in so doing they sometimes introduce another degree of uncertainty in the shape of capital controls, particularly in the case of smaller economies whose markets are most likely to be affected by portfolio flows. As noted above, for the first quarter-century after World War II, under the prevalent Bretton Woods regime, pegged exchange rates and varying restrictions on capital flows were the norm rather than the exception. Even in 2010 many capital accounts in the emerging world are controlled – most visibly in the case of China, whose currency is pegged loosely to the dollar. More recently, controls were partially reintroduced in a number of countries in the wake of the Asian crises of the mid/late 1990s; at the time of writing, they were being considered again by a number of emerging economies concerned at the scale of recent inflows of funds. Some indication of the variability of some of the most important interest rates, commodity prices and exchange rates in the last half century or so can be gauged from the charts below.

US government bond yields, 1951-2009 (%)

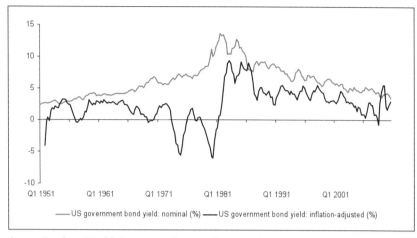

Source: Barclays Wealth Investment Strategy Group, national sources

Commodity prices, indexed: 1950-2010

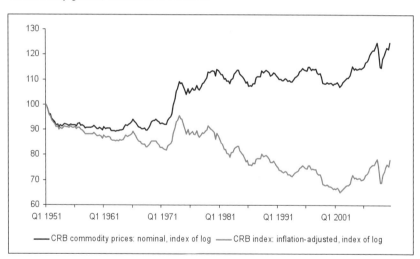

Source: Barclays Wealth Investment Strategy Group, CRB

Selected dollar exchange rates, 1957-2010

Source: Barclays Wealth Investment Strategy Group, Thomson Financial Datastream

Crude oil prices, 1973-2010

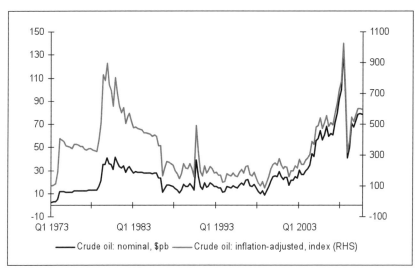

Source: Barclays Wealth Investment Strategy Group, Thomson Financial Datastream

Lastly, the workings of financial markets themselves can be a further independent source of considerable macro uncertainty. Nevertheless, the objective nature of the risk is perhaps questionable here, as is discussed a little more carefully below. The collective behaviour of investors and banks, in the context of largely deregulated but highly complex and interconnected markets, can foster extrapolative expectations and asset price bubbles. When those bubbles burst, the impact on balance sheets, discount rates and the financial system at large can be potent, as has

been seen on several occasions in recent history. This was the case, for example, in:
- the long-term capital management crisis in late 1998;
- the technology boom and bust in 2000; and
- most obviously, of course, the near collapse of the financial system in 2008-2009.

In the latter case, a genuine liquidity trap saw central banks lose their influence over interbank lending rates and companies across the world saw their access to working capital frozen. This triggered dramatic falls in trade and sharp rises in unemployment, as a result of a crisis that had its origins in the subprime segment of the US mortgage market.

Carmen M Reinhart and Kenneth S Rogoff (2009) identify 18 instances of bank-centred financial crises in advanced economies between 1945 and 2003 – 60 if emerging economies and the experiences of 2007-2008 are taken into account. Most of them passed without major macroeconomic consequences, but the risk of systemic collapse is something of which planners should be aware – particularly if the project coincides with a local property boom. Banking failure and sudden illiquidity in capital markets do not need to go global again to cause considerable local or regional distress: at the time of writing, Greece and Ireland are grappling with the aftermath of localised bubbles (one largely driven by the public sector, the other by private banks) and the wider Eurozone is worried about possible contagion.

3. **Contextual risk**

Geopolitical, macroeconomic and financial volatility is very visible: when it happens, we know about it. However, planners of long-term, large-scale projects also face more subtle high-level risks that we might characterise loosely as contextual in nature, but that might nonetheless be material. Here it is not the objective macroeconomic background that is changing, but rather the ways in which policy makers and investors view it. The reason that such intangible shifts can matter is that the valuation of such projects – indeed, the value of most things – in a market economy is unavoidably influenced by subjective considerations.

Such risks might include:
- domestic and cross-border legal and regulatory risk of the sort discussed elsewhere in this book;
- a change in the way in which monetary and fiscal policy is framed;
- a change in the analytical toolkit being used by the planners;
- measurement error and ambiguity;
- altered contagion and tail risk; and
- the ever-present risk of a change in the crucial but invisible realm of consumer taste and confidence.

It may seem unusual to include an altered macroeconomic policy regime as contributing to project risk. However, the response of the economies concerned to a given objective response will depend partly on the prevailing policy paradigm. For example, the chances of a recession turning into a more prolonged downturn may be influenced by whether the policy makers of the day believe in using aggregate demand-management tools in a counter-cyclical fashion. More topically, the danger of a single large bank failure developing into a systemic threat to the global money supply can depend on whether key policy makers believe that some institutions are too big to fail. It is not necessary to believe in the ultimate efficacy of those demand management or lender-of-last-resort tools to recognise that some key elements of the project appraisal framework could differ materially according to whether they are utilised. In each of the above examples, for instance, the likely evolution of short and long-term discount rates could be very different according to the policy response chosen, even if the real economy were unaffected.

There have indeed been many different macroeconomic policy regimes in operation within the market-led economies during the past 60 years or so, often alongside each other, and with some being pursued more zealously than others. A stylised chronology for the United Kingdom, for example, which has arguably seen more experimentation than most other large market economies over this period, would show an initial post-war framework of Keynesian demand-management policy using fiscal and increasingly monetary tools. This led, after some disenchantment with the stop-go results, to a half-hearted flirtation with industrial planning, and with prices and incomes controls. These in turn engendered further disenchantment and, after several balance of payments and fixed and flexible exchange rate crises, an enforced adoption of some monetarist and, eventually, neoclassical ideas. A further experiment saw:

- the temporary re-adoption of a pegged exchange rate;
- an enforced pragmatism; and
- most recently, the adoption of inflation targeting under the auspices of an operationally independent central bank – a regime that, until the recent banking crisis, was showing signs of some durability.

It remains to be seen whether the crisis-induced revival in more active demand-management evolves into something longer lasting (initial signs are that it will not).

A further source of contextual risk for planners is the possibility of a wholesale change in the analytical toolkit being used to evaluate projects. Details relating to the refinement of a specific methodology – whether to present results in terms of an internal rate of return or as a net present value, for example, when framing a discounted cash-flow analysis – can hardly be termed 'macroeconomic' in scale. However, the decision to choose a particular methodology can be, and is, influenced by evolving views on what constitutes best practice.

Some proponents of value-based stockmarket investing, for example, argue that any forward-looking approach to valuing assets is misplaced, and that a lengthy moving average of trailing (historic) earnings offers the least risky way of gauging underlying value. Others advocate a comparison of the replacement value of a company's assets with the value of those assets implied by its current stock price. Of course, neither approach is directly relevant to the assessment of a new project, where there is no past stream of earnings or stock price to analyse; but scepticism about the validity of forward-looking cost-benefit analysis is widespread and might foster greater reliance on valuations derived from comparable projects already in place, for example. At a time when discount rates are historically low, the result could be a marked discrepancy between (higher) estimated values derived from projected discounted cash flows and those derived from projects already in place.

The basic principle underlying the dominant valuation methodology today (the forward-looking discounted cash-flow framework) has been around for a long time – indeed, for as long as interest rates have existed (which, as Homer (1963) notes, is a considerable time indeed). However, several important refinements are relatively recent developments. Modern portfolio theory, which formalised the idea that risk and return are related, dates from Harry Markowitz's work in the early 1950s. The closely related capital asset pricing model – which introduced the idea of diversifiable and non-diversifiable risk, and the possibility of varying costs of capital within a market or sector – dates from the early 1960s. As recently as the 1980s, it was felt necessary to remind project evaluators in particular that net present values needed to be judged explicitly in terms of the amount of capital used to generate them; this led to today's formulations focused on economic value added and the cash-flow return on invested capital. The recent seizures in interest rate and equity markets are already causing some components of the dominant framework to be questioned, and the scepticism noted above may yet spread more widely. For example, the notion of an equity risk premium is becoming harder to substantiate after a prolonged period in which developed stock markets have delivered worse returns than riskless assets; the notion that the relative riskiness (the beta) of a particular industry can be gauged by the historical correlation of its stock prices with wider market indices is looking even more fragile.

A related source of uncertainty in project appraisal is the difficulty of measuring accurately even the seemingly less ambiguous elements of the investment arithmetic. The riskless interest rate, for example, is a key component of discounted cash-flow analysis, but its quantification is not quite as straightforward as it would seem. Suppose, for example, that it is decided that the long-term rate of interest paid on government bonds represents the core, riskless discount rate. In the largest, most liquid long-term interest rate market in the world – namely, that for US government bonds – nominal 10-year yields have fluctuated between 8% and 2% in the last two decades. Nominal yields less the current annual rate of inflation (a popularly used proxy for real yields) have fluctuated between 5% and -2%. In the United Kingdom, where a liquid market in inflation-linked government bonds – arguably the closest market-based approximation to a genuine real interest rate – has existed throughout this period, yields have fluctuated between 5% and 1% (with forward rates occasionally even lower). These are large proportionate variations. When yields are as low as 2%, even small changes can have a material impact on net present values (even allowing for the addition of a risk premium to the government yield to reflect the riskiness of private-sector cash flows).

Not all markets are as liquid as that for US treasuries. In 2008-2009, the massive interbank loan and asset-backed securities markets in both the United States and Europe effectively froze for prolonged periods. In some emerging countries, money markets may not exist to begin with and when they do, yield curves can be poorly defined. Such illiquidity is not always a result of local shortcomings: their governments may simply have little need to borrow (most visibly, perhaps, in the case of oil-rich states in the Middle East).

A topical illustration of the difficulties encountered in selecting an appropriate long-term discount rate to use in a long-term cost-benefit analysis was the Stern Review on the Economics of Climate Change produced for the UK government in 2006. Stern used a real discount rate below most estimates of the likely growth in the economy's productive capacity, incorporating zero 'pure time preference'. The selection of such a low rate implied that the welfare of future uncertain generations count equally with today's and was markedly at odds with most market rates at the time, as well as those used in forward-looking private analyses. Such selection was seen as potentially pivotal for the cost-benefit analysis. For some commentators, it discredited the review completely.

One response to the ambiguities of conventional rational cost-benefit analysis has been to turn to behavioural finance, which in recent years has been growing in popularity generally as disenchantment with the conventional economic research programme has spread. Here we encounter several additional shortcomings in the current standard appraisal framework beyond the definitional and measurement difficulties noted above. On the subject of discount rates, for example, the behavioural finance literature points to a marked discrepancy between individuals' revealed time preferences and those priced into conventional yield curves – specifically, studies reveal a very strong present bias: high discount rates over the immediate future, and then much lower forward rates than those embodied in existing market yield curves. More generally, project planners are seen as likely to suffer from overconfidence and hubris in their ability to forecast. In particular, they are considered to be likely to underestimate the costs and time needed to deliver a project, and to overestimate the benefits deriving from it. A sudden shift towards a behaviourally informed analytical framework from a conventional project appraisal could feasibly make the difference between viability and non-viability – particularly at a time of low market discount rates.

A further source of contextual uncertainty is the way in which investor expectations can interact to create the possibility of systemic risk, the domino-like effect that can transform a localised problem into a bigger, system-wide failure. There is some overlap here with the more exogenous financial risks noted above. The existence of such risk requires certain objective conditions to be in place, including, for example, the scalability noted by Taleb in his discussion of black swans, but those conditions are not always fully visible and the catalyst for the event is often subjective in nature. For example, systemic banking risk would be small if banks limited their lending strictly to a small portion of their deposit base, refrained from real estate lending and made only conventional, transparent loans. But fractional reserve banking existed alongside substantial commercial and residential mortgage markets, with the latter often in securitised form, for several decades before the crisis of 2008. There were some new, complicating objective variables in the loan cycle that preceded the crisis, most notably the existence of complex credit derivatives and contingent liabilities that made it even more difficult than usual to value bank loan books. However, the ultimate trigger for the crisis was a loss of confidence on the part of investors in property values and bank balance sheets. More recently, the sovereign debt crisis in the Eurozone, as well as the systemic risk that it posed, was arguably not precipitated by new information about the levels of debt in Greece and Ireland. Instead, it was caused by altered investor perceptions regarding both the countries' ability to repay it and the willingness of partner governments to offer support. The broad magnitude of their indebtedness and, in the case of Ireland, the scale of the contingent liability posed by the local banking system had been visible for many months; the flawed fiscal architecture of the single currency project had been known since its inception.

Another contextual risk for project planners to keep in mind is the ever-present possibility of an unexpected change in demand for the commodity, utility or service being provided, not because of a sudden change in any objective macro variable or resource price, but because of a shift in subjective consumer tastes or preferences. The viability of many nuclear energy projects has been undercut by public wariness; that of some conventional energy supply projects could yet be undercut by an intensification of environmental concerns.

Macro risks facing planners of large-scale cross-border projects

Political/geophysical risks	Economic/financial risks	Contextual risks
War between sovereign states	Growth, inflation and interest rates (cyclical and secular variations)	Policy regime
Civil war/revolution	Exchange rate volatility	Investment analytics
Terrorism	Housing/banking	Measurement error
Industrial unrest	Sovereign credit	Systemic risk
Asset seizure, punitive taxation	Competitiveness	Customer tastes
Natural disasters	Demography/participation	
Environmental/climate change	Natural resources	
	Technology change	

4. Main macro risks identified

In light of the above discussion, we can perhaps identify some of the most visible objective macro risks facing project planners. In the table below they are grouped under two broad headings: those that might be labelled political or geophysical in nature, and those that can be described as being economic or financial. The table also illustrates some of what we have labelled contextual or more subjective risks. As noted above, the distinction between objectivity and subjectivity in a market economy is (by definition) not always clear.

5. Conclusion

The future is unknowable and the past can be only the loosest of guides to the potential macro risks facing planners of large-scale infrastructure projects. Nonetheless, this brief discussion has identified many sources of such risk, ranging from the objective threats posed by conflict and natural disasters, and by cyclical and secular economic volatility, to the more subjective or contextual risks arising from altered policy and analytical paradigms, and from changes in investor and consumer sentiment.

The range and scale of macro shocks seen in the past half century have been dramatic, and there are doubtless new unknown unknowns looming ahead. Genuine risk is not the smoothly varying, statistically tractable quantity found in textbooks, but the more profound uncertainty that encompasses the possibility of wrenching dislocations, systemic collapses and worse. The best advice to planners is that they should stress-test their calculations extensively and do their best to keep an open mind and stay nimble.

The somewhat one-sided nature of this conclusion is daunting, but likely realistic. Being human involves facing more potential bad stuff than good. We each of us die; the earth could be wiped out by an asteroid strike. But an awareness of risk does not mean that we are shackled by it. The bad stuff does not always happen and prosaic day-to-day success can persist for years at a stretch.

Received macroeconomic wisdom argues that the developed world is facing a prolonged period of low growth and poor investment returns as a result of accumulation of debt in the cycle just ended and the ongoing imbalances in the global economy. The idea has been popularised in the notion of the 'new normal', as noted above. However, optimistic projections are not the only ones that are vulnerable to contextual rethinks. It is possible that the turbulent period we have just lived through was first and foremost a financial farce rather than an economic tragedy, and that it marks the end of a long period of disappointing growth, not the start of one. The causes of long-term prosperity are our endowment with real resources – land, labour, capital – and our ability to innovate and to learn by doing. The past few years have changed neither that fact nor those endowments nor our capacity for invention and improvement. The remarkable economic and political gains of the past half century do not have to be repeated – but they might be. To quote the late Julian Simon: "This is my long-run forecast in brief. The material conditions of life will continue to get better for most people, in most countries, most of the time, indefinitely. Within a century or two, all nations and most of humanity will be at or above today's Western living standards. I also speculate, however, that many people will continue to think and say that the conditions of life are getting worse."

References

Ahamed, Liaquat, *Lords of Finance: 1929, The Great Depression, and the Bankers who Broke the World*, Penguin (2009).

Bernstein, Peter L, *Against The Gods: The Remarkable Story of Risk*, New York: John Wiley & Sons (1996).

Fukuyama, Francis, *The End of History and the Last Man*, Penguin (1993).

Homer, Sidney, *A History of Interest Rates*, John Wiley & Sons (1963).

Maddison, Angus, *Chinese Economic Performance in the Long Run*, OECD (1998).

Reinhart, Carmen M and Rogoff, Kenneth S, *This Time is Different: Eight Centuries of Financial Folly*, Princeton (2009).

Taleb, Nassim Nicholas, *The Black Swan: The Impact of the Highly Improbable*, Allen Lane (2007).

Political risk in large projects

Arve Thorvik
Thorvik International Consulting AS

1. Introduction

The management of political, non-commercial and non-technical risks in industry has traditionally been given only limited attention. This area of risk management has been seen as comprising elements of engineering, economics and insurance expertise. Politics is a discipline with a narrow scope for quantifiable measurement and a field in which everyone sees themselves as an expert and a citizen, as well as – at least in many countries – a voter. This chapter approaches the issue of political risk from the perspective of political science and builds on practical experience, primarily from the international oil and gas industry. The perspective is international – it considers business development efforts involving foreign direct investments and cross-national projects.

The chapter tries to provide some tools to foster a better understanding of what kinds of risk need to be tackled. It also examines approaches to problem solving or mitigation. Lastly, it makes two points not commonly made in such discussions:

- Political risks do not necessarily originate with politicians, as their root causes are often inside the project itself; and
- Western democracies pose political risks that are at least as numerous and as complicated as less developed political systems.

1.1 Definition of 'political risk'

The first issue to address when considering a modern, straightforward definition of 'political risk' is to understand that political risk arises not only in developing countries – it may also be present in developed nations. In addition, political risk is not limited to *coups d'état*, civil wars or riots.[1] The definition provided by the *Financial Times* is as follows: "Political risk is the risk of operating or investing in a country where political changes may have an adverse impact on earnings or returns. This concerns not only politically unstable countries, but also places where normal democratic procedures may bring about a change of government and thus a possible negative change in policy (e.g.: on tax, regulatory constraints; tariffs, etc.)."

Investopedia's definition is similar, providing that political risk is "the risk that

---

[1] The Marsh Insurance Company introduces its presentation of political risk in the following way: "War, terrorism, and factional violence continue to threaten oil and gas production. Moreover, oil-and-gas contract renegotiations and changes in operating rules underscore the potential for interference by host governments. Careful risk management that includes political risk insurance can enable companies to take advantage of opportunities in developing markets."

an investment's returns could suffer as a result of political changes or instability in a country. Instability affecting investment returns could stem from a change in government, legislative bodies, other foreign policy makers, or military control. Political risk is also known as 'geopolitical risk', and becomes more of a factor as the time horizon of an investment gets longer".[2]

The International Risk Management Institute gives a wider and more system-neutral definition: "The exercise of political power causes political risks in international business, and this power can affect a firm's value."[3]

These definitions present the following weaknesses:
- They are often limited to political change or instability. In the author's opinion, political risk also appears when there is no abrupt or institutional change: if a government does not play by its own rules and does not provide a level playing field among competing projects, it creates risk.
- The International Risk Management Institute seems to support this approach: "a dramatic political event may pose little risk to a multinational enterprise, while subtle policy changes can greatly impact a firm's performance. A student-led protest for political change may not change the investment climate at all, while a change in local tax law can erode a firm's profits very quickly."[4]
- They seem limited to processes that are external to the project. Many political risks in fact stem from poor planning or performance by the project itself.

Accordingly, the author offers the following definition: 'political risk' consists of any actions taken by legitimate or illegitimate political groups or authorities that cause an important project to be cancelled, stopped, seriously delayed or significantly altered in scope. This may happen despite the project having been initially approved by the authorities, being entitled to such approval or having every reason to expect to be approved, in accordance with the applicable law in effect at the time. The result, if not mitigated, would be a significant financial or reputation loss to the project's owners or developers.

1.2 Scope and limitations

This chapter concentrates on large-scale projects that have a significant physical presence in a society/community, and that have an equally significant impact on that community. Such projects typically relate to infrastructure (eg, sea and airports, highways, railways, dams, power plants, pipelines, waste treatment, storage or disposal), the development of natural resources (eg, oil and gas, mines, plantation-style farming) or large-scale industrial activity (eg, metal processing, paper and pulp, refineries).

Governments or authorities are deeply involved in such projects. Resources, including land, are often owned by governments at a regional or national level. The project may be:
- initiated by governments;

---

2   www.investopedia.com/terms/p/politicalrisk.asp.
3   Wagner, Daniel, "Defining Political Risk", AIG (www.irmi.com/expert/articles/2000/wagner10.aspx).
4   *Ibid.*

- contested for in a more or less transparent public bidding process;
- executed by large – often foreign – private companies using a number of subcontractors to do the work; and
- labour-intensive over a limited and defined period.

By definition, therefore, a 'project' spans the periods starting with the initial idea and concluding with the beginning of the operations or commissioning of the development.

The challenges covered by this chapter are those that will normally be subject to political regulation and decision making. They relate to the environment, cultural heritage, health and safety issues, and the economic contribution to society at large. This chapter aims to serve two purposes for practitioners in large-scale projects around the world. The first is to improve the understanding of what causes or increases political risk relative to the political environment in which a project operates. The table below is an attempt at such a classification.[5]

Table 1: Classification of political risk

		Origin of the political risk			
		Project	Stakeholders	Government	Unpredictable
Project environment	Dysfunctional societies				
	Autocratic regimes				
	Transitional regimes				
	Western-style democracies				

The table above is organised along two axes:
- the type of political environment (ie, the political system) in which the project operates; and
- the source of the political risk – namely:
  - the project itself;
  - the way that the project handles its relationship with stakeholders;
  - the relations between the project and the government(s); and

---

5   A company with shareholders and headquarters in Europe executing a project in Africa will have to focus on both environments.

- the events or changes that could not possibly have been predicted by any normal project organisation.

The second aim of this chapter is to offer suggestions as to how to prevent, reduce or mitigate political risk

2. A simple typology of political systems

The effects of political actions – or inactions – that may change or stop a project are essentially universal. The kinds of challenges and political players that a project faces, however, differ considerably according to the kind of society in which the project operates. This chapter does not attempt to create new typologies of political system, but considers a simple, four-step ladder of political development for the purpose of distinguishing between the types of player and challenge that a project will encounter – namely, dysfunctional societies, autocratic regimes, transitional regimes and Western-style democracies.

2.1 Dysfunctional societies

Dysfunctional societies are characterised by a weak or undermined central authority that does not control its entire territory and whose national laws are compromised by regional, religious or cultural differences. Foreign companies developing major projects will have difficulties finding credible and legitimate counterparts with which to negotiate and strike a deal.

Companies that venture into projects in such parts of the world must be ready to take on considerable risk, and have excellent systems in place to handle both physical and financial risk. Oil exploration and production in the Niger Delta is a strong example of this (see text box below).

> **Nigeria, 2007**
> "Insurgents protesting over poverty and pollution in the Niger Delta raised the stakes in their battle for fairer revenue distribution, a cleaner environment and better living conditions. In 2006, their campaign resulted in the shut-down of 25% of Nigeria's oil production and helped drive global crude prices to close to $80 a barrel. Who are these rebels? Can multinational firms cope with the threat they represent to operations, personnel and material assets?
>
> Community-based militias and rebel groups conducted 31 attacks on the petroleum industry last year: 70 international staff were abducted in 21 raids; four acts of sabotage on pipelines and flow stations were reported; three violent assaults were made on international staff and assets and, in December, three car bombs were set off in Warri and Port Harcourt, two of the main oil-towns of the delta. Rebel attacks have continued with undiminished strength in 2007."
> Excerpt from Nodland, Arild and Hjellestad, Odin, "Security in the Niger Delta", Bergen Risk Solutions (March 2007)[6]

---

6   www.bergenrisksolutions.com/index.php?dokument=23.

High and protracted levels of violence do not necessarily result from actions by the oil companies themselves; rather, they stem from the uneven distribution of public wealth generated by the oil industry. Thus, companies become the target of protests against a partly invisible federal government. Nigeria has remained a major global producer and exporter despite these serious challenges. To the outside observer, it seems that this might be at least as much the result of successful containment strategies by companies protecting and isolating themselves from the society as a result of engagement and integration into local societies.

Containment or engagement:[7] The prime responsibility for any project or industry in such circumstances is to protect its employees and properties. At the same time, any moves towards social responsibility easily result in the company taking over the government's responsibility to provide basic services to its citizens (eg, hospitals, schools, roads). Therefore, assuming social responsibility is not without problems and risks creating a new type of corporate feudalism, a system which Europe abandoned in the late 19th and early 20th centuries. When moving into such dysfunctional societies, a project or company must take a clear view on the balance between engagement and containment.

Two examples illustrate this necessity. When BP, Statoil and their partners built the Baku-Tbilisi-Ceyhan and South Caucasus pipelines in parallel, stakeholder engagement was a key element of the projects. All along the route, local communities were engaged and mobilised, and the projects used as much local labour as possible.

By contrast, when the Chinese built the gas pipeline from Turkmenistan to China, and soon afterwards from Kazakhstan to China, the model was the same as that which the Chinese had used in similar projects in Africa. In an approach reminiscent of the construction of US railroads about 150 years ago, the Chinese moved in all labour from China – from the cook to the project manager (see text box below).

China, 2010

"Ranked first for African market share, Chinese contractors – active in 53 out of Africa's 57 nations – are playing a leading role in the continent's development. Some 80 per cent of Chinese projects here are related to basic infrastructure, with many of the contracts received through initiatives from government agencies and financed by grants. Diplomatic channels, financial aid, and a reputation for quality have all contributed to Chinese contractors' expanded presence in Africa.

There are several key features that give Chinese contractors an edge over their predominantly Western competitors in developing countries. First and foremost is cost. Chinese labour could cost a mere 5 per cent of the compensation given to counterparts from the US and Japan. Additionally, Chinese firms are able to source inputs cheaply, such as generators, turbines and cement, from manufacturers at

---

[7] This section does not refer only to dysfunctional societies: These choices, and the examples used, apply equally to autocratic and possibly also to transitional societies.

home, to forward massive cost savings on to clients in developing nations.

The experience and adaptability of Chinese contractors confers another advantage in developing nations. Firms often bring many of their own workers on-site who are first trained at home prior to expatriation."

Excerpt from Avery, Charles, "Building by design: how China develops the developing world", The Beijing Axis[8]

2.2 Autocratic regimes

In the developing world, autocracy constitutes the most common form of political regime.[9] Although most of us have a pretty good understanding of how democracy works, we know much less about how autocracies behave and the extent of their institutional diversity.

Barbara Geddes introduced a threefold categorisation of autocracies into military, one-party and 'personalist' autocratic regimes.[10] In countries with autocratic regimes, civil society is either non-existent or fully controlled. If the regime is ever challenged, it is mostly by an armed opposition or possibly by intervention from abroad (Saddam Hussain effectively ruled oil-rich Iraq from 1968 until the US-led invasion in 2003).

Geddes[11] classifies dictatorships according to who controls access to political office:

In military regimes, a group of officers decides who will rule and exercises some influence on policy. In single-party regimes, one party dominates access to political office and control over policy, though other parties may exist and compete as minor players in elections.

Personalist regimes differ from both military and single-party in that access to office and the fruits of office depend much on the discretion of an individual leader.[12]

If a major project is to be developed in an oil, gas or mineral-rich country run autocratically, then political risk does not relate to stability or security, as is the case in dysfunctional societies. Research shows that oil wealth tends to make regimes more stable: "Oil wealth is robustly associated with increased regime durability, even when controlling for repression, and with lower likelihoods of civil war and anti-state protest ... In short, oil wealth has generally increased the durability of regimes, and repression does not account for this effect."[13]

Therefore, political challenges to large projects in such countries are unlikely to come from instability or from civil society. They are more likely to come from the prevalence of corruption (see section below) or from the mere fact that many of these

---

8	knowledge.insead.edu/economy-china-development-100608.cfm.
9	www.stanford.edu/~magaloni/magalonifrontiersduke.vfin.pdf.
10	Geddes, Barbara, "What Do We Know About Democratization After Twenty Years"", *Annual Review of Political Science*, Vol 2: 115-144, June 1999.
11	Geddes, Barbara, "Paradigms and Sand Castles: Research Design in Comparative Politics", Ann Arbor: University of Michigan Press (2003), p 51.
12	Those in the oil and gas industry who have experience working in Turkmenistan, Kazakhstan or Azerbaijan will have an intimate understanding of the meaning of the term 'personalist regime'.
13	Smith, Benjamin, "Oil Wealth and Regime Survival in the Developing World, 1960–1999", *American Journal of Political Science*, Volume 48, n2 (April 2004), pp232-246 (plaza.ufl.edu/bbsmith/AJPS_OilRegimes.pdf).

regimes are ruled by rather unpredictable individuals who might abruptly change the framework conditions for business (see text box below).

Venezuela, 2010

"Chavez has nationalized a lot of businesses in Venezuela and he will nationalize more before the next presidential election in 2012. After 12 years in office, the former soldier shows no signs of slowing his drive to recreate Venezuela as a socialist state.

Since September, he has nationalized a fertilizer factory part-owned by U.S. giant Koch Industries and Italy's Eni; bottling factories belonging to Ohio-based Owens Illinois; a motor lubricants company; and the country's largest farm products company. More than 200 companies have passed into state hands this year.

Chavez says he has 'a little list' with more planned takeovers, although he has taken a breather in recent weeks."

Excerpt from Daniel, Frank Jack, "Factbox: Key political risks to watch in Venezuela", Reuters (December 10 2010)[14]

Corruption: All serious Western companies that are listed on a major stock exchange and are large enough to undertake cross-border projects will have in place a system for corporate governance that includes strong rules against engaging in corruption. Nevertheless, large projects are often implemented in countries where corruption is widespread. This is especially true for the energy sector.[15]

Corruption is no doubt one of the main political risks in such projects. At the practical level, the situation might be like this: a refusal to pay off a politician or official somewhere in the bureaucracy in the host country is likely to cause a significant delay in the project. If a bribe is paid, then the short-term political risk to the project is reduced. However, the medium to longer-term home country risk to the project owner increases enormously and could kill the company, as well as causing – at the very least – a significant fall in shareholder value, together with a complete change of management and board.

Engaging in corruption is therefore not an option for a serious company with a sustainable approach to business. To foster the right attitudes and internalise the ethical values in all staff is commonplace. More frequently overlooked is the need to implement common values and rules in project-specific joint ventures where companies with differing nationalities, company cultures and experiences work hand in hand.[16] Equally important – and often missed – is the need to signal the project or company's anti-corruption values and internal governance rules to the host government, local authorities, partners and subcontractors *before* the project

---

14 af.reuters.com/article/energyOilNews/idAFRISKVE20101201.
15 The following oil and gas-producing countries all rank below number 100 in the 2010 Transparency International Corruption Perception Index: Kazakhstan (105), Gabon (110), Vietnam (116), Ecuador (127), Azerbaijan (134), Nigeria (134), Iran (146), Libya (146), Russia (154), Venezuela (164), Equatorial Guinea (168), Angola (168), Sudan (172) and Turkmenistan (172).
16 For a broader discussion of this issue in the context of joint venture arrangements, see the chapter of this book titled "Investor risks and responsibilities: joint ventures and cross-border energy infrastructure".

starts. Doing this *post facto* is no easy task – especially in countries such as those listed in footnote 16 above.

2.3 Transitional societies

Ideally, countries should graduate from autocratic regimes (eg, the Soviet Union) to transitional systems (eg, Russia, Ukraine or other countries of the Commonwealth of Independent States), where political processes are partially transparent. The democratic institutions are weak, as are the legal structure and institutions (typically possessing a low level of bureaucratic capacity), and civil society is undeveloped. Political parties are frequently built around an individual former autocratic ruler. Such societies pose a whole array of political risks: unpredictable decisions, abrupt political changes, disorganised and semi-violent environmental groups, corruption and tilted playing fields between competitors, often controlled by the rulers.

> **Russia, 2008**
> "Since March, BP and TNK-BP have been subjected to a series of inquires, investigations, and raids at the hands of Russian authorities. BP's offices have been raided twice by the FSB, the state security body formerly known as the KGB. TNK-BP's offices have been inspected four times by the labour ministry. A former employee was arrested on charges of industrial espionage. Dudley was interrogated for six hours at the Ministry of Internal Affairs on tax-avoidance schemes engineered at TNK two years before he arrived.
>
> Last week, TNK-BP was forced to send the last of the BP secondees home amid a row over their visas. Khan, who is also responsible for government relations at TNK-BP, ignored a request from Dudley for 150 visa renewals for foreign staff. Instead, he requested just 63 from immigration authorities, in effect ending the employment of nearly half the company's crucial technical specialists. At one point, TNK-BP security men denied BP staff entry to TNK-BP headquarters."
> Excerpt from Danny Fortson, "Now it's war at BP-TNK", The Sunday Times *(July 27 2008)*[17]

> **China, 2010**
> "Foreign companies are losing market share in China across a broad range of industries because of discriminatory treatment by the government and regulators, according to the European Chamber of Commerce in China … 'Compulsory certification in excess of what is reasonable is being used to keep foreigners out of the market and business license requirements continue to exclude foreign companies from entire sectors,' the group said."
> Excerpt from the Financial Times *(September 2 2010)*[18]

2.4 Western-style democracies

Western-style democracies are characterised both by transparent political decision

---

17   business.timesonline.co.uk/tol/business/industry_sectors/natural_resources/article4406891.ece.
18   "Foreign companies 'losing out' in China", by Jamil Anderlini, *Financial Times*, September 2 2010

making, in accordance with the rule of law, and by a well-developed civil society. They include a large private sector and a fairly low level of corruption.[19] That does not mean that there are no political risks in these societies; rather, it means that the challenges are different.

Political risk is usually associated solely with non-Western, non-democratic, less developed societies. This would turn out to be true if the limited definition of 'political risk' – as discussed above – were applied. However, while challenges to projects normally take a less violent or dramatic form in Western-style democracies than in the other regimes discussed above, examples of major projects being politically challenged in these democracies still abound. Such projects include oil sands and shale gas in Canada and the United States, and a railway station in Germany (see text box below).

Germany, 2010
"Berlin – Demonstrations against plans to shift the German city of Stuttgart's railway station and main lines underground resumed Saturday, four days after mediation talks ended. The nationally televised negotiations under a mediator in Stuttgart's town hall had partly eased the bitter dispute between the government and conservationists. They ended with the mediator, Heiner Geissler, recommending on Tuesday that the project proceed. The hard core of the protesters said they would fight on against the project until they halted it."
Excerpt from "Demonstrations against Stuttgart rail plan resume after talks", Deutsche Presse-Agentur (December 4 2010)[20]

Many of the protests are motivated by environmental concerns, but take on a political momentum of their own when they arise outside the established channels. Ranging from peaceful demonstrations to violent actions by radical groups (so-called 'eco-terrorism'),[21] they present serious challenges to projects organisations, which are not normally designed to handle such issues.

National politics and geopolitical games: Even though a large project is in and by itself a political actor, there are definite limits to its ability to play. In particular, energy projects have a tendency to become embroiled in what is colloquially known as geopolitics – a game that transcends the projects themselves. To take an obvious example, any energy project in Iran (or any involvement with an Iranian company) will lead to a situation where nuclear proliferation issues and the human rights record of the regime in Tehran become more decisive to the fate of the project than the

---

19 There are, however, six EU member states ranking below number 50 on the Transparency International corruption Index: Hungary (50), Czech Republic (53), Italy (67), Romania (69), Bulgaria (73) and Greece (78).
20 www.monstersandcritics.com/news/europe/news/article_1603450.php/Demonstrations-against-Stuttgart-rail-plan-resume-after-talks.
21 'Ecoterrorism', also called 'ecological terrorism' or 'environmental terrorism', consists of the destruction or the threat of destruction of the environment by states, groups or individuals in order to intimidate or coerce governments or civilians. The term has also been applied to a variety of crimes committed against companies or government agencies and intended to prevent or to interfere with activities allegedly harmful to the environment (www.britannica.com/EBchecked/topic/765758/ecoterrorism).

merits of the projects itself. This was true even before the UN sanctions regime was clarified and strengthened. This example, together with many of a similar nature, is not publicly expressed by, or based upon, legal policies of the major powers in the West. However, companies are made to understand that there would be no future for them and that there will be no financing available if they continue to be involved.

As this chapter was being written, the jury was still out on which pipeline project would be chosen for the completion of the Southern Gas Corridor from the Caspian Sea to Europe. This issue at an early stage grew into what was termed a strategic issue for the European Union, as well as the United States. The publicly stated favoured option was Nabucco. That political support – intangible as it was – was enough to prevent competing projects from obtaining permits and putting financing in place. The situation may still evolve to allow the best commercial decision (which will cost the EU taxpayers the least money) to be made, but this is far from certain.

The history of gas pipelines is full of such examples. The very first gas pipelines from Norway to the European continent (from Troll) would not have been built were it not for the fact that President Reagan and his administration felt that Norwegian gas was needed to loosen the gas grip of the Soviet Union on Europe. The Baku-Tbilisi-Ceyhan and the South Caucasus oil and gas pipelines from Azerbaijan to Turkey would most likely not have been built either were it not for the intervention of President Clinton.

The European Union is now walking a tightrope between political objectives and its own free market philosophy; it is indeed picking winners among projects – not least in the energy sector – by selecting priority projects in the so-called 'Trans-European Network Programme',[22] and a 10-year development plan for infrastructure.

While the industry is not at all comfortable with such processes, at least they are fairly predictable in most cases: those who enter the competition without being explicitly on the priority list so with open eyes. The difficulties arise when priorities shift and abrupt changes in direction occur.

So if a project (particularly an energy project) assumes a regional or global relevance, its developers should do the following:

- Display at least as much competence in political processes as in technical and commercial issues.
- Pay attention to the major powers in the project's region (eg, the European Union and the United States), be well represented in these countries' capitals and keep these countries' authorities well informed about the project.
- Ensure support or, at the very least, the absence of opposition from the host governments to enable the project to progress. No matter how much governments claim to have fair and free markets, they tend to hinder projects that do not fulfil their own political objectives. To develop close and positive relations with the host governments – provided that they are not hostile to the major powers – is of paramount importance to the success of any large project.

---

22   This project was subsequently scrapped by the owners. The author makes no evaluation of whether the environmental impact work on this project was done to a satisfactory standard. The point is that it was not perceived by stakeholders to be good enough, causing the government to kill the project.

3. Root causes of political risk

Industry people are inclined to blame the external world for challenges that are not commercial or technical in nature. The truth is, however, that a large part of political risk can be foreseen and prevented, reduced or eliminated altogether by the project itself. A majority of what is considered political risk is risk of a different nature that continues to grow to reach the political sphere and significantly damage the project. Approaching this risk in a systematic way may help in understanding and handling it. There has been a tendency both in literature and among industry representatives to consider as political risk only actions taken by politicians, and not those taken by project developers. The author believes that this approach is flawed. Major projects – as defined above – take up a large place in any society. They automatically receive a lot of attention. Action – or lack of action – by the project developers can easily result in political processes that can backfire on the project. The first step in handling political risk is thus to recognise that the project itself is a political actor, a participant in processes and an object of political debates.

Table 2 below summarises the author's approach to classifying political risk. It makes sense to look at three categories of root causes of political risks: those that are economic or financial in nature, those that fall within the concept of sustainability and those that relate to the project's external communication.

Table 2: Classification of types and origins of political risk

			Proximity to project
			Internal Local National Regional Global →
Root cause of risk		Economic/commercial	————————→
	Sustainability	Health	——————→
		Safety	——————→
		Environment	————————————→
		Corporate social responsibility	————→
	Communication and external relations		——————————→
	Unpredictable		——————————→

3.1 Economic and commercial causes of political risk

A number of common economic and commercial aspects of a project may easily turn into political issues if not handled or communicated correctly. Examples include:

- the location of the project company's headquarters (eg, in Switzerland for tax reasons, at the sponsor's headquarters for recruitment, travel and cost reasons, or in a host country to allow for easy communication with host authorities);
- the development and use of local suppliers and contractors; and
- whether to take on a local partner (if not mandated by law).

The project and the host governments will find themselves on opposite sides of a negotiating table on a number of such issues. While one cannot offer generalisations on this subject, the main avenue for effective project management is to be ready to live by the law of the land, and to exchange open and frank information at all stages of the project in order to avoid taking the authorities by surprise and thereby creating risk.

3.2 Approach to sustainability issues

Quite often, political project risks in Western societies are connected with the environment, health or safety. Such issues, whether based in reality or in mere perception, are handled by the project developers according to three parameters: law and regulation, requirements from the financial community (particularly the World Bank, the European Bank for Reconstruction and Development and the European Investment Bank) and corporate standards.

This will prompt project developers to undertake environmental and social impact assessments, and hopefully make necessary adjustments to the project based on the assessments' findings. If the impact assessment is handled in what the stakeholders see as an unsatisfactory way, such handling can become a significant political issue and a showstopper (see text box below).

> **Italy, 2007**
>
> "The Italian government has suspended a decree that permitted Britain's BG Group (BG.L) to build a liquified natural gas (LNG) terminal in the southern port of Brindisi, Italy's Environment Ministry said. The ministry said in a statement late on Friday it had agreed with the Economic Development Ministry to suspend the authorisation of BG's long-contested project until a study of the environmental impact of the terminal is completed.
>
> There was no immediate comment from BG, which has said in the past it had all valid authorisations for the 500 million euro ($705.4 million) project. The Environment Ministry said it would push for scrapping the previous authorisation altogether if BG's Italian unit Brindisi LNG did not agree to carry out the environmental impact study – which usually takes months.
>
> BG's plan to build an 8 billion-cubic-metre-a-year LNG terminal in Brindisi has been blocked for years by opposition from local green groups and politicians, despite backing from Italian and British political leaders."
>
> *"Italy suspends BG's Brindisi LNG project", Reuters United Kingdom, October 6 2007*[23]

---

23   http://uk.reuters.com/article/2007/10/06/bg-brindisi-idUKL0662089420071006.

The development of the Nordstream gas pipeline from Russia to Germany was significantly delayed – first because of problems with munitions dumps and archaeological finds in the Baltic Sea, and second because the governments of Finland, Sweden and Denmark hesitated for a long time before giving the necessary environmental project permits. Difficulties in finding an acceptable landfall point for gas pipelines have been a major issue in many existing and planned developments. The first Europipe gas pipeline from Norway to Emden in Germany is a good case in point. Environmental concerns and protests necessitated a complex solution to the landfall, including the drilling of a tunnel to cross the Wadden Sea National Park.[24]

Few areas in project management carry a greater chance of turning into significant political challenges in either the host or home country. Thus, the development of oil sands resources has caused greater problems for some oil companies (eg, Statoil) at home than it does in the host country, Canada. The protection of rainforests is a bigger issue in Europe than in Brazil or Indonesia. Protests are emerging against the development of shale gas resources in North America. If these protests advance and take a strong hold on the political system, regardless of whether the environmental arguments are true, they may alter the outlook for gas supply worldwide.

While sustainability performance is clearly becoming a field of competition, it is also an area that lends itself well to industry cooperation, such as setting industry standards.[25] This is helpful not just in order to build an industry image, but also to protect the individual company against risk. The actions of one black sheep in an industry might create significant problems for the whole industry. The handling of the oil spill by BP/Halliburton in the Gulf of Mexico in 2010, which caused political outrage across party lines in the United States, resulted in losses for the entire industry through a drilling moratorium on all companies operating in the Gulf of Mexico and new, tougher legislation placing restrictions on other projects and operators in the same geographical area.

While health, safety and environmental issues are pretty well regulated by (most) national governments, lenders and corporations themselves, corporate social responsibility – which is equally critical to the successful completion of any large project – is a much more diffuse and diverse field to work in. The following Wikipedia description of what characterises a business that focuses on corporate social responsibility (with the exclusion of philanthropic work) covers the concept rather well: "CSR-focused businesses would proactively promote the public interest by encouraging community growth and development, and voluntarily eliminating practices that harm the public sphere, regardless of legality. CSR is the deliberate inclusion of public interest into corporate decision-making, and the honoring of a triple bottom line: people, planet, profit."[26]

24	Henning Grann: "Europipe Development Project: Managing a Pipeline Project in a Complex and Sensitive Environment", The Industrial Green Game 1997, pp 154–164. Washington, DC: National Academy Press.
25	The Extractive Industries Transparency Initiative has developed in this direction.
26	en.wikipedia.org/wiki/Corporate_social_responsibility.

3.3 Communication and external relations

Modern global companies know that they can succeed in communications policy only if they are as open and transparent as commercially possible, if they are proactive rather than reactive and if internal and external communication goes hand in hand. In post-Wikileaks times, executives understand that only the most sensitive commercial information should be kept secret, and that this must be protected far beyond what is common today. Openness must be the rule.

It is common to use the term 'stakeholder management'. This indicates – willingly or unwillingly – that stakeholders can or should be managed. That is, of course, not a viable course to take with local communities, neighbours and environmental non-governmental organisations, among others. Stakeholders will, to varying degrees, have their own objectives and agendas, which the project developers cannot expect to manage. Stakeholders have to be engaged – hopefully positively. The approach taken to such engagement also varies greatly. Thus, some projects call townhall meetings to inform stakeholders about decisions made, how those decisions will influence them and, possibly, how they will be compensated for any negative effects. Such an approach might create a political risk, rather than mitigate it. By contrast, other projects call stakeholders at the earliest stage of the project, using an independent facilitator and giving stakeholders an opportunity to voice opinions, to engage with the project and to propose changes. While the author does not advocate giving control of the project away to stakeholders, non-governmental organisations and townhall meetings, he is certain that specific elements of the second approach will significantly reduce risk to the project.

Stakeholders and local communities: Any large-scale project development plan will end up on the desk of the highest levels of governments. In addition, such projects span a long period from conception to completion – long enough to extend beyond the electoral cycle in countries where elections take place. This means that any well-managed project should keep at least some measure of contact with those who may become the next occupants of cabinet offices. Conversely, any project would be well advised to avoid becoming the pet project of the current rulers – this could cost dearly in many regimes, including Western democracies (see text box below).

Bulgaria, 2010
"Ever since the center-right government of Bulgarian Prime Minister Boyko Borisov took office in the summer of 2009, it has been balking at the construction of the Burgas-Alexandroupolis oil pipeline, which had been promoted vigorously by the former Socialist-led Stanishev Cabinet and the Socialist President of Bulgaria, Georgi Parvanov. It has also been met with staunch resistance along Bulgaria's southern Black Sea coast over environmental concerns."
Excerpt from "Bulgaria Might Kill Burgas-Alexandroupolis Pipeline by Not Paying Dues", Novinite Ltd (December 16 2010)[27]

---

27   www.novinite.com/view_news.php?id=123256.

Sometimes governments are project owners. Despite their own area of expertise, politicians also make serious mistakes when it comes to political risks. In their eagerness to create political compromises and move to action, they overlook local stakeholders and face serious opposition, which they had not counted upon. One eample is a high-voltage land line in Norway, which had received all construction permits and was ready to go when local communities along the planned line staged major protests, causing the government to order a temporary halt of construction and an investigation of alternative routings, including sea cables.

Equally surprising to a government unprepared for the intensity of the opposition was the dispute between local residents and the German federal and state authorities over the Stuttgart railway station (see text box below).

Germany, 2010
"The protesters believe the station is too expensive and was planned without enough prior public consultation. After demolition work on parts of the 80-year-old station began in the summer, tens of thousands of demonstrators turned out week upon week calling for the project to be abandoned."
Excerpt from Der Spiegel *Online, January 20 2010*[28]

3.4 Unpredictable causes

There will always remain, in any classification of political phenomena, a box for those abrupt changes that could not be expected, over which the project itself has absolutely no influence and which could ruin any perfectly run project. These events are the political equivalent of acts of God. The insurance industry defines the coverage of political risk in the following way: "Political risk insurance is coverage for business firms operating abroad to insure them against loss due to political upheavals including war, revolution, confiscation, incontrovertibility of currency, and other such losses."[29]

Venezuela, 2010
"Venezuela will nationalize a fleet of oil rigs belonging to U.S. company Helmerich and Payne, the latest takeover in a push to socialism as President Hugo Chavez struggles with lower oil output and a recession.

A former soldier inspired by Cuba's Fidel Castro, Chavez has made energy nationalization the linchpin in his 'revolution'. He has also taken over assets in telecommunications, power, steel and banking.

The 11 drilling rigs have been idle for months following a dispute over pending payments by the OPEC member's state oil company PDVSA. Oil Minister Rafael Ramirez said on Wednesday the rigs, the Oklahoma-based company's entire Venezuelan fleet, were being nationalized to bring them back into production."
Excerpt from Daniel, Frank Jack, *"Venezuela to nationalize U.S. firm's oil rigs"*, Reuters (June 24 2010)[30]

---

28 http://www.spiegel.de/international/germany/0,1518,732228,00.html.
29 www.allbusiness.com/political-risk-insurance/4971893-1.html.
30 www.reuters.com/article/idUSTRE65N0UM20100624.

Except for preparing world-class emergency plans to protect people and property, and taking out insurance, there is little the project developers can do.[31] The worst headaches for any project manager are caused by such risks: events that are completely unexpected and that could hardly be planned for. Those range from military *coups* to peaceful demonstrations. One of the world's most oil-rich countries, Nigeria, saw approximately 10 *coups* and *coup* attempts between 1964 and 2000.[32] *Coups* and *coup* attempts have been more widespread than most people think and they are not a thing of the past in some regions of the world (see Table 3).

Table 3: *Coups d'état* and attempted *coups* 1960-2009

	Asia	Sub-Saharan Africa	North Africa and Middle East	Latin America	Europe, Commonwealth of Independent States and Turkey
1960-1969	6	15	7	10	4
1970-1979	9	11	3	6	5
1980-1989	6	13	1	10	2
1990-1999	2	7	3	9	5
2000-2009	10	16	0	5	1

Author's own compilation, based on data from Wikipedia[33]

At the other end of the political regime spectrum, in Western-style democracies the fortune and future of a project that is favoured and advocated by a government that completely misjudges the sentiments of its own population might be just as dramatically affected if citizens oppose that project. For example, the European Union decided to make carbon capture and storage (CCS) a priority measure to reduce carbon dioxide emissions. A major energy company in Germany decided to avail of the available subsidies and to act before mandatory regulations were implemented, and commenced a project to build a large coal-fired power plant with CCS. However, the project eventually had to be cancelled, since no local government was willing to issue permits for storage of carbon dioxide, despite all scientific evidence of safety.

---

31 This article focuses on such events where project management can make some difference, although admittedly the unexpected events are at the core of what has been commonly known as political risk.
32 The exact number depends on definition. Failed *coups* and assassination of the ruler, among other things, have been included here.
33 en.wikipedia.org/wiki/List_of_coups_d'%C3%A9tat_and_coup_attempts.

For the project sponsor referred to above, financial loss was limited. In addition, there could hardly be any reputational consequences, because the sponsor was prevented from developing an environmental project favoured by the European Union. The cancellation is also highly unlikely to harm the energy company's relationship with the German government – it may possibly even have boosted the authorities' sympathy for that company. However, a situation where a government cannot enforce its own decisions creates uncertainty and difficulties for the industry, and future investments in such projects become unlikely.

4. Preventing and mitigating political risk

The main focus of this chapter is the description of political risks and their causes. What project developers can and should do to mitigate or prevent such risks is, of course, a key consideration. However, there is room for only a superficial description of such consideration in this chapter.

Many of the large-scale, cross-national projects on which this chapter focuses are, by necessity, joint ventures. The need for a modicum of competence and experience, financial strength and relationship with governments dictates this. The partners in the joint venture may all have a very solid basis for their own business operations in the form of values and attitude, as well as operational procedures both for communicating with the external world and for quick actions to solve problems. However, a majority of joint ventures fail. This is because they typically give insufficient consideration to their common approach to handling challenges. Therefore, project joint ventures should at the outset discuss, and agree on, common values, organisational practices and requirements, relationships with the outside world and exit strategies.

4.1 Values and attitudes

As demonstrated above, political challenges and risks for a project frequently stem from the project itself. A necessary element in avoiding political risk is to build project-specific values and attitudes among management and staff at all levels. A project where values such as openness, respect and honesty take centre stage and where dialogue and cross-cultural communication are commonplace will be in a good position when talking to the outside world, and when trying to avoid mistakes in analysing or handling political issues. Whether the issue is dealing with corruption or dialoguing with environmental non-governmental organisations, the values distilled to project staff will be decisive in keeping problems under control.

4.2 Organisation and competence

The chief executive officer of a global oil company once said: "There are two core competences in my company: geology and politics." Gas pipelines are said to comprise only two ingredients: 10% steel and 90% politics. Unfortunately, few companies take any notice of this saying when developing their knowledge base. While engineering, finance and law are considered tangible sciences and professions, political science, government relations and stakeholder management do not enjoy the same position with corporate decision makers.

Thus, while the recruitment of technical and commercial professionals is considered to warrant an intent consideration of the applicants' qualifications, many managers believe that everyone can handle political issues as this is seen as requiring only common sense. What many project developers learn the hard way is that the role of the softer side of project management ((health, safety and environment, corporate governance, corporate social responsibility, government affairs, stakeholder relations, media relations) is crucial. Expertise, experience, reputation and seniority in these disciplines are key elements to the success of all large-scale projects.

4.3   **External communication**

The importance of external relations can be expressed as simply as this: a project that does not enjoy easy access to, and trust from, government, stakeholders and the media will fail. Regardless of whether one calls this lobbying, public affairs, diplomacy or government relations, it is a key component. Project managers need to possess the values, competence, willingness and, not least, backing and mandate from the project owners to engage in such activities. Unfortunately, the centralised management style in fashion in many global companies causes them to leave little room for manoeuvre for project management: external communications and profiling are kept under the strict control of corporate headquarters and very close to the chief executive officer – and even more so in joint ventures. This means that many project managers go into political battle with a handicap, unless the project ranks at the top of the corporate agenda.

A final word on flexibility: most projects will at one stage face pressure to change their scope. Few pipelines and roads end up exactly where the line was first drawn on the map. Industrial plant designs are often altered to account for neighbours' complaints. Airport runways are moved because of noise. This chapter provides a number of such examples. The challenge for project managers is therefore to strike a balance between intertwined interests: to fulfil the task given by the owners and, at the same time, to exercise sufficient flexibility and willingness to listen to political signals at an early date. The author therefore strongly advocates giving the managers of large projects – and particularly those that are joint ventures – a separate and clear identity and profile, and a mandate empowering them to fight for their task – that is, to complete their own project on time, on budget and on specifications.

Multilateral and bilateral investment agreements

Tom Cummins
Ronnie King
Ashurst LLP

1. Introduction

The infrastructure necessary to ensure the production, transport and use of energy products requires enormous commitments of capital. Facilities are typically long term and immovable. Creation of a stable legal regime for investors is essential for the financing and development of projects.

One of the pervasive themes of this book is the distance between the source and use of energy products. Connecting areas of production with centres of demand constitutes a pressing geopolitical issue, especially in Western Europe, where it has been estimated that flagship gas pipeline projects will require €20 billion of commercial debt.[1] Commitment of financing to these projects will require a high degree of confidence on the part of investors and funders that their investments will be secure over the duration of the pipelines' operations. This need for security of investments is heightened by the strategic significance of energy resources and the political pressures that governments in regions rich in such resources may bring to bear upon investors.

Transboundary issues can arise in the energy sector in any circumstances where infrastructure and transit routes span several states. Problems may also develop where resources are exploited in areas subject to claims to sovereignty by more than one state. Looking further ahead, proposals for electricity supergrids to transport solar and wind-sourced energy between nations emphasise that transboundary issues will long endure in our evolving energy economy.

The nature of transboundary energy transportation and transmission gives rise to concerns that do not occur in other energy infrastructure projects. An investor in a power plant in an overseas state has only one host government to deal with. The power plant operator is also typically concerned only with the tariff rate that it receives and the tax regime in that state. The investor will wish to receive a return on its investment. The host state will want power to be supplied to industrial and commercial consumers. Although the host state's view of the bargain struck with the investor may change once the upfront construction risks have been overcome and production has commenced, there is a commonality of interest in ensuring that power production continues.

By contrast, a transboundary pipeline project may involve states that have no direct interest in the use of the infrastructure. A transit state may gain benefit from

---

1 "Financing Europe's future pipelines", *Infrastructure Journal* (August 4 2010).

the pipeline only through transit fees. Subject to different political and commercial imperatives from states that supply and use the transported product, the transit state may be more willing to exert pressure on investors to obtain more favourable commercial terms during the lifetime of the project. Issues may also arise as a result of political tensions between neighbours that are not present in a one-state infrastructure investment.

In this chapter we consider the protections available against governmental interference to investors in transboundary energy infrastructure. We explain the network of multilateral and bilateral investment treaties that exists and the means for the resolution of investment disputes.

2. **Options available to investors**

In broad terms, states offer four options to protect investments: investment protection legislation, investment contracts, bilateral investment treaties and multilateral investment treaties.

Legislation adopted to incentivise investment may provide exemptions for investors from paying certain taxes. A dedicated fiscal regime for investors in a particular industry might be offered – for example, for the exploitation of oil reserves. The concern with reliance on such legislation is that it may be only temporary. A government committed to encouraging investment and that enacts legislation for this purpose may be replaced by a government with different priorities.

Investors may enter into investment contracts with states. Concession agreements or production-sharing contracts may contain investor protection provisions such as stabilisation clauses intended to protect investors from actions by the state that adversely affect the value of their investments. The effectiveness of these clauses, which frequently seek to freeze the legislative position at the date of the contract, is variable. A government may still pass legislation in breach of a stabilisation clause. This may leave the investor with limited avenues of redress.

In the context of energy infrastructure projects, host government agreements and intergovernmental agreements may be entered into. These provide for a stable legal regime for a specific project. By way of example, an intergovernmental agreement providing for the development of the Nabucco gas pipeline from Asia to Europe was signed in July 2009. It guarantees political support from the governments of countries through which the pipeline will transit and is valid for 50 years.

Bilateral investment treaties are agreements between two states to protect and promote investments by investors of each state in each other's jurisdictions. Multilateral investment treaties involve more than two states entering into such agreements.

As states compete for foreign direct investment, these instruments have become increasingly popular. The offer of a suite of protections for investments makes host states more attractive as destinations for capital. For energy investors, these instruments may provide long-term certainty throughout project concept selection and planning, construction and operation and, lastly (but importantly), decommissioning.

Host government agreements and intergovernmental agreements are the subject of a later chapter in this book. This chapter focuses on bilateral investment treaties and multilateral investment treaties.

3. Bilateral investment treaties

3.1 Introduction

The law and practice of bilateral investment treaties developed gradually as international trade developed. Agreements to facilitate trade and commercial relationships between nationals of different countries have existed for hundreds of years. As global trade increased and became more sophisticated, treaties providing protections for investments (rather than simply encouraging trade) became common. At the same time, arbitration gained in popularity as a means of settling international disputes in a neutral forum.

In 1966 the International Centre for Settlement of Investment Dispute (ICSID) Convention, a multilateral treaty promoted by the World Bank, came into force. The convention is the culmination of efforts by investor states in the wake of the highly publicised nationalisation of BP's investments in Iran during the 1950s. It established the ICSID, which is dedicated to the conciliation and arbitration of international investment disputes. Following the establishment of the ICSID, bilateral investment treaties providing for arbitration directly between investors and host states multiplied and are now the norm. The convention has thus been a significant factor in the huge increase in bilateral investment treaties in recent decades.

There are now over 2,600 bilateral investment treaties and more than 270 other types of bilateral investment agreement, such as economic partnership agreements or free trade agreements (variations on bilateral investment treaties containing many of the same protections). All countries are party to at least one bilateral investment treaty or similar agreement.[2] The leading signatories to bilateral investment treaties are set out in the graph overleaf.

3.2 An overview of bilateral investment treaties

Many bilateral investment treaties offer substantially the same protections. They are usually short and contain 10 or so key provisions. Some states have published model bilateral investment treaties that form the basis of the treaties into which they enter, such as the 2004 US Model Bilateral Investment Treaty.

Claims arising under bilateral investment treaties have produced a significant body of international arbitration decisions. Key trends that emerge from those decisions are identified below and briefly discussed. The precise wording of the treaties does differ, however. In addition, awards of investment tribunals are not binding on subsequent tribunals and varying approaches of tribunals are a feature of the decisions, perhaps reflecting the political nature of the issues that they are dealing with. Investors should appreciate that bilateral investment treaties are not a

---

2   UN Commission on Trade and Development, "The Role of International Investment Agreements in Attracting Foreign Direct Investment to Developing Countries" (2009) (www.unctad.org/Templates/Download.asp?docid=12543&lang=1&intItemID=2068). The only known exception is Monaco.

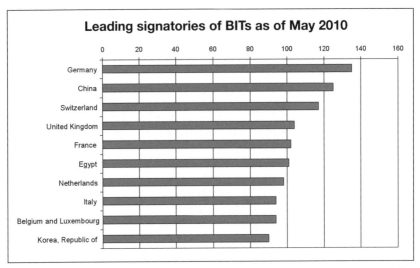

Source: World Investment Report 2010 Investing in a low-carbon economy, UN Commission on Trade and Development, 2010

panacea. Investors carry the burden of proving the value of their investments in cases where they allege host state interference. In circumstances where contracts are in the pre-operational stage, it may prove difficult to discharge the burden of proof to secure compensation for the substantial sums already invested.

What follows is a review of some standard provisions found in bilateral investment treaties. Example wording from bilateral investment treaties is used to illustrate the legal issues that may arise.

(a) **Preamble**

The Government of the United States of America and the Government of [Country] (hereinafter the "Parties");

Desiring to promote greater economic cooperation between them with respect to investment by nationals and enterprises of one Party in the territory of the other Party;

Recognizing that agreement on the treatment to be accorded such investment will stimulate the flow of private capital and the economic development of the Parties;

Agreeing that a stable framework for investment will maximize effective utilization of economic resources and improve living standards;

Recognizing the importance of providing effective means of asserting claims and enforcing rights with respect to investment under national law as well as through international arbitration;

Desiring to achieve these objectives in a manner consistent with the protection of health, safety, and the environment, and the promotion of internationally recognized labor rights;

Having resolved to conclude a Treaty concerning the encouragement and reciprocal protection of investment;

Have agreed as follows:[3]

Most bilateral investment treaties commence with a preamble. This performs a similar role to the recitals in a contract between private parties. It sets out the broad aims of the document and the factors that the parties have taken into consideration.

Investors considering bilateral investment treaties often do not dwell on the wording of the preamble because it does not set out substantive rights or obligations. However, it is not irrelevant; it may be used to interpret the substantive provisions.[4] Tribunals have frequently used preambles for the purposes of interpretation, seeking to strike a balance between giving effect to the strict wording of the substantive provisions in the treaty and the intent expressed in the preambles.[5]

The excerpt from the US Model Bilateral Investment Treaty above refers to the achievement of investment protection "in a manner consistent with the protection of health, safety, and the environment".[6] If an investor acts in a way that is inconsistent with, for example, the protection of the environment, a host state may argue that a substantive protection must be qualified by the preamble. In light of the criticisms often levelled on environmental grounds at large-scale energy infrastructure projects such as pipelines, the wording of the preamble may assume considerable significance. Investors should be conscious that bilateral investment treaties may also expressly exempt actions taken for environmental or public health reasons from the protections provided. Tribunals may be called on to determine whether a state action, ostensibly taken for reasons of environmental protection, is in reality a form of expropriation designed to undermine the value of the foreign investment.

(b) Definition of 'investment'

> *(a) "investment" means every kind of asset and in particular, though not exclusively, includes:*
>
> > *(i) movable and immovable property and any other property rights such as mortgages, liens or pledges;*
> >
> > *(ii) shares in and stock and debentures of a company and any other form of participation in a company;*
> >
> > *(iii) claims to money or to any performance under contract having a financial value;*
> >
> > *(iv) intellectual property rights, goodwill, technical processes and know-how;*
> >
> > *(v) business concessions conferred by law or under contract, including concessions to search for, cultivate, extract, or exploit natural resources.*[7]

---

3 Treaty between the Government of the United States of America and the Government of [ ] Concerning the Encouragement and Reciprocal Protection of Investment, 2004 US Model Bilateral Investment Treaty (www.unctad.org/sections/dite/iia/docs/Compendium//en/model_USA.pdf).

4 See Article 31 of the Vienna Convention on the Law of Treaties 1969 (untreaty.un.org/ilc/texts/instruments/english/conventions/1_1_1969.pdf).

5 See, for example, *Plama Consortium Limited v Republic of Bulgaria* (decision on jurisdiction, February 8 2005).

6 Indeed, Annex B to the US Model Bilateral Investment Treaty provides that "except in rare circumstances, non-discriminatory regulatory actions by a Party that are designed and applied to protect legitimate public welfare objectives, such as public health, safety and the environment, do not constitute indirect expropriations".

7 Agreement between the Government of the United Kingdom of Great Britain and Northern Ireland and the Government of the Azerbaijan Republic for the Promotion and Protection of Investments, January 4 1996 (www.unctad.org/sections/dite/iia/docs/bits/uk_azerbaijan.pdf).

'Investment' is often defined broadly in bilateral investment treaties. The UK-Azerbaijan Treaty excerpted above is an example of this. It provides protection for "every kind of asset". This includes both the more obvious categories of asset, such as physical property, and contractual rights, such as those for the exploration and development of oil and gas reserves. Many bilateral investment treaties entered into during the course of the past decade or so have adopted a similar approach.

Other treaties define investments more restrictively and seek to limit claims that may be brought by investors. So, for example, the Sweden-Latvia Treaty expressly requires that an investment be "made in accordance with the laws and regulations of the other Contracting Party" (ie, the host state).[8] The US Model Bilateral Investment Treaty requires that an investment be one that "an investor owns or controls, directly or indirectly, that has the characteristics of an investment, including such characteristics as the commitment of capital or other resources, the expectation of gain or profit, or the assumption of risk".

Tribunals have sought to distinguish between investments and ordinary commercial transactions that do not benefit from protection under a bilateral investment treaty.[9] In doing so they have adopted similar tests to those used to determine whether an investment exists for the purposes of the ICSID Convention (see section 4.2 below). However, any investment under a bilateral treaty must be considered by reference to the precise definition contained in the treaty.

Certain decisions of tribunals may have special relevance to investors in energy infrastructure. In *Mihaly International Corporation v Democratic Socialist Republic of Sri Lanka* (ICSID Case ARB/00/2), expenditure incurred *prior* to an investment in the development of a power station in Sri Lanka was held not to be an investment. The power station development never progressed and the tribunal stated that it could find no evidence in investment treaty practice that "pre-investment and development expenditures ... could automatically be admitted as 'investment' in the absence of the consent of the host State to the implementation of the project".

In *Romak SA v Republic of Uzbekistan* (PCA Case AA280), the tribunal found that "rights ... embodied in and [that] arise out of a sales contract, a one-off commercial transaction pursuant to which [the claimant] undertook to deliver wheat against a price to be paid by the Uzbek parties" were not an investment. This decision emphasises that tribunals are alive to the possibility of claimants seeking to take advantage of protections under bilateral investment treaties in circumstances in which they can have no application, such as where claims are brought in respect of commercial transactions rather than investments. Accordingly, rights under a contract that does not itself meet recognisable characteristics of an investment do not qualify for protection.

One should be aware that unless a bilateral investment treaty expressly provides otherwise, the definition of 'investment' extends not only to investments in the form of contracts entered into directly between a host state and an investor, but also

---

8   Sweden and Latvia Agreement on the Promotion and Reciprocal Protection of Investments, September 19 1994 (www.unctad.org/sections/dite/iia/docs/bits/sweden_latvia.pdf).

9   See, for example, the decision of an *ad hoc* committee (a body convened to review an award) in *Mitchell v Democratic Republic of Congo* (ICSID Case ARB/99/7 (decision on annulment)).

to contracts between private parties, such as a contract between a power company and a contractor for the construction of a pipeline. If the host state interferes with the performance of the contract – for example, by confiscating the contractor's equipment – then the parties may have claims under a bilateral investment treaty.

(c)   *Definition of 'investor'*

> The term "investor" shall comprise with regard to either Contracting Party:
>
> (i) natural persons having the citizenship or nationality of that Contracting Party in accordance with its laws;
>
> (ii) any corporations, companies, firms, enterprises, organisations and associations incorporated or constituted under the law in force in the territory of that Contracting Party;
>
> provided that that natural person, corporation, company, firm, enterprise, organisation or association is competent, in accordance with the laws of that Contracting Party, to make investments in the territory of the other Contracting Party.[10]

Bilateral investment treaties apply only to investments made by investors of one party to the treaty in the territory of the other party to the treaty. The issue of whether an investor falls within the definition provided in the relevant treaty has given rise to a large number of disputes. As indicated above in the UK-Russia Treaty, the requirements for investors to constitute 'investors' for the purposes of the treaty differ depending on whether the investor is a natural person or a company.

The criteria for determining whether an investor is protected by a bilateral investment treaty differ between treaties. The definition in the UK-Russia Treaty is broad. Other treaties adopt more restrictive language. The France-Nigeria Treaty, by way of example, defines an 'investor' as "any national or any legal person constituted in the territory of one Contracting Party in accordance with the legislation of that Party, having its head office on the territory of that Party, or controlled directly or indirectly by the nationals of one Contracting Party, or by legal persons having their head office in the territory of one Contracting Party and constituted in accordance with the legislation of that Party".[11] Other treaties require that an investor carry out "effective economic activities" in the territory of the state whose nationality it claims.[12]

The issue of control or ownership, in the context of defining an 'investor', is an important one. Project companies are often set up in host states in order to fulfil a requirement imposed by the host state. They therefore have the same nationality as the host state. This means that they would not be entitled to the protection of the

---

10   Agreement between the Government of the United Kingdom of Great Britain and Northern Ireland and the Government of the Union of Soviet Socialist Republics (now Russian Federation) for the Promotion and Reciprocal Protection of Investments, April 6 1989 (www.unctad.org/sections/dite/iia/docs/bits/uk_ussr.pdf).

11   Agreement on the Reciprocal Promotion and Protection of Investments between France and Nigeria, February 27 1990 (untreaty.un.org/unts/60001_120000/30/8/00058384.pdf).

12   See, for example, Agreement between the Government of the Republic of Korea and the Government of the Republic of Chile on the Reciprocal Promotion and Protection of Investments, September 6 1996 (www.unctad.org/sections/dite/iia/docs/bits/chile_korea.pdf).

bilateral investment treaty, in the absence of express provision. For this reason many treaties refer expressly to direct or indirect control by nationals of one contracting party as sufficient for a company to qualify as an investor. If so, a project company that is incorporated in the state in which infrastructure is to be constructed and that is owned by overseas investors would be entitled to the protection of any relevant bilateral investment treaty.

Many bilateral investment treaties contain so-called 'denial of benefits' provisions. These operate to permit a host state to deny the benefits of a treaty to a company that is controlled by investors of a non-party to the treaty. These provisions are aimed at preventing so-called 'letterbox companies', which have no substantial business activities in the contracting state, from taking advantage of treaty provisions.

Structuring investments so that vehicles are incorporated in specific jurisdictions for the purposes of obtaining treaty protections is not unusual. It is often done for taxation purposes. It is equally important that treaty protections be considered at the same time.

(d) *Specific investor protections*

The paragraphs below deal with the substantive protections that are typically provided in investment treaties and that form the basis of an individual claim. As a preface to this section, it should be noted that many of these protections will be seen to overlap. Investment treaty jurisprudence has commonly found certain state acts to be breaches of several categories of protection.

Fair and equitable treatment:

Investments of nationals or companies of either Contracting Party shall at all times be accorded fair and equitable treatment and shall enjoy full protection and security in the territory of the other Contracting Party. Neither Contracting Party shall in any way impair by unreasonable or discriminatory measures the management, maintenance, use, enjoyment or disposal of investments in its territory of nationals or companies of the other Contracting Party. Each Contracting Party shall observe any obligation it may have entered into with regard to investments of nationals or companies of the other Contracting Party.[13]

Providing fair and equitable treatment to investors is one of the core substantive protections found in bilateral investment treaties. However, it is hard to define precisely what 'fair and equitable treatment' is. For this reason, investors often include claims for breaches of the duty of fair and equitable treatment in disputes under bilateral investment treaties in addition to claims brought under more precisely formulated provisions. The duty is something of a catch-all protection.

Claims for breach of the obligation to provide fair and equitable treatment often arise in circumstances where investors consider that the courts of a host state have treated them unfairly. This may involve courts failing to consider claims brought by

---

13   Agreement for the Promotion and Protection of Investments between the United Kingdom of Great Britain and Northern Ireland and Malaysia, May 21 1981 (untreaty.un.org/unts/60001_120000/27/6/00052272.pdf).

investors or dispensing justice in a discriminatory way. In *Mondev International Ltd v United States of America* (ICSID Case ARB(AF)99/2), a claim brought by Canadian investors under the North American Free Trade Agreement, the tribunal stated that:

> *the test is not whether a particular result is surprising, but whether the shock or surprise occasioned to an impartial tribunal leads, on reflection, to justified concerns as to the judicial propriety of the outcome ... In the end the question is whether, at an international level and having regard to generally accepted standards of the administration of justice, a tribunal can conclude in the light of all the available facts that the impugned decision was clearly improper and discreditable, with the result that the investment has been subjected to unfair and inequitable treatment.*

Claims have also been brought where host state authorities have made unfair administrative decisions. This may be because the legitimate expectations of investors have not been met or the host state has acted inconsistently. In *Saluka Investments BV v The Czech Republic* (UN Commission on International Trade Law (UNCITRAL) arbitration), the tribunal held that the protection of legitimate expectations was the "dominant element" of fair and equitable treatment. However, the tribunal cautioned that "no investor may reasonably expect that the circumstances prevailing at the time the investment is made remain totally unchanged. In order to determine whether frustration of the foreign investor's expectations was justified and reasonable, the host State's legitimate right subsequently to regulate domestic matters in the public interest must be taken into consideration as well".

It is recognised that host states may be entitled to make changes to the business environment within which an investor operates. However, if specific representations have been made to investors, legislative or regulatory changes should not be made contrary to these representations.

Most favoured nation/national treatment: The obligation to provide fair and equitable treatment places an absolute obligation on the host state. Bilateral investment treaties also often provide for relative standards of treatment for investors. A most favoured nation provision ensures that an investor is treated no less favourably than other external investors. National treatment provisions ensure that investors are treated no less favourably than nationals of the host state. As in the excerpt below from the US Model Bilateral Investment Treaty, these protections are frequently contained in the same provision:

> *With respect to the establishment, acquisition, expansion, management, conduct, operation and sale or other disposition of covered investments, each Party shall accord treatment no less favorable than that it accords, in like situations, to investments in its territory of its own nationals or companies (hereinafter "national treatment") or to investments in its territory of nationals or companies of a third country (hereinafter "most favored nation treatment"), whichever is most favorable (hereinafter "national most favored nation treatment"). Each Party shall ensure that its state enterprises, in the provision of their goods or services, accord national and most favored nation treatment to covered investments.*

The most favoured nation provision is a powerful tool for investors. It enables them to take advantage of favourable provisions in other bilateral investment treaties entered into by the host state. In *MTD Equity Sdn Bhd v Republic of Chile* (ICSID Case ARB/01/7), Malaysian investors incurred significant expenditure seeking to establish a planned community in Chile. Following the refusal of permits for the community, the investors relied on most favoured nation provisions in the Chile-Croatia and Chile-Denmark bilateral investment treaties that were more extensive than those found in the Chile-Malaysia Treaty.

In *Maffezini v the Kingdom of Spain* (ICSID Case ARB/97/7), an Argentine investor failed to comply with a complicated dispute resolution clause in the Argentina-Spain treaty. The clause required disputes to be submitted to Spanish courts prior to arbitration. The investor did not do so and relied on the simpler procedure in the Chile-Spain treaty. The tribunal held that he was entitled to do so. This decision has been criticised and tribunals have adopted differing approaches to whether procedural rights, as distinct from substantive rights, in other bilateral investment treaties can be invoked under most favoured nation provisions.

Expropriation:
> *Investments of nationals or companies of either Contracting Party shall not be nationalised, expropriated or subjected to measures having effect equivalent to nationalisation or expropriation (hereinafter referred to as "expropriation") in the territory of the other Contracting Party except for a public purpose related to the internal needs of that Party on a non-discriminatory basis and against prompt, adequate and effective compensation.*[14]

It is a misconception that bilateral investment treaties do not allow expropriation. In fact, they typically recognise that expropriation may occur and provide for the levels of compensation to which the investor is then entitled. As the excerpt above illustrates, expropriation will often be permissible if it is justified and non-discriminatory, and if prompt, adequate and effective compensation is paid to the investor.

Expropriation takes different forms. Direct expropriation might involve the seizure of an asset by the government – for example, gas processing facilities. Indirect expropriation might involve measures by a host state depriving the investor of the economic benefit of its investment. Another example of indirect expropriation might be amendments to the fiscal terms in order to seek a larger share of transit fees from a pipeline. Attention has been paid in recent years to so-called 'creeping expropriation', where host states gradually introduce measures that have the effect of depriving the investor of the benefit of its investment over time.

Compensation for expropriation is available only, in principle, if the investor has been substantially deprived of its rights (eg, *Saipem SpA v The People's Republic of Bangladesh* (ICSID Case ARB/05/7)). Whether a substantial deprivation has occurred will be fact specific.

---

14 Agreement between the Government of the United Kingdom and the Government of Turkmenistan for the Promotion and Protection of Investments, February 9 1995 (www.unctad.org/sections/dite/iia/docs/bits/uk_turkmenistan.pdf).

In *Generation Ukraine Inc v Ukraine* (ICSID Case ARB/00/9), the tribunal stated that "creeping expropriation is a form of indirect expropriation with a distinctive temporal quality in the sense that it encapsulates the situation whereby a series of acts attributable to the State over a period of time culminate in the expropriatory taking of such property". In *Walter Bau AG v The Kingdom of Thailand* (UNCITRAL arbitration), the tribunal stated that "a strong interference with contractual rights needs to be shown". It found that no creeping expropriation had occurred in a claim brought by a shareholder in a tollroad, noting that the host state's "alleged misdeeds ... were inaction rather than affirmative action".

Protection of investments: Bilateral investment treaties usually require host states to offer full security and protection to investors. An example of such wording is provided in the fair and equitable treatment clause excerpted above. Although the precise formulations differ, tribunals have found that host states are obliged to exercise due diligence and to be vigilant with regard to the physical protection of investments. The obligation to provide physical protection for the assets of foreigners is the most historic of the rights recognised in bilateral investment treaties and long pre-dates them. In *Asian Agriculture Products Ltd v Democratic Republic of Sri Lanka* (ICSID Case ARB/87/3), the Sri Lankan government was found liable for failing to take precautionary measures before launching an attack that destroyed the investor's farm.

Tribunals have taken different approaches as to whether a protective obligation extends to legal protection, as opposed to physical safeguarding. In *Azurix Corporation v The Argentine Republic* (ICSID Case ARB/01/12), the tribunal noted that full protection and security "is not only a matter of physical security; the stability afforded by a secure investment environment is as important from an investor's point of view". It found that Argentina had failed to protect adequately Azurix's investment in water and sewerage assets.

Free transfer of funds:

Each Contracting Party shall in respect of investments guarantee to nationals or companies of the other Contracting Party the unrestricted transfer of their investments and returns. Transfers shall be effected without delay in the convertible currency in which the capital was originally invested or in any other convertible currency agreed by the investor and the Contracting Party concerned. Unless otherwise agreed by the investor transfers shall be made at the rate of exchange applicable on the date of transfer pursuant to the exchange regulations in force.[15]

Investors are often anxious that they will be unable to transfer returns made from their investments to bank accounts outside the host state. Most bilateral investment treaties contain provision for this. In some treaties the protection provided is qualified by the application of the laws of the host state for the purposes of

---

15 Agreement between the Government of the United Kingdom and the Government of Turkmenistan for the Promotion and Protection of Investments, February 9 1995 (www.unctad.org/sections/dite/iia/docs/bits/uk_turkmenistan.pdf).

protecting creditors' rights (eg, on insolvency of an investor), enforcement and satisfaction of judgments and the collection of taxes.

Umbrella clauses:
Either Contracting Party shall observe any other obligation it may have entered into with regard to investments in its territory by investors of the other Contracting Party.[16]

Under 'umbrella' (or 'observance of undertaking') clauses, the host state agrees to observe *any* obligation that it has entered into with regard to investments in its territory. Such clauses can radically broaden the scope of what can be claimed under a bilateral investment treaty. If a host state has entered into an agreement with an investor and breaches that agreement, an umbrella clause may have the effect of elevating a contractual claim into a bilateral investment treaty claim. The United Nations has estimated that around 40% of bilateral investment treaties in force contain umbrella clauses or clauses with equivalent effect.[17]

One of the most discussed issues in investment treaty arbitration is whether umbrella clauses elevate all breaches of contract by a state into breaches of bilateral investment treaties. Investors frequently argue that they do. The approach of tribunals has been inconsistent. In *SGS Société Générale de Surveillance SA v Islamic Republic of Pakistan* (ICSID Case ARB/01/13), a tribunal refused to elevate a breach of contract into a breach of the relevant bilateral investment treaty because to do so would have consequences "so far-reaching in scope, and so automatic and unqualified and sweeping in their operation [and] so burdensome in their potential impact upon a Contracting Party". The tribunal held that "clear and convincing evidence" would have to be adduced to indicate that this was the intention of the parties to the bilateral investment treaty.[18]

Conversely, in *SGS Société Générale de Surveillance SA v Republic of the Philippines* (ICSID Case ARB/02/6), the tribunal took the contrary position: it considered that the wording of the bilateral investment treaty was sufficiently clear for it to take the view that a failure by the Philippines to observe a contractual commitment amounted to a breach of the treaty.[19]

Subsequent tribunals have followed the approach taken in both *SGS* cases. However, some commentators and practitioners consider that interpreting the clear meaning of umbrella clauses, in the manner of *SGS/Philippines*, is the more appropriate course. *SGS/Pakistan* has been criticised as being overly restrictive. A credible counterargument is that typically the state will be alleged to have breached an obligation under a contract

---

16 Agreement between the Government of the Republic of Korea and the Government of the Republic of South Africa on the Promotion and Protection of Investments, July 7 1995 (www.unctad.org/sections/dite/iia/docs/bits/korea_southafrica.pdf).
17 UN Commission on Trade and Development, "Bilateral Investment Treaties 1995-2006: Trends in Investment Rulemaking" (2007) (www.unctad.org/en/docs/iteiia20065_en.pdf).
18 The relevant wording in the Switzerland-Pakistan Treaty was: "Either Contracting Party shall constantly guarantee the observance of the commitments it has entered into with respect to the investments of the investors of the other Contracting Party."
19 The relevant wording in the Switzerland-Philippines Treaty was: "Each Contracting Party shall observe any obligation it has assumed with regard to specific investments in its territory by investors of the other Contracting Party."

that contains its own dispute resolution mechanism (typically, arbitration). The dispute resolution clause in the underlying contract invariably stipulates that all disputes arising under the contract be resolved by following the contractual dispute resolution mechanism. Therefore, a bilateral investment treaty claim that bases itself on an assertion that the state has broken the underlying contract may be considered to breach the dispute resolution procedure under that contract. This has led to the development of a view that only breaches which go beyond alleged failure of contractual performance by the state by involving state action directed at depriving the investor of the value of its investment justify a bilateral investment treaty claim.

In *Sempra Energy International v The Argentine Republic* (ICSID Case ARB/02/16), the tribunal "distinguished breaches of contract from Treaty breaches on the basis of whether the breach has arisen from the conduct of an ordinary contract party, or rather involves a kind of conduct that only a sovereign State function or power could effect". The tribunal noted that the distinction may not always be clear.

Bilateral investment treaties may contain other protections in addition to those discussed above. As previously noted, they also provide for dispute resolution, usually by means of international arbitration. We consider this below.

4. Dispute resolution under investment treaties

4.1 Introduction

One of the most important elements of investment treaties is their provision for arbitration in a neutral setting between an investor and a host state. Investment treaties often offer a number of different arbitration options, usually including arbitration under the ICSID rules,[20] which are tailored to investment disputes. Any award made by an ICSID tribunal is enforceable in a state that has ratified the ICSID Convention as if it were a judgment of a local court. It should be appreciated, however, that successful enforcement frequently requires the cooperation of the host state; this tends to be forthcoming only if the host state has a pressing need to secure international assistance in the form of either loans or the occurrence of some political development that necessitates the satisfaction of treaty awards against the state.

Over 140 states have ratified the ICSID Convention.[21] Since the first case in 1972, more than 300 cases have been registered with the ICSID, with a marked increase in new cases since 2000. Over 60% of those cases are bilateral investment treaty claims. Energy Charter Treaty claims make up another 4% (see below). The popularity of ICSID arbitration in energy investment disputes is reflected in the fact that 39% of registered cases have arisen in the oil, gas, mining, power and other energy sectors.[22]

---

20 Arbitrations under the rules of UNCITRAL, the Stockholm Chamber of Commerce and the International Chamber of Commerce are also popular.
21 Note the important distinction between signature and ratification. A significant number of states have signed the International Centre for Settlement of Investment Dispute Convention but have not ratified it. Perhaps the most high profile is Canada, the only member of the G8 and one of only three members of the Organisation for Economic Cooperation and Development not to have ratified the convention (the others being Poland and Mexico, which have not signed the convention).
22 ICSID, "The ICSID Caseload – Statistics (2010-2)" (icsid.worldbank.org/ICSID/FrontServlet?requestType=ICSIDDocRH&actionVal=CaseLoadStatistics).

Prior to commencing arbitration, bilateral investment treaties often require parties to participate in good-faith negotiations as part of a minimum defined period of consultation. Some treaties contain a so-called 'fork in the road' provision, enabling a party to commence a claim in the local courts of the host state in preference to international arbitration. By doing so, the party waives its right to submit the same claim to arbitration.

4.2 The requirements for ICSID arbitration

The principal requirements for ICSID arbitration are that:
- the parties consent to ICSID arbitration in writing; and
- there exists a legal dispute arising from an investment between two ICSID contracting states (ie, states that have ratified the ICSID Convention).

Commonly, a host state consents to arbitration in its bilateral investment treaties. An investor then consents in writing by commencing a claim under the ICSID Rules.

The questions of what constitutes an investment and who is a national of another contracting state have proven controversial.

The ICSID Convention does not define an 'investment'. Tribunals have tended to use as a starting point the decision in *Salini Costruttori SpA v Kingdom of Morocco* (ICSID Case ARB/00/4), a claim arising from the construction of a Moroccan highway. Noting the lack of a definition of 'investment' in the convention, it was held that an investment must involve contributions by the investor, a certain duration of performance of the contract, a participation in the risks of the transaction and a contribution to the development of the host state. However, in recent years tribunals have resisted following an exclusive set of criteria. In *Biwater Gauff (Tanzania) Ltd v United Republic of Tanzania* (ICSID Case ARB/05/22), a case involving sanitation facilities in Tanzania, the tribunal advocated "a flexible and pragmatic approach".

In *Phoenix Action v The Czech Republic* (ICSID Case ARB/06/5), the tribunal reconsidered the so-called '*Salini* criteria' and noted that an investment made either in bad faith or illegally could not fall within the scope of the ICSID. However, in *Saba Fakes v Republic of Turkey* (ICSID Case ARB/07/20), which involved alleged expropriation of telecommunication assets, the tribunal considered the approaches adopted by tribunals and noted that some tribunals had added, to the four Salini criteria, the "requirement of a regularity of profit and return". The tribunal concluded that "the criteria of (i) a contribution, (ii) a certain duration, and (iii) an element of risk, are both necessary and sufficient to define an investment within the framework of the ICSID Convention". In the context of energy infrastructure projects, it might seem that it would be clear whether an investment exists. However, it may be that investors are entitled to ICSID arbitration as a result of investments in host states more subtle than physical infrastructure (eg, contractual rights).

ICSID arbitration is available only to nationals of a contracting state other than the contracting state in which the investment is made. However, this rule is qualified by the fact that parties to a dispute can agree that a national of a contracting state in which an investment is made be considered a national of another contracting state by virtue of foreign control. This provision is intended to apply where a state insists that a project company is incorporated within its jurisdiction. If the project company is controlled by nationals of a different contracting state, the parties can agree that it will be entitled to participate in ICSID arbitration.

The approaches of tribunals to the ICSID nationality requirement have varied. Thus, in *TSA Spectrum de Argentina SA v Argentine Republic* (ICSID Case ARB/05/5), a case concerning the privatisation of the Argentinian radio spectrum, the claimant argued that it should be treated as Dutch because its immediate parent company was Dutch. The tribunal concluded that even though the claimant's immediate parent company was Dutch, the ultimate owner was an Argentine national. Jurisdiction was denied. However, in *Tokios Tokelės v Ukraine* (ICSID Case ARB/02/18), the tribunal took a different approach. Even though nationals of Ukraine owned 99% of the claimant's shares and made up two-thirds of its management, the tribunal considered the claimant to be Lithuanian because it was incorporated there.

This variety of approach is reflected in other cases. As a result, it is difficult to ascertain principles upon which investors can rely as to when an investor will be considered to satisfy the nationality test. The paradigm case of an investment being made by a national of a contracting state whose nationality is that of his birth or of a corporate whose state of incorporation is also its place of management and major production centre remains the safest course.

4.3 Arbitration in the context of the investor-state relationship

Investors commonly use the threat of international arbitration proceedings as a means of renegotiating contracts with host states in circumstances where their rights have been affected. Unlike arbitration proceedings between private companies or individuals, investment treaty arbitration is often public and the subject of extensive commentary. The spectre of ICSID proceedings and their association with the World Bank often act as an incentive for host states to engage with investors, particularly if they fear that adverse publicity will affect their ability to obtain foreign investment.

5. Multilateral investment treaties

Multinational investment treaties are frequently concluded on a regional basis and are intended to promote trade and mutual investment. A number contain similar protections to those provided in bilateral investment treaties. Often, free trade agreements between states also contain investment protections. Important examples of multilateral investment treaties/free trade agreements include:
- the North American Free Trade Agreement, between Canada, Mexico and the United States;
- the Association of Southeast Asian Nations (ASEAN) Agreement for the Promotion and Protection of Investments, between Brunei, Cambodia, Indonesia, Laos, Malaysia, Myanmar, the Philippines, Singapore, Thailand and Vietnam;
- the Protocol of Colonia for the Reciprocal Promotion and Protection of Investment in MERCOSUR[23], between Argentina, Brazil, Paraguay and

---

23 Southern Common Market ('*Mercado Común del Sur*' in Spanish).

Uruguay; and
- the Cartagena Free Trade Agreement, between Columbia, Mexico and Venezuela.

Of greatest significance for investors in energy infrastructure, however, is the Energy Charter Treaty.

6. The Energy Charter Treaty

6.1 Introduction

The Energy Charter Treaty is the only multilateral investment treaty devoted to the energy sector. It has its origins in the early 1990s, following the collapse of the Soviet Union, when consideration was given to encouraging energy cooperation and the opening up of energy markets in Eurasia. Opened for signature in 1994, the treaty came into legal effect in April 1998. It covers:
- the promotion and protection of investments in energy;
- trade in energy, energy products and energy-related equipment;
- freedom of energy transit; and
- improvement of energy efficiency.

It also contains provisions for the resolution of disputes through international arbitration.

As at October 30 2010 51 countries had signed the treaty. It has also been signed collectively by the European Union and the European Atomic Energy Community. Signatory states that have ratified the treaty are referred to as 'contracting parties'. Of the signatories, only Australia, Belarus, Iceland and Norway have failed to ratify the treaty. Belarus applies the treaty provisionally; notably, so did Russia until October 18 2009, when it terminated its provisional application of the treaty and gave notification that it did not intend to become a full contracting party. The significance of Russia's actions and the meaning of provisional application are discussed further below.

In addition, a number of states, including the United States, are 'observer states'. This means that they are entitled to participate in certain aspects of the treaty without having any binding obligations thereunder or requirement to contribute to the treaty's budget. The map opposite illustrates the relationship of countries with the treaty. Signatory states are dark grey. Observer states are light grey. The observer states include the member states of ASEAN, which itself has observer status.

6.2 Investment protection under the Energy Charter Treaty

The Energy Charter Secretariat states that "the fundamental aim of the Energy Charter Treaty is to strengthen the rule of law on energy issues, by creating a level playing field of rules to be observed by all participating governments, thereby mitigating risks associated with energy-related investment and trade".[24]

The investment protection regime is contained in Part III of the treaty. It seeks to

---

24 www.encharter.org/index.php?id=7.

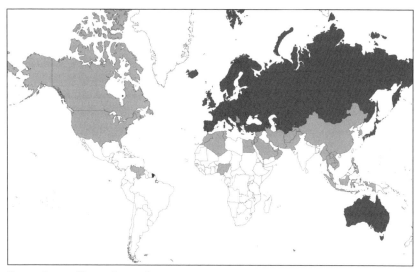

Source: Energy Charter Secretariat

ensure the protection of investments based on the principle of non-discrimination. During the treaty's preparation, its drafters did not wish to reinvent the wheel with regard to investment protection.[25] Accordingly, they adopted many of the protections contained in bilateral investment treaties, the effect and application of which have been considered above. The Energy Charter Treaty contains provisions for:

- fair and equitable treatment;
- protection from unreasonable or discriminatory treatment;
- national treatment and most favoured nation treatment;
- provision for compensation in the event of expropriation;
- transfer of capital; and
- full protection and security.

The obligation on contracting parties to avoid unreasonable and discriminatory treatment has application only once investments have been made. During the negotiation of the treaty, there were attempts to ensure that such treatment was afforded to potential investors during the pre-investment phase. This was seen as desirable in order that investors applying for licences or participating in tender processes could not be discriminated against in favour of indigenous competitors. Although agreement on this principle could not be reached, the treaty contains a mechanism to encourage such non-discrimination. The treaty draws a distinction between 'investments' (which are afforded full protection) and the 'making of investments' (defined as "establishing new Investments, acquiring all or part of existing Investments or moving into different fields of Investment activity"), in relation to which contracting parties are obliged to endeavour to accord to investors non-discriminatory treatment.

---

25   See Energy Charter Secretariat, "The Energy Charter Treaty – A Reader's Guide" (2002), section C.

The fact that treaty protection is not extended to the pre-investment phase reflects the ongoing debate as to what rights/expenditure should be protected under an investment treaty. As a generalisation, states that are typically investor states adopt a very wide approach, while host states adopt a narrower one. Furthermore, the Energy Charter Treaty obliges contracting parties to endeavour to keep discriminatory practices to a minimum and progressively to remove existing restrictions affecting investors in this way. The Energy Charter Treaty envisages the entry into, in due course, of a supplemental treaty that will extend full protections to the pre-investment phase.

The definitions of 'investor' and 'investment' under the Energy Charter Treaty are broad. The treaty sets out a non-exhaustive list of investment categories that again recalls wording found in many bilateral investment treaties:

"Investment" means every kind of asset, owned or controlled directly or indirectly by an Investor and includes:

(a) tangible and intangible, and movable and immovable, property, and any property rights such as leases, mortgages, liens, and pledges;

(b) a company or business enterprise, or shares, stock, or other forms of equity participation in a company or business enterprise, and bonds and other debt of a company or business enterprise;

(c) claims to money and claims to performance pursuant to contract having an economic value and associated with an Investment;

(d) Intellectual Property;

(e) Returns;

(f) any right conferred by law or contract or by virtue of any licences and permits granted pursuant to law to undertake any Economic Activity in the Energy Sector.

In essence, protected investments are those associated with economic activity in the energy sector relating to certain energy materials and products. The definition of 'economic activity in the energy sector' comprises any economic activity concerning the exploration, extraction, refining, production, storage, land transport, transmission, distribution, trade marketing or sale of 'energy materials and products'. This latter phrase encompasses materials and products in the nuclear, coal, natural gas, petroleum and electrical industries.

Importantly, the definition of 'investment' expressly extends to those investments made prior to the entry into force of the Energy Charter Treaty in 1998, although investors are entitled to protections under the treaty only in respect of such investments from that date.

Consistent with bilateral investment treaty practice, investors under the Energy Charter Treaty can be both natural persons and companies. Investors can incorporate companies in Energy Charter Treaty contracting parties and invest through them, thus gaining access to the protections of the treaty. However, a 'denial of benefits' provision in the treaty entitles a contracting party to deny advantages to a company that has no substantial business activities in the treaty contracting party through which it claims the jurisdiction of the Energy Charter Treaty.

The treaty also obliges contracting parties to ensure that any state enterprises which they maintain or establish to conduct their activities in relation to the sale or provision of goods and services shall observe the investment protection provisions.

6.3 Transit under the Energy Charter Treaty

The Energy Charter Treaty also provides for a special regime for the transit of energy goods through contracting parties. Such states are required to take necessary measures to facilitate the transit of energy materials insofar as this is consistent with the principle of freedom of transit. Measures must be taken without discrimination on the basis of the origin, destination or ownership of the energy materials or products. The treaty also imposes a prohibition on variations on pricing that may be based on such discrimination. Delays, restrictions and charges on the transit of energy materials or products are also forbidden.

Prior to the treaty, the international rules governing the transit of energy goods (and goods in general) were found in the General Agreement on Tariffs and Trade, a multilateral instrument first entered into in the 1940s for the purposes of facilitating and liberalising trade. The Energy Charter Treaty took as its basis the non-discrimination principles of the agreement and developed them for specific application to the energy industry.

Following the entry into force of the Energy Charter Treaty, discussions took place concerning whether the treaty should be supplemented by additional rules governing transit of energy products that reflected the evolving energy economy of Eurasia. A working group was established in 1998. A draft Protocol on Transit has been prepared that contemplates uninterrupted energy flows between contracting parties, access to available transit capacity, non-discriminatory tariffs and other similar protections. The protocol has, however, been in draft for a number of years, with agreement on its final form no closer to being reached.

6.4 Dispute resolution

The Energy Charter Treaty provides for international arbitration, offering investors the choice of ICSID arbitration (see section 4 above), *ad hoc* arbitration under the UNCITRAL rules or arbitration under the rules of the Arbitration Institute of the Stockholm Chamber of Commerce. Not all provisions of the Energy Charter Treaty entitle investors to commence arbitration proceedings against defaulting states. Although breach of any of the investment protections in Part III of the treaty entitles an investor to commence such proceedings, breach of the transit and trade provisions is to be resolved at the contracting party level. Thus, an investor that is adversely affected by a transit-related issue that does not constitute a breach of the investment protection provisions will be unable to commence arbitration under the Energy Charter Treaty to enforce its rights.

The first arbitration proceedings under the Energy Charter Treaty were commenced in 2001. They related to rights under an electricity sales agreement.[26] Since then, nearly 30 claims have been started. Perhaps the most significant of these

---

26   *AES Summit Generation Ltd v Hungary* (ICSID Case ARB/01/4).

are a trio of disputes arising from the collapse of the Russian oil company OJSC Yukos Oil Corporation in 2006.

These cases (*Hulley Enterprises Limited v The Russian Federation* (PCA Case AA 226), *Yukos Universal Limited v The Russian Federation* (PCA Case AA 227) and *Veteran Petroleum Trust v The Russian Federation* (PCA Case AA 228)) were brought against Russia by shareholders in Yukos which claimed that their investments had been expropriated. They contended that they had suffered considerable losses, estimated to be in the region of $100 billion.

Russia objected to the jurisdiction of the arbitral tribunal in these parallel proceedings on a number of grounds. One of these was that as it was subject only to the provisional application of the Energy Charter Treaty, it was not liable for a failure to provide investment protections to the claimants. This was because provisional application under the treaty has effect only where such application is not inconsistent with the constitution, laws or regulations of the contracting party. Russia argued that submission to international arbitration was inconsistent with its domestic laws which required sovereign acts to be referred to the Russian courts.

The tribunal disagreed. It found that there was nothing in Russian law that prevented the provisional application of treaties. This was the key issue – not whether arbitration under the treaty was contrary to Russian law. The tribunal also took the view that Russia's termination of provisional application of the treaty did not preclude the investors' claims. Under the relevant provision of the treaty, parties which had made investments prior to the termination date would continue to benefit from the investment protection regime for a further 20 years from termination. Accordingly, the tribunal ruled that it had jurisdiction to consider the merits of the investors' cases.

The *Yukos* decision is potentially highly significant. It means that investments made in Russia prior to October 18 2009, when provisional application was terminated, will benefit from protection under the Energy Charter Treaty until 2029. Given the importance of Russia to European energy markets, this ruling greatly strengthens the force of the treaty as a mechanism for energy sector investment protection.

6.5 Model agreements

Although host government and intergovernmental agreements are not the subject of this chapter, it should be noted that the Energy Charter Secretariat has developed model forms of these agreements for transboundary pipeline developments. They are intended to form the starting point for negotiations between potential participants in infrastructure projects.

7. Structuring a transaction to enhance investment protections

Investors in transboundary energy infrastructure should consider the following (non-exclusive) list of issues at the outset of any project to enhance the likelihood of benefiting from investment protections:
- Consider the options available to the investor – are there bilateral investment treaties that may offer substantive protections for the investment? Does the Energy Charter Treaty or another multilateral investment treaty apply?

Would the investor constitute an 'investor' for the purposes of any investment treaty? Is the investment of a type that would be protected by any investment treaty?

- Consider adopting a contractual structure that increases the available protections – could the investment be made by a group company in a state with a favourable bilateral or multilateral investment treaty? Alternatively, incorporating a holding company in a country that is a party to a bilateral or multilateral investment treaty with the host state will provide a further avenue for redress if the project is compromised. However, if such company does not have substantial activities in the jurisdiction, it will be necessary to consider the application of any denial of benefits provisions in the relevant treaty.
- If there are options as to which treaties may be applicable, check the substantive provisions in each – as is clear from the discussion of provisions commonly found in bilateral investment treaties, there is some variety as to the protection on offer. It is worthwhile considering at the outset of a transaction which treaties are best suited to protecting the investment.
- Consider the dispute resolution provisions available – is there an efficient mechanism for the determination of disputes? If the bilateral investment treaty provides for ICSID arbitration, would the investor and investment qualify for such proceedings?
- Check whether it is feasible to seek a waiver of sovereign immunity – the law of sovereign immunity, in part, acts to entitle states to resist enforcement of arbitral awards against their assets. This means that a host state which has had an award made against it arising from a bilateral investment treaty will be able to resist payment of the award to the investor. In circumstances where a significant investment is planned in a host state, it is sensible to seek a waiver of sovereign immunity by the host state.
- Consider the availability of other mechanisms to mitigate the risk of investing – outside the framework of investment treaties, investors should consider other ways to reduce the risk of suffering the effects of adverse governmental actions. One approach would be to obtain finance from a multilateral institution such as the World Bank, the European Bank for Reconstruction and Development or the European Investment Bank. There is evidence that the presence of such institutions in a project reduces the willingness of a host state to interfere, as this would damage its relationship with the institutions and reduce its likelihood of obtaining financing in the future. Another possible avenue would be to obtain political risk insurance for the project from a private insurance company. The cost of such insurance is likely to be lower if investment treaty protection is in place. Indeed, the existence of a bilateral or multilateral investment treaty may be a precondition to the provision of insurance.

8. Conclusion

Sometimes overlooked by investors in international projects, and more frequently inadequately understood, the extensive network of bilateral and multilateral investment treaties available affords a host of valuable protections. The nature of transboundary energy infrastructure is such that investment protection should be at the forefront of pre-investment planning. Careful structuring of investments in order to maximise the protections available can be invaluable should relationships between investors and host states sour.

It remains to be seen how the law of bilateral and multilateral investment treaties will evolve. In recent years many observers have written about a backlash against investment arbitration. Academics have issued a public statement criticising the investment treaty regime for hampering the ability of governments to act for their people in response to the concerns of human development and environmental sustainability.

There has also been criticism of the perceived investor-friendly approach adopted by tribunals.[27] Some observers perceive a structural weakness of the ICSID system – namely, the process for selection of arbitrators. In a three-person tribunal, each party has a right of nomination, with the chair being appointed by the ICSID from a fixed list of candidates who may be included on the list only as nominees of contracting states. Certain investors have complained of pro-government sympathy on the part of chairs.

Recently, Ecuador and Bolivia have withdrawn from the ICSID. Other states have forcefully and, at times, successfully argued that the protections that they offer to investors are subject to the defence of necessity, that certain breaches of the terms of their bilateral investment treaties may be forced upon them given the prevailing economic and social circumstances, and that for this reason they should not be liable to investors. As emerging economies develop and increase in confidence, they may be less willing to respect the protections contained in their investment treaties. The notion of permitting arbitration tribunals to adjudicate disputes with high significance for national wellbeing may become increasingly controversial.

Intergovernmental risk, however, will always be present. The existing regime, despite the criticisms levelled at it, remains the best means available to investors to mitigate the risks considered in this chapter.

References
Cameron, Peter, *International Energy Investment Law: The Pursuit of Stability*, Oxford University Press (2010).
Energy Charter Secretariat, "The Energy Charter Treaty – A Reader's Guide" (2002).
McIlwrath, Michael and Savage, John, *International Arbitration and Mediation: A Practical Guide*, Kluwer Law International (2010).
Mclachlan, Campbell, Shore, Laurence and Weiniger, Matthew, *International Investment Arbitration: Substantive Principles*, Oxford University Press (2007).
Newcombe, Andrew and Paradell, Lluís, *Law and Practice of Investment Treaties: Standards of Treatment*, Kluwer Law International (2009).
UN Commission on Trade and Development, "Bilateral Investment Treaties 1995–2006: Trends in Investment Rulemaking" (2007).

27　See Public Statement on the International Investment Regime (August 31 2010).

Intergovernmental agreements

Katie Baehl
R Coleson Bruce
George F Goolsby
Baker Botts LLP

1. **Introduction**

    A theme of this volume is risk allocation and mitigation among contracting parties. It is axiomatic that the most efficient approach allocates a risk to the party best positioned to manage the risk, and that proper risk apportionment best places the parties to avoid or mitigate risks generally. While this principle is simple, applying it in complex transactions often is not, particularly given the multitude of players involved in such transactions. Further complicating the principle's application is the fact that risk takes many forms in the context of multilayered contracting required, for example, in the design, construction and operation of energy transmission and transportation systems across several jurisdictions. Such projects involve several suites of agreements, including those with or involving the host states, those involving construction, those dealing with ownership and operation, and those for financing and insuring the project. The stakes are increased when, as is often the case, the project spans multiple jurisdictions that, on their own or in concert, create high expectations for local development and present numerous legal, commercial, regulatory, political and social uncertainties.

    This chapter discusses the risks associated with such uncertainties and how they may be addressed through the use of a tailored, project-specific treaty – an intergovernmental agreement. Such treaties provide the benefits of:
    - ensuring host state support for the project;
    - providing or confirming the grant of particular rights, benefits or exemptions for the project; and
    - creating an adequate legal framework to provide the necessary predictability and stability to support long-term investment in an energy infrastructure project.

    This chapter begins by discussing the interests of the various stakeholders and the risks of particular concern to parties involved in such projects; it then describes how intergovernmental agreements address these concerns and relate to other project agreements.

    A few observations should be made at the outset. First, while this chapter speaks primarily in terms of cross-border oil and gas pipeline projects, intergovernmental agreements are also useful in other cross-border energy transmission projects such as large-scale telecommunications and power transmission projects. Second, while this chapter focuses on intergovernmental agreements, it is not intended to suggest that intergovernmental agreements are the sole or superior way to achieve project-specific risk mitigation. While some other options are noted, they are the subject of other chapters and therefore are not explored in depth here.

Lastly, while intergovernmental agreements exist in form versions[1] and while similarities in intergovernmental agreements are noted, an intergovernmental agreement signed for one project may be substantially different from an agreement prepared for a different project. The very nature of an intergovernmental agreement as a project-specific, state-to-state undertaking means that the final features of the agreement will be a function of the needs of, and risk management challenges posed by, a particular project in a specific location at a given time. Nonetheless, and despite the probable variability between any two intergovernmental agreements, certain commonalities are identifiable and discussed below.

2. Prominent concerns of project participants

Fundamental to project planning is a thorough review of local laws to assess the state of the law, its compatibility with project needs and how local laws respond to key project risks in a cross-border context. Typically, a dialogue between private interests and government interests follows such an evaluation, with discussion focusing on what benefits, obligations, risks and expectations need to be addressed, what laws are deficient or inapposite, and where there simply is no law in a critical area. What each stakeholder requires in order to participate in the project may vary, but often there is sufficient common ground to set an agenda and proceed. In this process the host states and private sector stakeholders[2] will each view their interests through the lens of their unique role in the project. Much of the discussion often centres on fiscal deal points and government grants, followed quickly by confirmation of the stability of what has been agreed.

This section discusses some principal concerns that the various stakeholders may ultimately wish to see addressed through an intergovernmental agreement. While this section contrasts various stakeholder concerns, the project participants' interests often overlap, are complementary or are amenable to compromise.

2.1 Host state concerns

It is useful to begin by discussing the typical concerns of the host states, because as a treaty, the intergovernmental agreement is the product of negotiations between or among the host states and is enforceable as a state-to-state agreement. Private sector stakeholders will wish to put forward an agenda for the intergovernmental agreement that provides a sound basis for protecting their investment in the project. However,

---

1   The form versions include the first and second edition of the Model Intergovernmental Agreement for Cross-Border Pipelines put forth by the Energy Charter Secretariat. The second edition is available at www.encharter.org (click on 'Trade & Transit' and then 'Model Agreements').
2   Private sector stakeholders include project sponsors, equity investors, contractors, lenders and insurers.

progress towards that goal during the negotiation of the intergovernmental agreement will require private sector stakeholders to understand, critically assess and ultimately appreciate the strategic interests of the host states in relation to the project.

Broadly speaking, a host state's concerns will be of two varieties:
- defining and protecting the various commercial benefits generated by hosting the project; and
- protecting and preserving national interests that may be affected by the host state's involvement in the project.

In the context of an international pipeline project, a host state's commercial concerns vary depending on what role the state plays in the project. One or more nations may be the source of the energy (ie, producing states); other states may be solely involved as the conduit to market (ie, transit states); with still other states both hosting the transportation or transit and receiving energy for local consumption (ie, consuming states). A producing state has within its boundaries a domestic supply of hydrocarbons that it and its private investors seek to monetise by gaining access to markets outside its boundaries. From such a state's perspective, the development of the state's hydrocarbon resources is paramount, with the pipeline being essential, but ancillary, to upstream production activities. Thus, a group of producing states has a common interest in creating a reliable, long-term transit system to or through its neighbouring countries to market; the group also wishes to assure that its transit state partner will not interrupt or impede that transit, thereby limiting access to market. By contrast, a consuming state with insufficient domestic resources to meet its energy demand will be keen to participate in a cross-border project to acquire a steady supply of hydrocarbons to sustain its economy. Meanwhile, a transit state gains neither the benefits of hosting the production nor the benefits of a consumer. Instead, it tends to focus on the benefits of hosting the transit infrastructure, such as receiving revenues from transit fees and taxes, being involved in the construction and operation of the in-state facilities, and seeking the benefits of domestic training, job creation and employment. While these benefits can be substantial, the transit state will be aware that it enjoys a key locational advantage; it may seek to parlay that advantage into increased or additional benefits. Given these varied interests, each host state to a multi-jurisdictional energy infrastructure project will negotiate project agreements with its particular situation, interests and goals in mind.

Notions of mutuality also enter the picture as multi-jurisdictional infrastructure projects are developing. Each state will have an interest in the terms and conditions of the various project agreements negotiated by the other host states. In certain situations, each state will wish to be sure that its grants are not disproportionate to those of the others. This is not to say that all terms must be the same. For example, the producing and consuming states have an interest in ensuring that each side performs its obligations as agreed. Because their roles differ, their commitments will differ as well. Both the producing and consuming states have an interest in gaining the cooperation of the transit state. Given that the producing state gains benefits from the production itself and the consuming state derives substantial advantages from reliable supplies of energy, those states' taxes and transit fees for the pipeline may be less than those insisted upon by the transit state. Moreover, the multiple elements of the project within their jurisdiction provide numerous opportunities to exact government take. Regardless of the fiscal terms initially agreed, the balance of benefits may change over time, potentially resulting in commercial asymmetries between host states and a desire by a state to alter the allocation of benefits, either at the state-to-state level or, more likely, at the state-to-investor level.

Aside from the strictly commercial concerns of host states will be concerns of national interests affected by the host states' participation in the project. Key among these interests are geopolitical and regional interests, national security and the impact of the project on constituent populations (eg, job growth or displacement of citizenry due to pipeline location). Perceived project complications may also include fears about loss of habitats or traditional local livelihoods, effects on regional political capital and risks on the environment and the health and safety of local communities. In addition, because the commodity being transported is necessary for industry and even military operations, project success and the risk of detrimental action by other parties (including other states) interrupting project operations may have significant national security and economic implications. If the project is large scale, the economy of an entire region may be both positively and negatively affected by the project's location, operations and ultimate success. All of these concerns will result in the host states requiring stability of more than the purely commercial aspects of the project; they will also demand stability of the project's legal framework in the other participant nations, particularly with respect to laws affecting the national security, environment, health and labour of citizens and other national interests.

At bottom, all host states will wish to ensure that:
- their own state receives its fair share of the commercial benefits associated with the cross-border endeavour;
- significant national interests are accounted for; and
- the other states are sufficiently committed to justify the state's own expenditure of time and resources and its commercial and political investments in the project.

The cross-border nature of these projects makes it essential that each participating state uphold its commitments to the project because detrimental action by any one state harms all others. As a result, each state will want to see that the other participant states honour their commitments through a reliable state-to-state legal mechanism by which it can enforce the terms of key project agreements against the other states. The same dynamic applies at the state-to-investor level, where each state expects the investors to perform as agreed – even in other states. An investor's breach in any state may imperil the whole of the cross-border project. These concerns will tend to result in each host state's desire for transparency and enforceability with respect to the rights and duties of the other project participants and host states, with attendant notice, communication and enforcement mechanisms available should project complications arise.

2.2 Private sector stakeholder concerns

Private sector stakeholders' overarching concern in relation to the host states is twofold:

- to secure the necessary grant of rights, privileges and exemptions from the states as are necessary to make the cross-border project viable; and
- to put in place the necessary investment protections, including requiring a legal framework offering long-term certainty and predictability of laws and standards.[3]

That framework includes state-to-state elements, as well as individual state-by-state to investor elements, including providing to the project and the project participants certain benefits and protections in relation to the project. In the absence of such grants and a viable legal framework, the perception of non-support and related instability may cause project investors to walk away or, if the project goes forward, risks left unaddressed (and therefore often allocated, by default, to the project sponsors) in this area may imperil the project's success over the long term.

Much of the private sector focus is on project economics and the protection, over the life of the project, of the economic basis on which the investment was made. In all cases, however, private sector stakeholders' overarching goals will be to do everything reasonably practicable:

- to define clearly the laws and standards applicable to the project;
- to ensure that those laws and standards remain applicable throughout the life of the project; and
- to make those laws and standards enforceable through both incentives for compliance and disincentives for non-compliance.

Achieving this goal will not always be easy in the context of cross-border projects. The amount of work to be done to achieve an agreed stable framework will depend on the exact landscape of the existing domestic and international laws and treaties implicated by a particular project. Private sector stakeholders may encounter domestic legal regimes that are highly complex or, alternatively, significantly underdeveloped. Additionally, they may encounter unclear or conflicting jurisdictional lines within a state, as is the case in Iraq, where the Kurdistan regional authorities assert significant control over developments in the region. Similarly, the issue of international boundaries and border disputes, and the question of what states have or assert jurisdiction in particular circumstances, are significant state-level risks about which private stakeholders have concerns. Boundary delimitation claims exist around the world. The ongoing debate about which littoral states of the Caspian must be consulted or have approval rights in respect of activities on, in and under the Caspian Sea is a prime example. Such disputes are often longstanding and unlikely to be resolved prior to commencement of the project. Accordingly, private

---

3   For a discussion of the risks inherent in transnational gas project and how such concerns may be addressed by a range of possible project agreements and treaties, including the Energy Charter Treaty, see Griffin, Paul, *Transnational Gas Projects and their Agreements: A Practical and Legal Guide*, Herbert Smith (2002).

sector stakeholders need state-level assurance that the project is nevertheless welcome by all disputants and project activities will not be disrupted as a result of these disagreements.

Private sector stakeholders are also likely to be concerned about changing domestic laws and changes in governments, given the long lifespan of many cross-border energy infrastructure projects. Because of the expense of building this infrastructure, energy transmission projects will remain in place for long periods of time – often for 40 or more years (typically, at least 20). In 40 years a nation may experience multiple changes in political control (legal or otherwise), and, in the absence of long-term commitments under an intergovernmental agreement, such changes in control would otherwise bring significantly new policies affecting the project. In this respect, project sponsors must be concerned with:

- small-scale legal changes, such as changes in legislation that would have an impact on the overall function of the project (including tax and regulatory laws and laws governing the ability to do business in the jurisdiction); and
- more drastic legal changes in the organisation and policies of a country resulting from political changes or otherwise.

None of this is to overlook perhaps the most fundamental risk that private investors are concerned about when they focus on the host state: the risk of nationalisation or expropriation of project assets – whether such nationalisation or expropriation occurs all at once by clear and direct action, or is more creeping and occurs over time as either various project value elements are taken or their further enjoyment is denied without compensation or other remedy. In this regard, while a host state retains its sovereignty, including the right to take drastic action in respect of assets and investments within its borders, a key expectation of private sector investors involved in these projects is that the state will not exercise this inherent power and, if the state must nationalise or expropriate, it will do so in a manner that is consistent with recognised international standards and procedures. This includes affording the project participants fair and equitable treatment and prompt, adequate and effective compensation. It is rare for major cross-border projects to proceed without some form of express commitments and protections regarding expropriation and nationalisation.

While existing investment treaties of general application may address expropriation and the stabilisation of laws, other key issues to private sector investors are unlikely to be sufficiently addressed by existing domestic laws or treaties. For example, private sector investors will wish to achieve tax efficiencies, including in particular avoiding double or multiple taxation by the host states. Closely related is the issue of efficiencies in so-called 'home country taxes' – that is, the taxes paid by the private sector investors in their home countries as a result of revenues earned through a foreign-based project. In relation to the host states, tax concerns also include the issue of 'host country take', which refers to the taxes, transit fees, local hiring and training requirements, as well as other fiscal or commercial terms by which project revenues are payable, directly or indirectly, to the state. These concerns will manifest themselves in any project negotiations as each state compares its investment incentive and investment protection package against those of the others.

In addition to concerns regarding the amount and form of taxation and other issues relating to host country take, private sector stakeholders will focus on creating a body of rights and regulations governing the conduct of the project. This includes establishing uniform technical and regulatory specifications across the host states and consistency of application in all host jurisdictions. It is far superior to have a uniform set of standards across three states than to have three pipeline segments, each subject to substantially different standards. Regulations may address issues such as safety, construction, and environmental and technical standards; they may also include other types of regulation, such as anti-monopoly and utility regulation. Certainly, the ability to fly the line and attend to telecommunications over the entire length of the pipeline requires cross-border agreement on regular access for operations. In addition, in the cross-border context there are concerns about project interruptions and the free transit of goods based on differing standards and regulations. Lastly, private sector stakeholders must be sensitive to the issues of state-level insistence that:

- the project and project operations be conducted in a manner consistent with those standards finally agreed; and
- mechanisms be put in place to allow both state-to-state and state-to-investor dispute resolution, and the ability to enforce remedies against an offending party at each level.

All told, a private sector stakeholder's degree of enthusiasm in making an investment will vary with the ability:

- to secure the essential economic benefits package;
- to define the law and standards that will apply to the project sufficiently; and
- to assure that those laws and standards are in place throughout the duration of the project, and that they create sufficient enforcement incentives and mechanisms which will function for many years across multiple jurisdictions.

3. Intergovernmental agreements

The fundamental concerns faced by host states and private sector stakeholders may, in certain circumstances, be addressed by intergovernmental agreements. In short, intergovernmental agreements offer great flexibility and may be used in conjunction with host government agreements[4] to define and provide for enforcement of the law of a project (eg, to craft the specific laws and regulations that will apply to the particular project) with treaty-level protections, thereby providing the kind of long-term legal predictability and stability necessary to the success of cross-border energy infrastructure projects.

---

4   Host government agreements can have a variety of names. However, as a category, they are project-related agreements between investors and a host state (or possibly states – or even national oil companies) that specify the government's support for, and involvement in, the project by, for example, granting certain rights, benefits and exemptions in favour of the project. These agreements typically also set forth rules and regulations for the investor. The relationship between host government agreements and intergovernmental agreements is discussed in more detail in section 3.2 of this chapter.

3.1 The nature of intergovernmental agreements

Central to a discussion about the usefulness of an intergovernmental agreement in a cross-border project is emphasising that such agreements are distinct from other project agreements because intergovernmental agreements operate at the treaty level. As international agreements concluded between states in written form and governed by international law, intergovernmental agreements meet the definition of a 'treaty' under Article 2 of the Vienna Convention on the Law of Treaties. As a result, intergovernmental agreements are fully enforceable under international law. Another aspect of intergovernmental agreements is how they relate to the domestic laws of the participating states. Under many legal systems, for example, treaties also become part of the domestic law of the state and may even prevail over other domestic law (excepting only the state's constitution) in the local law hierarchy. This may provide an opportunity to deal with conflicting domestic laws through use of an intergovernmental agreement.

The use of a treaty to address risks associated with a cross-border project offers the advantage of heightened enforceability. At the international law level, states retain their sovereign right to denounce or withdraw from treaties; however, the standards set forth in a treaty, such as an intergovernmental agreement, can be enforced under international law by the other signatories unless and until the state officially denounces the treaty in accordance with the Vienna Convention or the treaty itself. Due to the need for project stability, one seldom sees intergovernmental agreements expressly providing for denouncement in a project-specific context. As a result, any denouncement[5] would itself be a violation of the basic commitments of the intergovernmental agreement and would be expected to trigger remedies at both the state-to-state and state-to-investor level.

There are other advantages in the use of a treaty to address the risk of a party reneging on the deal. For example, while breaching another project agreement would undoubtedly be a significant issue for investors, the simultaneous violation of a treaty may gain more international attention and condemnation.[6] This remains true when the breach results from a change by the state of its domestic law in contravention of the host government agreement. Moreover, treaties are governed by an international set of rules of interpretation and enforcement, providing a stable basis for interpretation of intergovernmental agreements and denying the ability, or making it difficult, for the state to alter the deal by changing the law – an important feature in states where domestic laws are in flux, are difficult to find and follow, or are otherwise unfamiliar or silent about common international norms that the counterparties may take for granted (eg, principles governing the proper

---

5 Unless such denouncement is otherwise made in accordance with the Vienna Convention on the Law of Treaties.
6 This is certainly not the case where the domestic law involves the nationalisation of an industry, but is likely to be true if the domestic law involves a matter of indirect expropriation, such as through new or additional taxation or regulation beyond that agreed.
7 Note that project agreements other than the host government agreement may be incorporated into the intergovernmental agreement by reference. For example, the turnkey agreement was appended to the intergovernmental agreement governing the Baku-Tbilisi-Ceyhan pipeline (discussed in section 3.5 of this chapter).

interpretation of contracts or the principle of good faith). As a result, securing a treaty-level commitment by which each state is bound to honour its host government agreement and other domestic law-related commitments[7] serves to move these agreements beyond merely contractual or local law promises and makes it more difficult for the state to avoid its obligations.

Treaties are the ideal legal framework to manage many of the state-level risks and reaffirm grants made by each state to the project participants. As state-to-state agreements governed by principles of international law, treaties offer an attractive degree of clarity and stability to the legal framework of a project. Treaty-level agreements assure that the laws and standards of the project will apply, to the extent provided by international laws, throughout the duration of the project, with treaty-level enforcement incentives and mechanisms functioning across multiple jurisdictions. And because intergovernmental agreements are negotiated by the host states, a degree of state-to-state transparency and enforceability is inherent in the project, thus providing a conduit through which many host state concerns may be addressed.

However, a few words of caution are warranted. First, intergovernmental agreements do not exist in a vacuum. Host states and project sponsors should be sensitive to the implications of entering into a treaty, including considering its effects on existing treaties and the application of international norms. By proposing to put in place a set of prevailing laws and standards that apply to a particular project, exclusive of other laws, standards and regulations, it will be necessary to consider how the particular intergovernmental agreement will operate alongside, or be consistent with, other existing treaties. Host states should be clear that intergovernmental agreement primacy is not in derogation of treaties to which the state is a party that establish international norms relating to human rights, health and safety and the environment. Similarly, care should be taken to assure consistency with constitutional limits, including constitutional provisions regarding the hierarchy of laws, how treaties affect or are embodied in domestic law, how domestic law may be changed and how grants of rights are made effective. Lastly, because treaties are negotiated by states, political concerns (both domestic and international) beyond those strictly related to the project will play a role in the conclusion of the intergovernmental agreement and may well affect the agreement's substance. Project sponsors may be clear about project needs, but in these state-to-state negotiations only the states will be privy to their negotiations and only the states will agree on the final terms.

3.2   **Working alongside host government agreements**
Another crucial facet of intergovernmental agreements is their relationship to host government agreements or other project-level agreements. In addition to including project-specific clauses designed to grant (or recognise the grant of) specific benefits for the project, or to eliminate or reduce either risks or impediments that unnecessarily burden the project, intergovernmental agreements also recognise (and may even incorporate) the underlying host government agreements. In certain instances, as state-to-investor and investor-to-state undertakings, host government

agreements are primarily contractual, and thus interpreted and enforced as binding contracts – albeit that by decree, parliamentary action or other means, the host government agreement is approved as a governmental obligation fully enforceable under applicable domestic law. On other occasions the text of the host government agreement may be fully enacted and become part of the applicable domestic law, in which event it is enforceable both contractually and as domestic law.[8] In yet other circumstances, if the text of the host government agreement is made an integral part of the treaty-level intergovernmental agreement, the terms of the host government agreement may become domestic law that prevails over other domestic law, because in many legal systems treaty-level obligations trump conflicting domestic law other than the constitution.[9]

When the host government agreement has been incorporated into the treaty, the host states can enforce the more detailed terms of the host government agreement against the other host states. This provides the host states with additional rights to protect their domestic interest, as well as giving the investors the potential for political support in the event that a host state fails to abide by the terms of the intergovernmental/host government agreement package. Thus, while intergovernmental agreements are separate from host government agreements and do not detract from their importance to the project, intergovernmental agreements may be used in tandem with host government agreements to make necessary grants and to mitigate and allocate risks in a way that a host government agreement cannot accomplish on its own.[10]

The various ways in which an intergovernmental agreement may complement a host government agreement is illustrated by how, in a cross-border pipeline project, intergovernmental agreements address the need for uniform technical, environmental and health and safety standards, as well as regulatory requirements (or at least requirements similar enough to allow the pipeline or transmission line to function efficiently across the entire system). While much of the detail can be provided and heavy lifting accomplished by the individual host government agreements, an intergovernmental agreement is useful in two ways. First, it sets forth (or affirms, if the detail is in the host government agreement) specific technical,

---

8   In fact, many host government agreements call for their terms to be enacted into local law in whatever manner is required under local law. Thus, the requisite approval of the host government agreement itself may give the terms of the agreement the status of domestic law; however, the concern remains as to whether the host government agreement is the prevailing domestic law. For a brief discussion of the relationship between international and domestic law, see Brownlie, Ian, *Principles Of Public International Law*, Oxford University Press (2003), pp47-48.

9   When intergovernmental agreements are used more generically, as when neighbouring states create energy corridors to attract certain types of transportation or transmission project by designating certain zones and defining available benefits, particularised host government agreements may then be used in association with the corridor intergovernmental agreement for specific projects.

10   As a matter of negotiation – unlike an intergovernmental agreement, where the states themselves control the agenda and agree terms – the project sponsors are at the table with each state in the negotiation of the host government agreement. As a result, as part of agreeing to the necessary terms of the host government agreement(s), the project sponsors may raise the need for and purpose of an intergovernmental agreement, or the participating states (particularly the producing state or consuming state) may suggest that state-to-state support is in order. If the point is raised by the investors, they may even press for the terms of the host government agreements to be enacted into domestic law through the domestic law process or be incorporated into the intergovernmental agreement, as discussed in this section.

environmental and health and safety standards applicable to the entirety of the pipeline, as was done in the agreements relating to the Baku-Tbilisi-Ceyhan pipeline (a project traversing Azerbaijan, Georgia and Turkey). In that project, these standards vary from state to state and local agencies remain vested with regulatory control. However, the variations are within reasonable limits. Regulation is therefore focused and the pipeline operates as an integral system rather than as three pipelines patched together. The result is that even though there is room to adjust the standards to meet particular state requirements, the intergovernmental agreement serves to stabilise workable variations and provide essential uniformity. Again, in the case of the Baku-Tbilisi-Ceyhan pipeline, these standards are monitored and enforced in each state by the relevant domestic regulatory agency, thus allowing each state to regulate independently, but always in accordance with agreed standards.

Second, an intergovernmental agreement may call for uniform standards and an entirely new regulatory framework to be overseen by a central, intergovernmental regulatory agency in which all host states are involved, as was done in the West African gas pipeline project (traversing Benin, Ghana, Nigeria and Togo). In this project, the central authority oversees the pipeline project on behalf of all four host states. The intergovernmental agreement for this project also sets forth an appeals process for agency decisions. Thus, intergovernmental agreements and host government agreements may work in concert to bind the host states to a uniform 'law of the project' through treaty obligations.

3.3 Interacting with investment treaties

While treaty-level protections may be desirable in the context of a given project, an intergovernmental agreement or intergovernmental/host government agreement package is not the only strategy available to secure many of the rights and benefits necessary to a given project. A primary factor influencing the content of a given intergovernmental agreement is the existence and content of other investment treaties between or among the host states. Sometimes pre-existing treaty regimes may be relied upon to confer certain benefits and privileges, and to allocate and mitigate certain state-level risks in a multi-jurisdiction project. Pre-existing treaties may be generally applicable to transnational business investments and operations, such as the bilateral and multilateral investment treaties discussed elsewhere in this volume, or the Energy Charter Treaty (essentially, a multilateral investment treaty tailored to energy infrastructure).[11] To the extent that these pre-existing treaties apply to transnational business operations in general, they underscore the participating states' ongoing interest in hosting such projects and help define the benefits made available and the law applicable to a particular energy infrastructure project.

In some instances, the host states may be parties to an extensive regime of investment treaties. Such regime may define 'expropriation', require prompt, adequate and effective compensation in the event of expropriation and provide for fair and equitable treatment. It may also grant flexibility with regard to acceptable

---

11   For a more detailed discussion of investment treaties, see the chapter of this book on multilateral and bilateral investment agreements.

forms of business organisation for the free transfer of goods and money in and out of the country and with regard to the hiring of both local and foreign employees at the discretion of the participant. In these cases, the existence of extensive pre-existing investment treaties may eliminate the need to negotiate the extended list of project rights and privileges and provide the necessary promise of stability respecting the investment.

At the other end of the spectrum are those projects spanning jurisdictions that feature few or no pre-existing treaty-level agreements. The success of such projects will depend on securing necessary grants and benefits, as well as securing stabilisation and crafting a more robust legal framework to address – on a consistent basis (both at the treaty level and in the state-to-investor relationship) – the benefits, privileges and obligations associated with the project, and what law and standards apply to the particular project. In the absence of such consistency in applicable laws and standards, a truly integrated project – such as a hydrocarbon pipeline operating across several jurisdictions as one project – may be very difficult to achieve, or even impossible.

When host countries are members of applicable and available investment treaty regimes, but those regimes provide only some of the particularised benefits or the type or level of intergovernmental protections necessary to a particular energy infrastructure project, further action may be necessary. In these cases, the host states have pre-existing treaty-level agreements that cover many of the necessary features of doing business and address many of the risks associated with the investment itself; however, such agreements do not deal with certain other critically important points unique to the particular project. Rather than negotiating terms anew, intergovernmental agreements can augment and particularise treaties in various ways, including by incorporating their terms by reference (and perhaps even strengthening them by providing additional or better-tailored terms). The intergovernmental agreement can be used to fill gaps in the remaining terms required by the project or to extend benefits to investors or types of business organisation not included in the existing investment treaties.

Intergovernmental agreements may also be used to give access to existing bilateral or multilateral investment protections when existing treaties preclude such access. For example, in order to qualify for protection under certain general investment treaties, an investor may need to incorporate locally, do business only in a particular organisational form or locate its headquarters within the jurisdiction – all of which may be sensible for inbound investment wholly within a single state, but impossible or impracticable when the project is cross-border in nature. Indeed, some investment treaties establish a minimum threshold of in-state activities, such as holding board meetings. Meanwhile, the host state often requires that the project sponsors act through a subsidiary incorporated in that host state. While project organisational structures can be created to solve these problems, such structures may be complex and otherwise inefficient from an operational, financial or tax efficiency perspective. Thus, seeking access on agreed terms through a specific intergovernmental agreement may be the better option over the long term.

3.4 Addressing project-specific risks

An intergovernmental agreement may simply aim to address gaps or deal with a few particular concerns not covered by general investment treaties, or it may sit atop a more robust, project-specific, legal framework intended to provide something more akin to a uniform law of the project. In either case, the intergovernmental agreement may provide project participants with a useful tool to allocate and mitigate certain risks.[12]

For example, an intergovernmental agreement reduces enforceability risks by allowing the states themselves to address issues and resolve grievances relating to the project. Unlike bilateral investment treaties, a project-specific intergovernmental agreement recognises the mutual interdependency inherent in a cross-border transmission project. It provides an internal framework across the host states, allowing each state to enforce the commitments of the other states in favour of the project and requiring each state to perform lest the other states protest.[13] This enhancement of the rights of interested parties with a stake in project success is significant because energy transmission projects involve the construction of infrastructure (eg, pipelines or transmission lines) that is captive respecting location and cannot be moved from any of the host states. The result is not only that the project sponsors have the right of enforcement against each state under the individual host government agreements (and vice versa), but also that each host state also gains a right of enforcement against other host states under the intergovernmental agreement to the extent that the problem presents at the state level.

Intergovernmental agreements may also reduce risk by supplanting domestic laws that are unsuitable for international transportation or transit projects. For example, local anti-monopoly or utility regulatory laws intended to protect local markets and local users of energy may be inapposite to energy projects involving solely the cross-border transit of hydrocarbons or energy. As mentioned above, a crucial concern to private sector investors is tax efficiencies – including, in particular, avoiding double (or multiple) taxation by the host states and achieving efficiencies in home country taxes.[14] Key tax issues are unlikely to be sufficiently addressed by existing domestic laws (or treaties); however, they may be addressed in detail in an intergovernmental agreement.

Additionally, intergovernmental agreements may provide for project-specific commitments or regulations on the part of the host states in order to reduce a wide

---

12   Indeed, the very process of drafting an intergovernmental agreement mitigates risk by increasing transparency through encouraging the host states to share information about their individual commitments to assure a reasonable balance in commitments and to permit state-to-state action if those commitments are not honoured. This aspect enhances a common basis of agreement and avoids surprises. This process also provides a way for the host states to work through issues specific to the project, including the application and enforcement of regulations and technical specifications, as discussed in this section.
13   Because many investment treaties focus on investment in the single hosting states, unless the investors have as their home state one of the other host states, the existing general investment treaty framework may have no mechanisms for the hosting states to consult with each other about project matters and may not provide a clear basis for state-to-state dispute resolution.
14   For more information regarding tax risks and how to avoid them, please see the chapter of this volume on tax risks.

variety of risks. The agreements may include covenants to take any actions necessary to ensure that the project functions in accordance with applicable standards, and that any future inconsistent laws will not apply or will otherwise be subject to remedy. Streamlining of local law requirements could also include exemptions or government assistance with local legal compliance matters.[15] Similarly, an intergovernmental agreement might include a provision stating that the resolution of any boundary disputes will not affect the legitimacy or rights of the pipeline project. Lastly, the end goal of bringing energy to market means that the project needs for all participating states to reaffirm the right of free transit (a concept embodied in international treaties such as the Energy Charter Treaty) as to all volumes contractually committed to markets beyond the state(s).[16]

Another project-specific concern that creates risks for the host state and private sector stakeholders alike pertains to both the construction of the pipeline and displaced persons. The extensive construction and subsequent operations involved in these projects mean that the project may need state assistance in acquiring the necessary lands and rights of way. In turn, the host state will be concerned with how the project will respond to the economic dislocations (temporary and permanent) arising from the displacement of its residents. Thus, if the pipeline crosses agricultural land but is buried below plough depth, the issues focus on how the project obtains the right of way, how resident populations are compensated and how, after pipeline installations, agricultural activities resume in a manner compatible with pipeline operations. Closely related is how the project will be affected by future infrastructure projects, such as other pipelines, transmission lines or highways.

Beyond the project-specific provisions desired by the project sponsors or the host states, risks that are concerning to potential lenders and insurers may be taken into consideration when drafting an intergovernmental agreement. The financial community will be concerned about political risk and welcome a palpable showing of state support in the form of an intergovernmental agreement and host government agreements. More practically, the intergovernmental agreement may confirm the several states' acceptance of step-in rights, provide a reaffirmation of cooperation and further assurances, and reaffirm other promises made under the host government agreements. These provisions encourage financing by assuring available remedies and a high degree of state commitment.

In addition to these various concrete protections that intergovernmental agreements confer on stakeholders, these agreements may also be drafted to include more general expressions of support for the project. These include, among other things, broad statements of support, reaffirmations of the binding nature of the project documents and commitments to cooperate and act in good faith. Although these provisions may be difficult to enforce, they serve two important purposes. The

---

15　For example, while construction standards specified by local laws may be acceptable in a purely domestic situation, in order to secure financing the project may be required to meet more stringent international standards. In such case intergovernmental agreement sanctioning those higher standards helps clear away conflicting local standards.

16　For a discussion of free transit under the Energy Charter Treaty and the General Agreement on Tariffs and Trade, see Redgwell, Catherine, "International Energy Security", in *Energy Security: Managing Risk in a Dynamic Legal and Regulatory Environment*, Oxford University Press (2004), pp40-42.

first is requiring the host states to take a stance that makes it more difficult to renounce the project on the basis that it is inequitable or illegal. The second is that such statements of support help investors to establish that their expectations regarding legality and financial returns were reasonably based and legitimate – an important factor in obtaining compensation in the event of a claim for violation of the applicable fair and equitable treatment standard (whether under the applicable intergovernmental agreement or otherwise) or indirect expropriation.[17]

Although many of the risks discussed in this section are common in cross-border pipeline projects, a number of project-specific elements and circumstances will affect the provisions ultimately necessary in a given intergovernmental agreement. For instance, the existence of an extensive general investment treaty regime may eliminate the need to negotiate an extended list of project requirements at the level of the intergovernmental and host government agreements, as discussed above. Other considerations include:

- the socio-economic and political stability of the host states;
- the relative condition and content of domestic law (including tax policy);
- disparities in applicable standards;
- foreign policy concerns; and
- the number and particular goals of the states involved (especially whether the state is a transit, consumer or producing state).[18]

In addition to dynamics in, between and among the host states, domestic economic policies often play a role in the form that an intergovernmental agreement takes and the provisions that will ultimately be included.

Thus, while the various portions of an intergovernmental agreement just discussed will be common, no two intergovernmental agreements will be developed and negotiated in the same way or result in identical outcomes. After all, the entire purpose of having a project-specific treaty is that state-to-state commitments – both between or among the states and in relation to each state's commitments to project investors – must address the needs and risks particular to the parties and the project. Otherwise, an intergovernmental agreement would be no different from an investment treaty existing independently of the project. In that situation, instead of a project-specific intergovernmental agreement, perhaps the states should consider a broader, more generic investment treaty approach to attracting and encouraging investment across all sectors.

3.5 Possible structures

Given the variety of project-specific factors, it should come as no surprise that intergovernmental agreements come in an array of structures and may address the

---

17 For a discussion of the role of investor expectations in claims of indirect expropriation or a violation of the fair and equitable treatment standard, see Dolzer, Rudolph and Schreuer, Christoph, *Principles Of International Investment Law*, Oxford University Press (2008), pp104-106 and 133-147.
18 For additional information on the particular concerns and risks involved in transboundary pipeline projects, see Browning, William E and Dimitroff, Thomas J, "Transboundary pipeline development and risk mitigation", in Picton-Turbervill, Geoffrey (ed), *Oil and Gas: A Practical Handbook* (2009).

same risk differently. A comparison of the intergovernmental agreements governing the Baku-Tbilisi-Ceyhan pipeline and the West African gas pipeline illustrates the diversity of approaches possible with regard to such agreements.[19] In the Baku-Tbilisi-Ceyhan pipeline, many of the project-specific concerns are addressed by the host government agreements, rather than the intergovernmental agreement itself. The intergovernmental agreement includes, among other things, a commitment by each of the states that it will "fulfill and perform on a timely basis"[20] its obligations under the intergovernmental agreement and the other project agreements (the full text of which was attached to, and made part of, the intergovernmental agreement). The intergovernmental agreement also specifically ensures that the technical, safety and environmental regulations will be governed by the host government agreements "notwithstanding any standards and practices set forth in the domestic law of the respective State".[21] Meanwhile, under the intergovernmental agreement, taxes are based solely upon activity taking place within each state's own territory and per the state-specific tax regimes laid down in the project host government agreements.[22] As a result, the particular rules and regulations for the pipeline, while largely uniform, vary from state to state based on the terms of the host government agreement negotiated between the project sponsors and each host state.

By contrast, individual state-to-investor host government agreements were not used for the West African gas pipeline; instead, the host states entered into a single omnibus host government agreement (referred to as the 'international project agreement') in addition to an intergovernmental agreement. These agreements call for a new regulatory framework overseen by a central intergovernmental regulatory agency (the West African Gas Pipeline Agency), in which all host states are involved.[23] The intergovernmental agreement then ensures the efficacy of the single regulatory programme, limiting the applicability of laws external to the project agreements by specifying that the pipeline and the obligations of the various parties, including the agency, will be governed exclusively by specified instruments, rules and principles.[24] Thus, regulatory consistency is ensured by the delegation of each host state's right to regulate to a single international institution and by the exclusion of the application of other laws. The intergovernmental agreement and international project agreement also unified the tax scheme by applying a specific formula for taxation and setting forth a fiscal regime that has been agreed to by all states.[25]

The omnibus host government agreement approach taken in the West African pipeline project has the benefit of requiring states to agree on a uniform plan for taxation, regulatory and other matters. However, this approach requires that all states involved agree upon a uniform approach; it also imposes on the parties the

---

19	For a more detailed explanation of the Baku-Tbilisi-Ceyhan pipeline and related transaction documents, see Goolsby, George, *Cross Border Transactions: The Baku-Tbilisi-Ceyhan Crude Oil Pipeline As A Case Study*, Matthew Bender & Co (2001).
20	Article II(4).
21	Article IV.
22	Article V.
23	Article IV.
24	See Articles III and VII.
25	See Article V.

obligation to reach unanimity to amend or modify terms. Obtaining such consensus will require additional negotiation and may require further compromises on the part of the project sponsors. The advantage is that once the intergovernmental and international project agreements come into force, the project sponsors need only obtain consents and approvals from one regulatory agency. Meanwhile, the approach taken for the Baku-Tbilisi-Ceyhan pipeline requires merely that the various host government agreements be compatible – not identical. This means that the project sponsors will need to obtain consents from a regulatory agency in each state (at least for certain actions) throughout the life of the project; however, it may initially allow the project to move forward more quickly by reducing the number of parties which must agree upon the precise terms of the deal and/or any changes thereto. Ultimately, the differences between these two projects illustrate the flexibility that is available in designing and agreeing an intergovernmental/host government agreement package that is both responsive to private stakeholder needs and otherwise workable for and acceptable to the state authorities whose job will be to monitor and enforce the laws and standards embodied in the treaty, the project agreements and domestic legislation, as appropriate.

4. Conclusion

Creating a tailored intergovernmental agreement can be a difficult and time-consuming effort. However, when the project is of strategic importance and requires a high level of coordination, cooperation and support, a project-specific intergovernmental agreement may be particularly useful and should be carefully considered in project planning. Some projects are located in jurisdictions where the existing framework of protections is acceptable and the tailoring can be done in host government agreements or other project-level agreements. Other projects are not of sufficient size or strategic importance and simply do not justify use of an intergovernmental agreement. This may be particularly true where the participating states have favourable domestic laws or investor-friendly treaty regimes in place. So, while there may always be elements of risks and desires for uniformity that an intergovernmental agreement might address, the degree of risk reduction or the ability to achieve uniformity by other methods may preclude pursuit of an intergovernmental agreement. Lastly, the consequence of trying and failing must also be assessed, because the states may simply refuse to consider the idea for reasons totally unrelated to the merits of the project. Ultimately, intergovernmental agreements invite the participating states to become more active participants in the project by emphasising important state-to-state considerations. When intergovernmental agreements operate in tandem with host government agreements, they provide significant comfort to private sector stakeholders pursuing cross-border projects.

References

Agreement among the Azerbaijan Republic, Georgia and the Republic of Turkey Relating to the Transportation of Petroleum Via the Territories of the Azerbaijan Republic, Georgia and the Republic of Turkey Through the Baku-Tbilisi-Ceyhan Main

Export Pipeline, November 18 1999, *84 LNG/Gas Contracts World 1*, Barrows (March 2003).

Browning, William E and Dimitroff, Thomas J, "Transboundary pipeline development and risk mitigation", in Picton-Turbervill, Geoffrey (ed), *Oil and Gas: A Practical Handbook* (2009).

Brownlie, Ian, *Principles Of Public International Law*, Oxford University Press, (2003).

Dolzer, Rudolph and Schreuer, Christoph, *Principles Of International Investment Law*, Oxford University Press (2008).

Goolsby, George, *Cross Border Transactions: The Baku-Tbilisi-Ceyhan Crude Oil Pipeline As A Case Study*, Matthew Bender & Co (2001).

Griffin, Paul, *Transnational Gas Projects and their Agreements: A Practical and Legal Guide*, Herbert Smith (2002).

Model Intergovernmental and Host Government Agreements for Cross-Border Pipelines, Energy Charter Secretariat (2007).

Redgwell, Catherine, "International Energy Security", in *Energy Security: Managing Risk in a Dynamic Legal and Regulatory Environment*, Oxford University Press (2004).

Sornarajah, Muthucumaraswamy, *The International Law On Foreign Investment*, Cambridge University Press (2004).

Treaty of 31 January 2003 on the West African Gas Pipeline Project, *Basic Oil Laws & Concession Contracts*, Barrows (2008) (www.barrowscompany.com/Publications/Search.do).

Vienna Convention on the Law of Treaties, UNTS I-18232 (May 23 1969).

Trans-boundary energy projects and maritime transport risk

Glen Plant
Legal consultant

1. Introduction

Many energy projects involve long-distance transboundary movements of fossil or nuclear fuels, electricity or other forms of energy from source to market. Oceans, seas or international canals often lie *en route*.[1] Where they do, and where the carriage by ship of energy cargoes in large volumes is both technically feasible[2] and politically acceptable,[3] maritime means of transportation will often be more commercially competitive than terrestrial means.[4] Thus, the carriage of crude or refined petroleum oil by oil tanker, gas by liquefied natural gas or liquefied petroleum gas tanker, biofuel by chemical tanker or coal by dry-bulk carrier oil (together known as 'energy cargo ships') might well be more profitable than their pipeline, train or road vehicle equivalents. In such cases, maritime transport is likely to figure as the project's main – or at least a significant – mid-stream transportation solution. It might even be that a long maritime route will be preferred to a short terrestrial one.

How much more competitive maritime transportation is, if at all, will depend on a number of factors relevant to both maritime transport and terrestrial alternatives. Those relevant to maritime transport include such basic infrastructure matters as:

---

[1] With the exception of the international Suez and Panama Canals, this chapter does not deal with energy cargo movements by river or other waterways, across lakes or on those seas not generally considered to be open international waters subject to the general international law of the sea, such as the Caspian Sea. Although a number of each is internationalised for use by foreign ships, the author omits them for simplicity's sake, given the great variety of passage regimes involved.

[2] Electricity must generally be transported via electric cables.

[3] This is largely untrue of highly radioactive cargoes, such as used nuclear reactor fuel from Japan sent to France and the United Kingdom for reprocessing (between 1969 and 1990), and the recovered fissile materials (and associated waste) being shipped back to Japan (since 1995). Ships carrying such cargoes are heavily regulated, but enjoy the same international law rights of navigation as any other ship. Coastal states may require them only to carry special documents, observe internationally agreed special precautionary measures or confine their passage to designated sea lanes when in innocent passage though their territorial seas (Articles 22(2) and 23 of the UN Convention on the Law of the Sea 1982, 1833 UN Treaty Series 396). This notwithstanding, a number of states have prohibited the passage without prior authorisation of the ships *en route* to Japan through their territorial seas and even through their 200 nautical-mile exclusive economic zones. The ships have increasingly followed routes extended so as to keep them outside those waters to the extent possible. In this trade, ships are used less for competitive than for safety and security reasons. See further Plant, Glen, "Shipments of Irradiated Nuclear Fuel, Plutonium and High Level Radioactive Wastes in Flasks on Board Ships", *Journal of the Society of Underwater Technology* (July 2011, forthcoming).

[4] Aerial alternatives can generally be discounted, except for some high-value, low-volume nuclear energy cargoes, since the limited payloads of aircraft render the carriage of high-volume, low-per-unit-value energy cargoes cost prohibitive.

- the availability of suitable ports or terminals with adequate supply, storage and loading facilities;
- the depth and safety of the waters in question;
- the distances to be covered and chokepoints to be negotiated; and
- the major risks described in this chapter and their mitigation possibilities.

Shipping's commercial advantages arise largely from economies of scale and other efficiencies such as low fuel costs. They also arise from legal advantages, such as the prescription and enforcement of maritime regulatory standards (on such matters as safety, environmental protection and labour conditions), imposing a generally lower burden than their terrestrial equivalents. Of perhaps greatest importance is the fact that energy cargo ships enjoy liberal international rights of maritime navigation that make maritime transportation, unlike some transboundary terrestrial modes of transportation, a rent-free good. In general terms, ships of all states (and of all types and carrying all forms of cargo) enjoy in peacetime a freedom of navigation on the high seas and (in slightly more qualified terms) in foreign states' 200 nautical-mile exclusive economic zones.[5] Ships also enjoy a right of innocent passage in foreign states' territorial seas or archipelagic waters.[6] This is a more restricted right than the high seas freedom (and tankers may be required to confine their passage to designated sea lanes).[7] This right's exercise may, however, be neither hindered nor impaired by the coastal state,[8] nor suspended in straits used for international navigation lying within territorial waters (ie, international straits).[9] No state may charge foreign ships, including energy cargo ships, a tariff merely for navigating through its territorial sea – still less its exclusive economic zone or the high seas to seaward. Coastal and port states may charge only for services specifically rendered to ships.[10] This stands in contrast, for example, to tariffs charged by pipeline transit states. Special passage regimes, explained below, apply in many of the more important international straits, in part to offset their being high-risk chokepoints for energy cargo ships. In principle substantially the same, passage rights are enjoyed by

---

5   Articles 87 and 58 of the UN Convention on the Law of the Sea.
6   Ships are entitled to navigate in the waters of their own flag state as a matter of national, not international, law. In England, for example, a public right of navigation is enjoyed within territorial waters at common law, though not apparently in the 200 nautical-mile zone.
7   Article 22(2) of the UN Convention on the Law of the Sea. The convention rejects, however, any suggestion that the passage of certain ships could or should be regarded as inherently non-innocent by virtue of their cargo or size (a suggestion first made by Canada in 1970: "[p]assage by an oil tanker through the ice-bound territorial waters [of the North-West Passage] is inherently non-innocent"; see Wulf, N, 46 *Law and Contemp Prob* (1983), 155, 163); the influential US/USSR Joint Statement on Uniform Acceptance of Rules of International Law Governing Innocent Passage (September 23 1989) (*US Digest*, Ch 7, §2) confirms the then superpowers' opposition to the suggestion.
8   Part II, Section 3 and Article 52 of the UN Convention on the Law of the Sea.
9   Article 45(2), *ibid*.
10 See for example Article 26, *ibid*. This is also true of the international (chokepoint) straits discussed below. See, for example, Article 2 and Annex 1 of the Convention Regarding the Straits, Montreux (July 20 1936), 173 *League of Nations Treaty Series 213* (and Plant, Glen, "The Turkish Straits and Tanker Traffic: an Update", 24 *Marine Policy* (2000), 193, 212-13); and the Treaty for the Redemption of the Sound Dues, Copenhagen (March 14 1857), 116 *Consolidated Treaty Series* 357 (and the similar separate agreement with the United States), the very purpose of which was the remission in perpetuity of the Sound Dues formerly charged as the price of transit. International canals are, of course, exceptions, being artificial and costly to construct and maintain; a tariff is payable to transit both the Panama and Suez Canals.

neutral vessels in times of armed conflict.[11]

Maritime transport nevertheless involves risks – commercial, political, security and other. This chapter is concerned with legal aspects of the major risks, in particular those that are partly defined by law and are amenable to a degree of legal control. Such major risks can be roughly classified into those arising in the context of armed conflict and those arising in time of peace,[12] bearing in mind that in the past half-century, the concepts of peace and war have become blurred to the extent that it is not always possible to draw neat distinctions between the two.[13] They can also be classified, again roughly, into security risks, the risks of excessive law enforcement measures,[14] environmental protection law risks, safety risks, risk of waterway blockage and risk of being targeted by direct action protesters. The security risks considered in this chapter are few: belligerent acts, piracy, ship hijacking and armed robbery, and terrorist acts. Natalie Klein's list of such risks also extends to illicit trafficking in arms and weapons of mass destruction, illicit trafficking in drugs, smuggling and trafficking of persons by sea, illegal, unreported and unregulated fishing, and intentional and unlawful damage to the marine environment.[15] This chapter does not consider these in detail, however, because they refer to acts unlikely to be committed by, or to, energy cargo ships in the normal course of trade – that is, where the trade is not a mere front for maverick state regimes, terrorists or organised crime.

Where the level of one or more of these major risks is perceived to be high along a maritime route, or routes, putatively connected to an international energy project, this might be judged to outweigh normal commercial advantages of transportation by sea. One can expect to see mid-stream adjustments in favour of terrestrial means of transport when the overall identifiable risks associated with maritime passage outweigh, or at least substantially outweigh, both its commercial premium and the identifiable risks of use of those terrestrial means. Examples include the construction of oil pipelines to bypass the Turkish Straits, starting with the first use of the Baku-Tiblisi-Ceyhan pipeline in May 2006. When oil tankers trying to exercise in principle liberal passage rights through a chokepoint such as the Turkish Straits suffer delays (in this case for a combination of physical and human reasons), sometimes of many days (each day resulting in a demurrage charge calculated in tens of thousands of dollars), those rights cease to have the appearance of a rent-free good.[16] The high capital costs of constructing such bypass pipelines are at least partly offset, moreover, by their termination at deepwater ports or terminals: in the case of the Baku-Tiblisi-Ceyhan pipeline, crude (from the Caspian Sea) may be loaded in Ceyhan on board

---

11 See Part I, *San Remo Manual on International Law Applicable to Armed Conflicts at Sea*, available online at www.icrc.org/ihl.nsf/FULL/560?OpenDocument and with *Commentary*, Cambridge (1995). Belligerent vessels' navigation rights are modified under the law of armed conflict.
12 Control over peacetime risks is likely to be easier than that over wartime risks.
13 Paragraph 4(1) of the US Department of the Navy, Office of the Chief of Naval Operations, *Commander's Handbook of the Law of Naval Operations*, Naval Warfare Publication NWP1-14M, 1995- (*Commander's Handbook*), available online at usnwc.edu/getattachment/a9b8e92d-2c8d-4779-9925-0defea93325c/1-14M (July 2007 version).
14 All unjustified and excessive law enforcement actions in principle engage state responsibility under international law and are compensable. See, for example, Articles 106, 110(3), 111(8) and 232 of the UN Convention on the Law of the Sea. See also the safeguards concerning coastal state enforcement of environmental standards, in Part XII, Section 7 of that convention.
15 *Maritime Security and the Law of the Sea*, Oxford (2011), Cap1.

very or ultra large crude carriers far more cost efficient in carrying it to US markets than the smaller Suezmax tankers of the right dimensions to transit the Bosphorus.

Risks for maritime energy cargoes are indeed likely to be particularly high in maritime chokepoints – those narrows or canals that are:

- of strategic significance for energy cargo flows;
- impossible or expensive to avoid; and
- particularly vulnerable for physical, political or other reasons.[17]

"The blockage of a chokepoint, even temporarily, can lead to substantial increases in total energy costs. In addition, chokepoints leave oil tankers vulnerable to theft from pirates, terrorist attacks, and political unrest in the form of wars or hostilities as well as shipping accidents which can lead to disastrous oil spills."[18]

The two most important chokepoints are:

- the Strait of Hormuz, which saw the transit of between 16.5 and 17 million barrels per day of oil in 2008 – about 40% of all seaborne traded oil;[19] and
- the Straits of Malacca and Singapore (15 million barrels per day in 2006).[20]

Others of importance are:

- the Suez Canal (4.5 million barrels per day in 2006);[21]
- the Strait of Bab el-Mandeb (3.3 million barrels per day in 2006);[22]
- the Turkish Straits (2.9 million barrels per day in 2009 and likely to increase);[23]
- the Danish Straits (3.3 million barrels per day in 2009 and likely to increase);[24] and

---

16   Plant, Glen, *supra* Footnote 10, at 193-94; Elkind, J, "Economic Implications of the Baku-Tbilisi-Ceyhan Pipeline", in Starr, S and Cornell, S (eds), *The Baku-Tbilisi-Ceyhan Pipeline: Oil Window to the West*, Central Asia-Caucasus Institute & Silk Road Studies Program (2005), 39, pp46-47 (www.silkroadstudies.org/new/inside/publications/BTC.pdf). Bypass pipelines should be distinguished from parallel pipelines, which provide a supplementary service to energy cargo ships transiting chokepoints.

17   Although the Strait of Dover and Fair Isle Channel are, like certain other transit passage straits, important for energy cargo ships, they are not listed as chokepoints because energy cargo ships of all sizes face no more than normal navigational difficulties when transiting them in ordinary peacetime circumstances.

18   US Department of Energy, Energy Information Administration, "World Oil Transit Chokepoints" (www.eia.doe.gov/cabs/World_Oil_Transit_Chokepoints/Full.html).

19   *Ibid.* The volumes dropped to 15.5 million barrels per day during the 2009 recession (*The Economist*, February 5 2011, map on p72). This transit passage strait is the only maritime route into and out of the Persian Gulf for oil, liquefied natural gas and liquefied petroleum gas tankers serving markets worldwide and is capable of accommodating the largest tankers.

20   US Energy Information Administration, *ibid*. The volumes dropped to 13.6 million barrels per day during the 2009 recession (*The Economist*, *ibid*). The volumes of transiting oil are rising by about 3% per year (Roach, A, "Malacca, Straits of", *Max Planck Encyclopaedia of International Law*, 8th online edition (www.mpepil.com) (hereafter *Max Planck Encyclopaedia*).

21   In 2006 an estimated 4.5 million barrels per day of oil flowed through the Suez Canal, mostly northbound to EU or US markets (US Energy Information Administration, *ibid*). While oil volumes have recently declined, partly as a result of recession, Somali piracy (see *infra*) and political unrest, liquefied natural gas transits have increased (to 17.5 million metric tons in 2009) (www.eia.doe.gov/cabs/Egypt/Oil.html).

22   US Energy Information Administration, *ibid*. This transit passage strait is the only maritime access for energy cargo ships into and out of the Red Sea.

23   *The Economist*, *supra* footnote 19.

24   *The Economist, ibid*. Russia is exporting increasing volumes of crude oil via the Danish Straits.

25   *Ibid*.

- the Panama Canal (over 0.8 million barrels per day in 2009).²⁵

International law takes account of, and seeks to offset, the chokepoint effect of all of these international straits and canals.

In the listed, as in other, important international straits, foreign ships enjoy not a mere right of innocent passage, but either:
- a right of transit passage closer to the high seas freedom,²⁶ as for example in the Strait of Hormuz, the Strait of Bab el-Mandeb and the Straits of Malacca and Singapore (hereafter referred to as 'transit passage straits'); or
- a passage regime governed in whole or in part by a longstanding international convention (or conventions) in force specifically relating to such straits (see Article 35(c) of the UN Convention on the Law of the Sea), as with the Danish and Turkish Straits (hereafter referred to as 'Article 35(c) straits').

It does not follow that tankers of all sizes can use all of these straits. The reasons for the size restriction mentioned above on tankers carrying Russian and Caspian crude oil through the Turkish Straits are partly physical (the Bosphorus is particularly narrow and winding) and partly the result of bordering state regulations.²⁷ The main channel in the deepest of the three routes through the Danish Straits, the Great Belt, is dredged to a minimum depth of only 17 metres. There is no alternative maritime route for tankers in either case. Although, moreover, the Malacca Strait is a transit passage strait, an agreed interpretation of several provisions of the UN Convention on the Law of the Sea has the practical effect of imposing a 3.5 metre minimum under-keel clearance requirement on transiting foreign tankers,²⁸ which prevents transits by the largest tankers (the very large crude carriers of between 200,000 and 300,000 dead-weight tonnage and the ultra large crude carriers of over 300,000 dead-weight tonnage). Alternative tanker routes to Far Eastern markets through Indonesian straits are considerably longer.

Foreign energy cargo ships are also afforded broad rights of transit through the international canal chokepoints, subject to their obeying the relevant rules and regulations, which include the payment of tolls. Transit through the Suez Canal is allowed to vessels of all nations, subject to their complying with the conditions stated in the Canal Authority Rules of Navigation and other Egyptian rules and regulations.²⁹ Transit through the Panama Canal is allowed to vessels of all nations, subject to their complying with the Canal Authority's Regulation on Navigation in Panama Canal Waters.³⁰ Tankers transiting the canals are, however, restricted, depending on dimensions permitted by the Canal Authorities, respectively to

---

26 Part III, Section 2 and Article 53 of the UN Convention on the Law of the Sea.
27 Plant, Glen, *supra* footnote 10, and "Dardanelles and Bosphorus", *Max Planck Encyclopaedia*.
28 Letter from the Malaysian Government to the President of the Third UN Conference on the Law of the Sea, dated April 28 1982, UN Document A/CONF62/L145, XVI OR 250 (Malaysia), and /Adds 1-8, *ibid*, pp251-53 (Australia, France, Germany, Japan, the United Kingdom and the United States).
29 Article 1 of the Convention Respecting the Free Navigation of the Suez Maritime Canal, Constantinople (1888). The rules can be found at www.suezcanal.gov.eg/NR.aspx and www.lethsuez.com/SCA_Circulars/SCA_frames.asp?circ=circ012010.htm.

Suezmax size (120,000 to 200,000 dead-weight tonnage)[31] and Panamax size (50,000 to 80,000 dead-weight tonnage).[32] For vessels of and below these sizes, the canals are nonetheless chokepoints of choice in normal circumstances, since the alternative routes around Cape Horn and the Cape of Good Hope add respectively at least 8,000 nautical miles and 6,000 nautical miles to an energy cargo ship's voyage. The present Panama Canal expansion project will, when completed, permit larger (though still medium-sized) tankers to transit it; more importantly, it will open its use to liquefied natural gas tankers and coal bulkers from South America.

In all events, the majority of transboundary energy cargo movements (reckoned both by mass and by value) are by sea.[33] Some oil or other energy cargoes are still carried in tankers owned, or demise-chartered, by large (integrated) oil company fleets or the state of production. The vast majority is, however, carried by independent corporate operators, which own or charter vessels operating under the flags of a wide variety of states.

The flag state is primarily responsible for prescribing and enforcing standards for its ships. Shipping is a truly international business. While there must be some genuine link between ship and flag state, each state is free to fix the conditions for the granting of nationality, registration and flag to ships;[34] it is relatively cheap and easy to change nationality. The industry is therefore united in its desire to have uniform, globally agreed minimum standards – set out, for example, in safety and environmental protection instruments of the International Maritime Organisation and labour instruments of the International Labour Organisation, and incorporated into national laws and regulations. It follows in principle that where energy cargo ships face risks that arise out of states' legislation based on those global standards, those risks should be uniform wherever the ship sails.

Important in this respect is that the – in principle – exclusive jurisdiction of the flag state of an energy cargo ship is subject to a limited, though growing, number of exceptions permitting (but not requiring) coastal or port states to act. Those of greatest importance in this context are:

- the belligerent right of visit and search during armed conflict to ensure that an energy cargo ship is not contributing to the enemy's war effort;
- the universal jurisdiction, unique in peacetime, afforded to any state, under the customary international law of piracy *iure gentium*, to board and seize a pirate ship intended for use, or being used, for piratical acts against an energy cargo ship, to arrest the persons and seize the property on board and decide upon the penalties to be imposed;[35]

---

30 Articles 2 and 3 of the Treaty Governing the Permanent Neutrality and Operation of the Panama Canal, Washington DC (September 7 1977), 1161 *UNTS* 182. The rules can be found at www.pancanal.com/eng/legal/reglamentos/navigation-compilation.pdf.
31 Section II of the Rules of Navigation and Suez Canal Authority Circ 2/2010. It is common practice for northbound tankers too laden to enter the canal to lighten at and send part of their loads through the parallel Sumed pipeline before transiting the canal. See www.eia.doe.gov/cabs/Egypt/Oil.html.
32 There is a westbound parallel pipeline, the Trans-Panama.
33 "In 2007 total world oil production amounted to approximately 85 million barrels per day, and around one-half, or over 43 million barrels per day, of oil was moved by tankers on fixed maritime routes" (US Energy Information Administration, *supra* footnote 18).
34 Article 91 of the UN Convention on the Law of the Sea.

- the right of hot pursuit by a coastal state of foreign vessels reasonably believed to have breached that state's laws in its coastal waters, with a view to stopping and arresting those vessels. This is of potential utility in the apprehension, for example, of pirates or terrorists attacking energy cargo ships;[36]
- multilateral or bilateral consent agreements whereby the flag state permits another state to exercise jurisdiction – for example, against suspected terrorists (or other malefactors) – on board an energy cargo ship;
- a port state's right to place a wide range of conditions on foreign energy cargo ships' voluntary calls at its ports or offshore terminals, based on its sovereignty in territorial waters (into which those energy cargo ships have no right of entry). Such conditions normally include environmental protection, safety and labour (and increasingly security)[37] standards. These can in principle be applied to the ship well before it reaches the port state's territorial or jurisdictional waters. They cannot be applied, however, to energy cargo ships merely transiting those waters without calling at a port or terminal: it is far easier, for example, for the near continental-sized United States, which has a good deal of inbound but little transiting energy cargo traffic, to impose higher than globally agreed standards than it is for the 27 member states of the European Union. This is clear from the United States' refusal to join the International Maritime Organisation's global oil tanker cargo spill pollution liability compensation scheme and its establishment of a unilateral regime generally more onerous for foreign oil tankers;[38] and
- the UN Convention on the Law of the Sea according a degree of concurrent prescriptive and enforcement jurisdiction to coastal states over foreign ships in their territorial seas and, to a lesser degree, their exclusive economic zones as a result of the failure of some flag states properly to enforce the minimum standards – especially environmental ones – set by the International Maritime Organisation or the International Labour Organisation. That jurisdiction is, however, tied to those globally agreed standards, at least as regards construction, design, equipment and manning matters that travel with the ship and cannot easily be changed as it enters different jurisdictions.[39] The exception is therefore a limited one.

Given both the need to obtain global consensus on minimum shipping standards and the complex jurisdictional issues surrounding ships that might pass through a number of jurisdictions on any given voyage, ships might benefit from generally less

---

35 Articles 101 to 107 *ibid*. See further *infra* section 3.2.
36 Article 111 *ibid*.
37 Klein, Natalie, "Legal Implications of Australia's Maritime Identification System", 55 *ICLQ* (2006), p337. The zone's main purpose is in fact to enhance surveillance in protecting offshore petroleum facilities from terrorism.
38 BMT/Plant, Glen, "Final Report of the Study on the Economic, Legal, Environmental and Practical Implications of a European Union System to Reduce Ship Emissions of SO2 and NOx", No 3623 (April 2000) (ec.europa.eu/environment/enveco/taxation/ship_emissions/pdf/mainfinal.pdf) and Appendix 4 (on file with author).
39 See especially Articles 21, 41, 42, 54, 211 and 220 of the UN Convention on the Law of the Sea.

onerous environmental standards than those applicable to terrestrial modes of transport; indeed, they might even be relieved of the burden of some environmental standards altogether. International shipping enjoys a free ride, for example, in having no express international law duty to mitigate its greenhouse gas emissions. Ship greenhouse gas emissions are omitted, unlike terrestrial transportation emissions, from national greenhouse gas inventories. Consequently, those emissions are also omitted from mitigation obligations mandated by the 1992 UN Framework Convention on Climate Change[40] and its 1997 Kyoto Protocol[41] for Annex I states parties (ie, developed states or those with economies in transition). It has to date proved impossible to reconcile the uniform global standards approach of the International Maritime Organisation, which is charged by Article 2(2) of the Kyoto Protocol with negotiating greenhouse gas standards for ships, with a fundamental principle of the UN Framework Convention on Climate Change, of common but differentiated responsibility, that requires in the first instance binding commitments from Annex I states alone.[42] Mitigation of ship emissions was placed on the agenda of the UN Framework Convention on Climate Change Conference at Cancun in December 2010. It was, however, dropped "[w]hen it became clear that parties would be unable to overcome their differences ... the global nature of the industry mak[ing] it difficult to make decisions on jurisdiction".[43] Regional action by the European Union is possible, probably through the addition of ship greenhouse gas emissions to the EU Emissions Trading Scheme. This might, however, result in legal challenges by foreign operators faced with increased operating costs that they regard as subsidies paid to support the EU scheme; US operators have challenged the comparable inclusion of international aviation emissions in the scheme.[44]

The above-described choice between maritime and terrestrial transportation options is complicated by further possibilities.

First, there might be more than one possible maritime route. Where an international energy project has chosen to rely on maritime transportation, choice of one route and/or vessel characteristic (eg, size or speed) over another will take into account not merely commercial but also major risk factors, including chokepoint concerns.

Second, one terrestrial option can have maritime features: states enjoy a right under international law to lay pipelines (or electric cables) across or beneath the maritime seabed, including that of the high seas or foreign states' continental shelves. This justifies transboundary energy projects, when deciding between ship and pipeline options, in making comparison between international pipeline

---

40   New York (May 9 1992), 1771 *UNTS* 107.
41   Kyoto (December 11 1997), 37 ILM (1998) 22.
42   See Plant, Glen, "Legal issues surrounding Ship GHG Emission Reduction Policy Options", European Commission *Technical support for European Action to Reducing (sic) Greenhouse Gas Emissions from International Maritime Transport*, Contract 07010401/2008/507628/ATA/C3, 2008 (on file with author).
43   "Cancun Climate Summit Exceeds Low Expectations, but Sidesteps Trade Issues", International Centre for Trade and Sustainable Development, 14(44) *Bridges Weekly Trade News Digest* (December 22 2010) (www.ictsd.org/i/news/bridgesweekly/99004).
44   The Air Transport Association of America succeeded, in May 2010, in having the English High Court refer the validity of the relevant EU measures to the European Court of Justice for a preliminary ruling (www.airlines.org/News/Releases/Pages/news_5-27-10.aspx).

transportation rights under the law of the sea and energy cargo ships' rights of navigation,[45] and of the relative risks of relying on one or the other. In some such projects it also involves a comparison between the risks and advantages of a maritime pipeline route and those of a terrestrial pipeline route: the North Stream trans-Baltic gas pipeline from Russia to Germany, for example, which crosses the territorial or jurisdictional waters of five states,[46] is designed to bypass Ukrainian or Belarusian territory, through which the transit of Russian gas has resulted in frequent disputes in recent years.[47]

The major energy cargo ship risks are dealt with in detail below, in intuitively apparent descending order of risk. The author makes the caveats, however, that this order will not always reflect reality (eg, a terrorism incident might on occasion cause more harm than an act of war), and that there is potential for overlap in the senses that:

- a non-war security risk carries the potential to escalate into a war risk;
- categories might overlap (eg, terrorists funding their activities with organised crime or piracy); and
- categories might even on occasion be indistinguishable – for example, although some hold that a terrorist can never be a pirate (because he is motivated by political, not private) ends, an alternative view is that the relevant distinction is not political versus private, but public versus private, that any act of violence on the high seas not attributable to, or sanctioned by, a state (as a public act) is piracy (a private act).[48] At least one judicial decision seems to support this view.[49]

---

[45] Any state may exercise the high seas freedom to lay submarine cables and pipelines if it does so with due regard for the interests of other states in their exercise of their high seas freedoms (or deep seabed activities) and for cables and pipelines already in position and their reparability (Articles 87 and 112 of the UN Convention on the Law of the Sea; see also Articles 113 to 115)). A state also enjoys:
- in foreign exclusive economic zones "subject to the relevant provisions of [UN Convention on the Law of the Sea], the freedoms referred to in article 87 of the laying of submarine cables and pipelines", which must be exercised with due regard for the resource and marine environmental protection rights and laws of the coastal states in its exclusive economic zone (Article 58 of the convention);
- on foreign continental shelves extending beyond the exclusive economic zone, the entitlement to lay submarine cables and pipelines, which the coastal state may not impede, subject to its right to take reasonable exploration and exploitation and environmental protection measures and to consent to the precise delineation of their courses (Article 79 of the convention); and
- no rights at all in foreign territorial waters (Article 79(4) of the convention).

The first two bullet points are comparable to the freedom of navigation of energy cargo ships in the superjacent waters, but the third bears no comparison with the rights of innocent or transit passage in the superjacent waters.

[46] See www.nord-stream.com and especially www.nord-stream.com/en/press0/press-releases/press-release/article/pipe-laying-now-underway-in-the-waters-of-five-baltic-sea-countries.html?tx_ttnews [backPid]=.

[47] "Gas wars: the dispute between Russia and Ukraine shows that Europe must reduce its energy vulnerability", *The Economist* (January 8 2009) (www.economist.com/node/12899511). "Russia-Belarus gas dispute: Commission strongly concerned about gas cuts in Lithuania", European Commission Press Release IP/10/797, Brussels (June 23 2010) (europa.eu/rapid/pressReleasesAction.do?reference=IP/10/797&format=HTML&aged=0&language=EN&guiLanguage=en). But see now Olearchyk, R, "Russia and Ukraine settle dispute over gas", *Financial Times* (November 30 2010) (www.ft.com/cms/s/0/dd338a26-fcba-11df-bfdd-00144feab49a.html?ftcamp=crm/email/2010121/nbe/EnergyMining/product#axzz 16sntuVup).

[48] Guilfoyle, D, *Shipping Interdiction and the Law of the Sea*, Cambridge (2009), pp36-40.

[49] *John Castle v MV Mabeco*, Belgian Court of Cassation (1986, 77 *ILR* 537).

In the discussion, regard is given to the possibility of litigation arising in national courts (and governed by national laws) between private parties interested in an energy cargo ship venture over loss or damage arising out of risked events. Such are the complexities of typical large vessel ownership, operation and finance, as well as of contracts of carriage or charterparties, that the risk of liability might fall on a number of different parties. However, rather than regard the risk of litigation as an additional major risk particularly associated with shipping segments of transboundary energy projects, one should recall that complex arrangements concerning liability for potential risks also surround transboundary pipelines, and that in practical terms the ultimate risk in both instances normally falls on the insurance and reinsurance markets (sometimes supplemented by government support).

In analysing the private claim risks particularly associated with the shipping segment, therefore, it makes more sense to have regard to the insurability *vel non* on commercial markets of major risks to energy cargo ships and the desirability of taking out marine insurance policies against them – bearing in mind the levels of risk and the amounts and costs of available cover. Large oil companies or governments operating their own tanker fleets might be able to self-insure, but independent tanker operators are generally too small to do so.

Typical commercial shipping ventures require a variety of different forms of marine insurance, most of which is taken out in Europe.[50] The normal policies taken out cover hull and machinery, freight and cargo. Each country's marine insurance companies will offer cover, on a time or voyage basis, and providing full or (at a lower premium) lesser levels of cover, and subject to conditions and warranties, such as that relating to seaworthiness. This will usually reflect the standard clauses of the country's institute of underwriters or equivalent, and seek to comply with its national marine insurance law. In England, long a leading marine insurance centre, to which the author refers below for illustrative purposes,[51] the International Underwriting Association of London (successor to the Institute of London Underwriters) prepares and regularly updates standard clauses. In addition, the UK Marine Insurance Act 1906 makes binding provision or raises presumptions on matters not displaced by agreement of the parties. Shipowners generally seek cover

---

50 England, Norway, Germany, France and other European countries took 59% of world marine insurance premiums in 2009, Japan 10%, the United States 9% and the rest of the world 22% (*Global Financial Markets: Regional Trends* (November 2010) (www.thecityuk.com)).
51 Useful reference might also be made in particular to the German General Rules of Marine Insurance, the Norwegian Marine Insurance Plan and the American Institute Clauses.
52 Such risks might extend, for example, to:
 • life salvage and certain salvage expenses;
 • personal injury;
 • towage contract;
 • collision, pollution and wreck;
 • contract and indemnity;
 • cargo liabilities;
 • loss of personal effects;
 • fines;
 • legal costs;
 • war risks excluded from the standard policies; and
 • omnibus cover for new risks.
53 33 USC 2701.

for third-party liability risks[52] through a mutual policy with their P&I club (a cooperative insurance association that provides cover on a protection and indemnity basis), the cover found acceptable by the club being governed by the club's rulebook. Higher levels of cover are possible with the club's members of the International Group, which operates pooling arrangements.

Where the normal hull, cargo or freight policy clauses exclude unusual risks, such as war or terrorism, the shipping venture might (for an additional premium) obtain marine insurance cover underwritten normally in conformity with special International Underwriting Association (formerly Institute of London Underwriters) clauses, or under its protection and indemnity cover. In a worst case, if there is no government provision to supply a gap in the insurance market, the shipping venture might be left to bear the risk alone.

The highly competitive nature of the world marine insurance and protection and indemnity markets tends to underscore the desirability of uniform global regulatory standards. Just as energy cargo ships should face uniform risks arising out of legislation based on those global standards, so they should not face the risk of having to pay different insurance premiums with respect to the risks posed by those standards according to where they sail. Generally this is the case, notwithstanding differences between national laws and between international and national laws governing war, piracy, terrorism and other risks. There is, however, one major exception: the higher oil pollution liability risk faced by oil tankers trading to the United States. Indeed, the United States' unilateral regime, established by the federal Oil Pollution Act 1990,[53] can be said to add a fourth type of normal insurance to those tankers – pollution insurance. An oil tanker wishing to trade to the United States must, among other things, produce evidence of financial responsibility to meet the liability requirements of the act, which differ in limitation levels and ease of breaking them from the globally agreed standards.[54] The authorities may deny entry to US waters or ports, or detain a non-complying tanker, which (with its cargo) may be subject to seizure and forfeiture. Certificates of financial responsibility that differ from the global oil pollution liability certificates are required as proof (and, in effect, security) from a guarantor, which may be sued directly by a claimant. The high level of financial responsibility to be demonstrated is generally based on a 'worst-case discharge' estimate. The International Group of P&I Clubs was for several years not prepared to provide cover, fearful that the liability limits under the US Oil Pollution Act 1990 would be readily broken by the courts and that the act would force them to provide cover contrary to their own club rules.[55]

Eventually, four means of guarantee emerged, the main ones being:
- fixed premium company insurance (now with the Shipowners' Insurance and Guaranty Co Ltd, which covers over 65% of the market); and
- membership of a regional mutual P&I club, Shoreline Mutual, covering claims in excess of normal protection and indemnity cover for oil pollution

---

54 See *infra* footnotes 140 to 142 and accompanying text.
55 See, for example, McQuiston, R, "The Oil Pollution Act of 1990: a Review of Three Years of Intensive Rulemaking", *Butterworths Journal of International Banking and Financial Law* (June 1993), 275, p278.
56 See further www.cofr.com and www.shoreline.bm. The other methods are self-insurance (5% of the

($500 million), which only traders to the United States pay (with a 30% market share).[56]

This was, however, almost certainly only because of the economic and financial importance of the US market.

The European Union is also increasingly prepared to set regional environmental standards for ships where it views the global standards to be too low. This carries the risk of regionally increased costs for energy cargo ships calling at EU ports or terminals (or possibly merely transiting EU waters). The danger of differing standards should not be overstated, however, since EU threats to adopt regional standards have several times served as an inducement to the International Maritime Organisation to raise the global standards.[57]

Regard is also given to the fact that increased risks to energy cargo ships might lead directly to higher operating costs – for example, through the use of more fuel in a diversion or the payment of 'danger money' to crews. Seamen have often been paid more highly in war zones and might be paid more in other perilous waters. In 2008, for example, part of the Gulf of Aden was declared a high-risk area, in view of Somali piracy, entitling transiting seafarers covered under terms agreed by the International Bargaining Forum to double pay and death benefits, costing the shipping industry between $30 million and $40 million a year. In 2011 the arrangement was extended to the Arabian Sea and the northern Indian Ocean.[58]

2. **War risks**

An energy cargo ship runs the risk in certain sea areas of being affected or caught up in international armed conflict, with the possibility of expensive delay or diversion, collateral damage to it and its cargo, or even capture or destruction.

In principle, (lawful) belligerent acts may take place anywhere at sea, except in neutral territorial waters.[59] In practice, however, the limited size and range of most navies tend to restrict such acts to waters adjacent to one or more of the warring states. Even so, the impact on neutral energy cargo ship movements can be considerable. Although interference with, and attacks upon, merchant ships – and in particular neutral oil tankers – by both sides during the 1980-88 Iran-Iraq War were at first limited to the northern part of the Persian Gulf, they gradually crept further and further south and east to occur throughout the Persian Gulf, as well as in the Strait of Hormuz and Gulf of Oman.[60] Over 400 ships, mostly oil tankers, were damaged (80 sunk or declared total losses), over 200 seamen's lives were lost and numerous polluting oil spills occurred.[61]

---

	market share) and procuring a bond from an approved US surety company (rare).
57	Including in relation to the phase-out of single hull oil tankers and to raising oil pollution compensation limits; but see the *Intertanko Case* (C-308/06, European Court of Justice (Grand Chamber), June 3 2008).
58	Matthews S, "Employers agree to extend piracy high risk area: Agreement seeks protection of seafarers through the provision of increased security measures", *Lloyd's List* (March 28 2011).
59	*San Remo Manual, supra* footnote 11, Part II, especially Paragraphs 14-22.
60	Walker, G (ed), *The Tanker War, 1980-88: Law and Policy*, 74 International Law Studies Series, US Naval War College (2000), pp40-75.
61	*Ibid*, at 74-75.

Energy cargo ships might also fall foul of internal armed conflicts (where insurgents have succeeded in establishing control over a defined portion of a state's territory) or of lesser insurgencies, civil commotions or riots. While loss or damage occasioned to an energy cargo ship by such conflict might be serious, it is far more likely to occur in a port or at a terminal than at sea, since those conducting the uprising or riot are unlikely to have significant navies or access to armed vessels.

The discussion below concentrates on risks encountered at sea during international armed conflict, since while energy cargo ships will always be advised to avoid unsafe ports or terminals, they might find it commercially undesirable or impossible to avoid areas at sea rendered less safe by international armed conflict. Energy projects were certainly not going to halt oil shipments from the Persian Gulf during the 1980-88 war.

In times of armed conflict, "the fundamental role of [belligerent] navies [is] to establish control at sea or to deny it to the enemy, linking that control to broad political and economic issues ashore".[62] A navy's efforts at control will not be restricted to enemy warships and merchant vessels; they will extend to interdiction of neutral shipping suspected of carrying contraband cargo to the enemy or otherwise aiding its war effort. This clearly conflicts with the general international interest in freedom of passage and commerce reflected in the provisions on navigation of the UN Convention on the Law of the Sea. The international law of naval warfare has traditionally sought to achieve a balance between these two conflicting interests by permitting peacetime international navigation rights to continue to apply in time of armed conflict,[63] but with some modifications permitting the exercise of belligerent rights – notably:

- for warships to visit and search a neutral ship to ascertain whether it is subject to capture – for example, because it is transporting goods considered to be contraband;[64]
- to place restrictions on, or prohibit the entry of, neutral ships into the immediate area of operations;[65]
- to use naval mines in accordance with specific rules;[66] and
- to enforce an effective blockade of enemy ports or coastal waters by preventing neutral as well as enemy ships passing through the outer curtain of the blockade to deliver goods or personnel to the enemy.[67]

---

62 Roach, J Ashley, "The Law of Naval Warfare at the Turn of Two Centuries", 94 *Am J Int'l L* (2000) 64, p64. See also *Eastern Extension Telegraph Co Ltd v United States*, UK-US Arbitration (1923) 6 RIAA 112.
63 *San Remo Manual, supra* footnote 11, especially Paragraphs 23-37.
64 *Ibid*, Paragraphs 118-19 and 121. A neutral merchant ship and cargo may be captured as prize for adjudication – or exceptionally destroyed when military circumstances preclude such taking in for adjudication – if shown by visit and search to be carrying contraband or conducting certain other acts (*ibid*, Paragraphs 146-51). If a neutral merchant ship resists visit and search or commits other acts giving it an enemy character, it may even be attacked (*ibid*, Paragraphs 67-68). Some naval powers recognise an exception in favour of neutral vessels travelling in convoy under the protection of a neutral warship (*ibid*, Paragraph 120), but the United Kingdom for one does not (Ronzitti, "Naval Warfare", *Max Planck Encyclopaedia*, Paragraph 16). Unfortunately, the question of whether an energy cargo is contraband is not straightforward. Belligerents might fail to produce lists of what they consider to be contraband (as Iran failed to do in the 1980-88 war), and even the *San Remo Manual* provides no indicative list.
65 Von Heinegg, W, "War Zones", *Max Planck Encyclopaedia*, Paragraph 6.
66 *San Remo Manual, supra* footnote 11, Paragraphs 80-91.
67 *Ibid*, Paragraphs 93-104.

The international law of naval warfare is a mix of customary and treaty law. Many of the treaty provisions are over a century old[68] and do not adequately reflect modern naval technological developments, major changes to the law of the sea (notably in the 1982 UN Convention on the Law of the Sea)[69] or even the practice developed during the World Wars. Consequently, the most authoritative place to seek the law now is generally considered to be the *San Remo Manual on International Law Applicable to Armed Conflicts at Sea*, prepared in the 1990s by a group of international lawyers and naval experts.[70] It is usually referred to along with the most important national naval manuals.[71]

Another major development that might be argued to displace[72] or, more likely, blur[73] the present-day operation of the traditional rules of naval warfare is the role of the UN Security Council under the UN Charter in regulating the lawful use of force. In principle, the charter prohibits resort to force, including at sea, except in the exercise of a closely circumscribed right of individual or collective self-defence,[74] or unless authorised by the Security Council acting under its Chapter VII (having determined that there is a threat to the peace, a breach of the peace or an act of aggression). When the council so acts, there is in principle no room for neutrality; all states should cooperate in removing the threat to the peace.[75] In practice, however, political agreement in the council to authorise use of naval force against a state under Chapter VII has been the exception rather than the rule. The council has

---

68	Those still of some relevance in this context include The Hague Conventions VIII (Relative to the Laying of Automatic Submarine Contact Mines), XI (Relative to Certain Restrictions with Regard to the Exercise of the Right of Capture in Naval War) and XIII (Concerning the Rights and Duties of Neutral Powers in Naval War), The Hague (October 18 1907), 2 *Am J Int'l L (Supp)*, pp138, 167 and 202.
69	It has been suggested that the UN Convention on the Law of the Sea now replaces many of the rights and responsibilities drawn from the laws of naval warfare. See, for instance, Lowe, AV, "The Commander's Handbook on the Law of Naval Operations and the Contemporary Law of the Sea", in Robertson, H (ed), *The Law of Naval Operations*, 64 International Law Studies, US Naval War College, 1991, 111, pp130-133, available online at www.usnwc.edu/Research---Gaming/International-Law/RightsideLinks/Studies-Series/documents/Naval-War-College-vol-64.aspx. *Contra* Astley, J III and Schmitt, M, "The Law of the Sea and Naval Operations", 42 *Air Force L Rev* (1997) 119, p138.
70	This is operational in flavour, being addressed to the wagers of a conflict rather than to prize courts or neutrals, and so omits certain controversial issues and tends to treat the neutral rights relevant to, among other things, energy cargo ships as the bits left over after belligerent rights have been exercised. See, for example, Neff, S, *The Rights and Duties of Neutrals: A General History*, Manchester (2000), pp198-206.
71	Notably the US *Commander's Handbook*, *supra* footnote 13, for which there is an *Annotated Supplement*, Thomas, A and Duncan, J, 73 International Law Studies, US Naval War College (1999), available online at www.usnwc.edu/Research---Gaming/International-Law/RightsideLinks/Studies-Series/documents/Naval-War-College-vol-73.aspx. Others dealing extensively with law of naval warfare issues are: • Royal Australian Navy, *Manual of the Law of the Sea*, ABR 5179 (1983); • Canada National Defence, *The Law of Armed Conflict at the Operational and Tactical Levels* (2004), especially in Cap 8, available online at www.forces.gc.ca/jag/publications/Training-formation/LOAC-DDCA_2004-eng.pdf; and • Fleck, D (ed), *Handbook of International Humanitarian Law*, 2nd ed, Oxford (2008), an annotated translation of the German manual. See also UK Ministry of Defence, *The Manual of the Law of Armed Conflict*, Oxford (2004).
72	See Lowe, *supra* footnote 69.
73	See Von Heinegg, W, "The Protection of Navigation in Case of Armed Conflict", 18 *Int J Mar & Coastal L* (2003), 401, p402. See further Wendel, P, *State Responsibility for Interferences with the Freedom of Navigation*, pp234-36.
74	The necessity for that self-defence must be "instant, overwhelming, and leaving no choice of means, and no moment for deliberation" (Moore, *Digest of International Law*, II, 24-30, 409-14; VI, 261-62; and VII, 919-20).
75	See, for example, *San Remo Manual*, *supra* footnote 11, Paragraph 7.

specifically authorised its use only once, when calling upon the United Kingdom "to prevent, by the use of force if necessary" shipments of oil to the unlawful and racist Smith regime after it declared unilateral declaration of independence in Southern Rhodesia;[76] this led to extensive interdiction by the Royal Navy of oil tankers in the Mozambique Channel in 1966.

In reality, the laws of naval warfare continue to apply, in a form modified to take into account a mixture of UN practices in reaction to conflict situations.

Energy cargo ships should be aware of belligerent states' practice of declaring exclusion or war zones, and of the difference between the two.

Exclusion zones, in which the same rules of the law of armed conflict apply as outside the zone, that do not exceed what is reasonably required by military necessity or the need to safeguard protected persons or objects, that are properly notified and that do not cut off neutral territory or waters are lawful under international law according to the *San Remo Manual*.[77]

War zones[78] are, on the contrary, unlawful – at least if they are as indiscriminate as those employed by both sides in the World Wars and in the Iran-Iraq War.[79] Energy cargo ships take risks if they enter these zones, which are established by a belligerent with the purpose of relieving itself of the rules of naval warfare requiring it to discriminate between legitimate targets and illegitimate ones (including neutral ships). At the beginning of the Iran-Iraq War, for example, Iran declared all Iranian waters a war zone, prohibited all transportation of cargo to Iraqi ports and absolved itself of responsibility for ships not following routes designated by it. Iraq responded by declaring all Persian Gulf waters north of 29°30′N to be a prohibited zone in which it would attack all vessels, including those docking at Iran's Kharg Island oil terminal. In the years following, Iraq's zone was converted to a general maritime exclusion zone, which was further extended in scope. Eventually, attacks by both sides began to occur even outside the declared zones. In 1984 neutral vessel damage by sea mines as far away as the Gulf of Suez and Strait of Bab el-Mandeb was generally attributed either to Iran or to Libya.[80] Neutral states became increasingly concerned by the impacts on energy cargo ships and their navigation rights, and fearful of possible extreme impacts, such as the blocking of the Straits of Hormuz chokepoint to tankers.

---

76 UN Security Council Resolution 221 (April 9 1966), Paragraph 5. This authorisation has been regarded as anomalous, since it preceded any council decision on a mandatory sanctions regime, normally the precursor to resort to force. The direct reference to naval force has not been used in equivalent Security Council resolutions since, for example, those concerning:
  • Iraq/Kuwait (UN Security Council Resolution 665 – August 25 1990);
  • Sierra Leone (UN Security Council Resolution 1132 – October 8 1997);
  • former Yugoslavia (UN Security Council Resolutions 787 – November 16 1992 – and 820 – April 17 1993);
  • Haiti (UN Security Council Resolutions 940 – July 31 1994 – and 1529 – February 29 2004);
  • North Korea (UN Security Council Resolution 1718 – October 14 2006); and
  • Libya (UN Security Council Resolution 1973 – March 17 2011).
77 *Supra* footnote 11, Paragraphs 105-06.
78 Otherwise known as 'danger zones', 'barred areas', 'total blockades', 'exclusion zones' or 'operational areas': Von Heinegg, *supra* footnote 73, at Paragraph 1.
79 Von Heinegg, *ibid*, Paragraphs 29 and 37; Ronzitti, *supra* footnote 64, Paragraph 20 (who adds "except perhaps where established in waters adjacent to the declaring State").
80 Walker, *supra* footnote 60, at 54.

The risks encountered by neutral energy cargo ships navigating through war zones (or other high seas areas where they are subject to unlawful visit or attack) might be mitigated by diplomatic and naval action of their flag states or states of beneficial ownership. Of course, the belligerents are unlikely to attack neutral merchant ships openly in the presence of protecting warships. They might, however, conduct clandestine attacks or otherwise seek to deny energy cargo ships their international navigation rights by means short of armed attack – for example, by harassing their navigation. In such case a protecting state is entitled to respond with proportionate countermeasures, such as a show of force by warships at action stations.[81] However, it is unlikely to be able to invoke the right of self-defence to justify an armed response.[82]

Diplomatic and naval assistance during the Iran-Iraq War included the following:

- the maintenance by several neutral states of naval patrols in the waters affected to give close support to flag or beneficially owned vessels, or even, in the United States' case, to escort convoys of such vessels.[83] In practice, naval patrols on occasion offered distress assistance outside the war zones to innocent vessels of friendly, neutral states;[84]
- the interposition of naval vessels between a belligerent ship unlawfully seeking to visit a protected vessel; and
- mine countermeasures.

Assistance even extended to encouraging the reflagging of some 14 tankers belonging to the Kuwaiti state oil company to the US or UK flag so as to afford them such protections against Iranian attack or interference. By 1987 there was strong interest among tanker owners in arming their vessels for self-defence, but also strong government discouragement.[85]

Armed conflict affecting an energy cargo might, of course, lead to multiple litigation risks. Various clauses in contracts of carriage or affreightment potentially come into play so as to cause private party compensation claims.[86] Under a voyage or time charterparty, moreover, the effects of an armed conflict might lead to disputes concerning safe ports clauses, deviation or even frustration of the contract. Litigation might arise, moreover, as a result of mariners on an energy cargo ship suffering illness, injury or death. Either the carrier or the shipper, depending on the type and

---

81 *Corfu Channel Case*, ICJ Rep (1949) 1, 30.
82 See the *Oil Platforms Case (Iran v United States)*, ICJ Rep (2003), concerning a US attack on an Iranian offshore oil platform from which attacks were alleged to have been launched against US ships during the Iran-Iraq War.
83 Over 100 convoys in all (Walker, *supra* footnote 60, at 72-73).
84 *Dep't St Bull* July 1988, at 61.
85 Plant, Glen, "The Legal Implications of Defending Merchant Ships", keynote paper to the Mediterranean Marine Consultants' Conference on Shipping under Fire: Reassessing Safety and Security at Sea, Athens (January 20-21 1987). One reason for avoiding carrying even light weapons for self-defence is that some states, including the United States, allow for attack on the principle that so-called 'defensive' weapons are difficult to distinguish from offensive ones (*Commander's Handbook*, *supra* footnote 13).
86 Such as fire, perils of the sea, acts of war, acts of public enemies, arrest or restraint of princes or deviation clauses. Of course, the contract can to some extent provide in advance for foreseeably higher than usual risks – for example, by providing (usually for additional consideration) for a degree of flexibility in respect of time of delivery, which would normally be of the essence.

terms of contract between them, will have sought to insure against such risks, thus raising the prospect of insurance claims or subrogation to an insured's claim.

War risks are excluded from standard hull and machinery, freight and cargo insurance policies, so that either reliance is placed on P&I war risks cover (see below) or an additional premium paid for an extension of cover to war risks. In England, for example, cover may be written under the International War and Strikes Clauses (Hull) and/or the International War Clause (Cargo) (or older equivalent Institute) terms, either on International Underwriting Association or similar Lloyds forms. These cover loss of, or damage to, the vessel or cargo caused by (for relevant purposes) "war civil war revolution rebellion insurrection, or civil strife arising therefrom, or any hostile act by or against a belligerent power; capture seizure arrest restraint or detainment and the consequences thereof or any attempt thereof; derelict mines torpedoes bombs or other derelict weapons of war; [and] any terrorist or any person acting maliciously of from a political motive". The clauses include exclusions; they exclude as too high, among others, nuclear risks and risks arising from a war involving any of the five permanent members of the UN Security Council (China, France, Russia, the United Kingdom and the United States). In respect, moreover, of cover taken out for vessels engaged in worldwide trade, the clauses include trading warranties requiring notice to be given of operation (which must be innocent and prudent)[87] within the territorial waters of listed states likely to be the seat of hostilities or unrest and the payment of an additional premium at the insurer's discretion where risks are very high.[88]

Indeed, some insurers may choose not to write certain war risks at all.[89] Where war risks insurance is not available on reasonable terms and conditions, states sometimes provide by legislation for their beneficially owned ships to receive adequate insurance or reinsurance provision.[90]

P&I clubs generally exclude insurance for P&I liabilities arising as a consequence of belligerent acts, but cover is largely reinstated by the War Risk Extension clause offered to all mutual members. This clause[91] sets out terms of cover that are somewhat different from the normal P&I Rules,[92] and are also conditional, like hull and cargo cover, on innocent and prudent trading.[93]

3. Peacetime security risk

As suggested above, complex factors have created various links between peacetime security risks, including:

---

87 Imprudent trading, including by carrying contraband or blockade running, is likely to exclude cover.
88 This was the practice for tankers trading to the Persian Gulf during the Iran-Iraq War. The London War Risks Rating Committee, for example, raised rates in 1982 and again in 1984. Walker, *supra* footnote 60, at 44; Miller, MD, *Marine War Risks* (1992), pp18-22 and 270-72. Premiums can be very high.
89 For example, Section 15-9 of the Norwegian Marine Insurance Plan (2010) (www.norwegianplan.no/eng/index.htm).
90 For example, the United States (42 USC app § 1282, 1287).
91 For example, Rule 21 of the Steamship Mutual P&I Club Rules (available at www.simsl.com/Rules-and-Cover/rules-class-i—-protection-and-indemnity-21-25.html).
92 It is limited to $500 million, any one event, and to an excess of either the proper value of the entered ship (up to a maximum $100 million), or the amount recoverable in respect of the claim under any other policy of insurance.
93 For example, Rule 24 of the Steamship Mutual Rules, *supra* footnote 91.

- acts of maverick regimes;
- proliferation of weapons of mass destruction;
- smuggling of arms, people and narcotics;
- organised crime;
- terrorism; and
- piracy and hijacking of, and armed robbery against, ships.[94]

Moreover, any of these, alone or in combination, could create a major blockage of a chokepoint or a marine pollution incident. The discussion below nevertheless deals only with those risks particularly important for innocent energy cargo ships engaged in normal trade; the US-led Proliferation Security Initiative is not discussed, for example, since energy cargo ships are unlikely to be subjected to Proliferation Security Initiative interdiction on suspicion of smuggling weapons of mass destruction.[95] It also deals with the risks separately, as there are no *inevitable* connections. The worst pirates operate off Somalia, for example, with little evidence of cooperation in terrorist activities.

3.1 Terrorism risks

The 9/11 attacks have made the United States the lead state in the war against terrorism. The terrorist threat at sea takes two main forms of interest to energy cargo ship operations:

- attacks by terrorists against an oil, liquefied natural gas or liquefied petroleum gas tanker at sea, using either conventional weapons or a small boat packed with explosives to serve as a weapon, as for example when Al Qaeda holed the double-hulled French oil tanker Limburg a few miles off Yemen as it approached an oil terminal on December 6 2002, causing a 90,000 ton spill;[96] and
- using a vessel, hijacked or otherwise,[97] as a weapon or to transport a weapon such as a radioactive dirty bomb or armed terrorists against a naval or commercial port, population centre or strategic waterway.

A great concern to transboundary energy projects is likely to be a terrorist incident resulting in blockage of a chokepoint. In January 1996, for example, nine pro-Chechen gunmen hijacked a Turkish ferry just north of the Bosphorus and kept passengers and crew hostage for three days, threatening to blow up the vessel and their hostages.[98] The Turkish authorities had reason to believe that the terrorists had considered blowing up one of the Bosphorus bridges so as to block the Turkish Straits.[99]

---

94 See, for example, Köknar, AM, "Maritime Terrorism: A New Challenge for NATO", *Energy Security*, Institute for the Analysis of Global Security (January 24 2005) (www.iags.org/n0124051.htm#14).
95 Concerning the initiative, see www.state.gov/t/isn/c10390.htm.
96 Her Majesty's Royal Navy, "The Nature and Trends of Global Maritime Security" (www.royalnavy.mod.uk/linkedfiles/upload/pdf/the_nature_and_trends_of_global_maritime_security.pdf).
97 "It has been reported that al Qaeda owns or controls about 15 cargo ships that could be used as floating bombs against ... high interest vessels, or to smuggle explosives, chemical or biological weapons, such as a radioactive dirty bomb into a US port, or to transport al Qaeda members into a third country", Roach, J Ashley, "Initiatives to Enhance Maritime Security at Sea", 28 *Marine Policy* (2004) 41, p42.
98 Köknar, *supra* footnote 94.

Another concern to transboundary energy projects is a terrorist action resulting in a major pollution incident,[100] even though it might be hard for claimants to establish liability and make recovery from the shipowner, operator or other responsible party for the consequences of a tanker spill caused by terrorists.[101] According to the Royal Navy, "oil is one of the prime targets for the terrorists... concentrated in the Middle-East, specifically in and around the Arabian Gulf". Since the attack on the Limburg, there have been a number of speedboat attacks on tankers anchored off Iraq's Al-Basra Oil Terminal. In April 2004, for example, terrorists detonated explosive-laden boats near a tanker moored alongside it.[102]

The blurred distinction between war/armed conflict and peace might have been rendered even less clear by the arguable widening of the right of self-defence under international law following the 9/11 attacks to include forceful reaction to certain acts perpetrated by non-state entities. Whether this widening has in fact occurred is unclear. Two UN Security Council resolutions[103] passed in the wake of the 9/11 attacks make broad-reaching provision for action against international terrorism declared by them to be a "threat to international peace and security". Although these resolutions do not expressly authorise the United States to use force against Al Qaeda, they refer in their preambles to the right of self-defence. In addition, the Declaration of the North Atlantic Treaty Organisation (NATO) Council of September 12 2001 stated that if it were determined (as NATO later confirmed that it was) that the terrorist action on 9/11 "was directed from abroad against the United States, it shall be regarded as an action covered by Article 5 of the [NATO] Treaty, which states that an armed attack against one or more of the Allies in Europe or North America shall be considered an attack against them all".[104] It might be inferred from these decisions that the legal concept of self-defence has been expanded to cover defensive measures taken following and in reaction to certain terrorist acts of terrorist groups. Majorities in two International Court of Justice cases have, however, resisted this trend, despite several strongly argued and convincing dissents:[105]

- *Legal Consequences of the Construction of a Wall in the Occupied Palestinian Territory (Advisory Opinion)*;[106] and
- *Armed Activities on the Territory of the Congo (Democratic Republic of Congo v Uganda)*.[107]

Arguments for the legality of anticipatory self-defensive measures against non-

---

99   Ibid.
100 "A Synopsis of the Terrorist Threat Facing the O&G Industry", *Oil and Gas Industry Terrorism Monitor* (2007) (www.ogi-tm.com/ogi_threats_st.php).
101 Concerning the legal difficulties in the Limburg case, see Olyslager, D, "The Limburg Terrorist Attack", Paper to the IUMI Singapore Conference (September 12-15 2004) (www.iumi.com/index.cfm?id=7199). In the event an *ex gratia* $1 million payment was made to Yemen in respect of the spill.
102 Royal Navy, *supra* footnote 96.
103 UN Security Council Resolutions 1368 (September 12 2001) and 1373 (September 28 2001).
104 40 ILM (2001) 1268. See also Murphy, S (ed), "Contemporary Practice of the United States Relating to International Law", 96 *Am J Int L* (2002), 237, at 244.
105 See Zemanek, K, "Armed Attack", *Max Planck Encyclopaedia*, Paragraphs 14 *et seq* and Sir Watts, Arthur, "Israeli Wall Advisory Opinion", *ibid*, Paragraph 42.
106 ICJ Rep 2004, at Paragraph 139.
107 ICJ Rep 2005, at Paragraphs 146 and 160.

state terrorist actors (and certainly for pre-emptive measures) are likely to be even more strongly resisted by the majority of the court.

Be this as it may, after the 9/11 attacks the United States began boarding vessels in the Indian Ocean, the Red Sea and the Strait of Hormuz looking for Osama bin Laden and his Al Qaeda associates, even though the United States was not engaged in armed conflict with any state. While consent from the ships' masters was generally sought for these inspections, the United States notified the maritime industry that it would compel boarding of a foreign vessel suspected of transporting terrorist suspects and from which such consent was not forthcoming. According to a Russian report, the specific legal basis for this action was never explicitly articulated, but President Bush referred in general to self-defence in the context of response to the attacks by Al Qaeda.[108] If self-defence, an exceptional right, was indeed the unarticulated justification, it was used to interfere with normal peacetime rights of navigation and so must be seen as an extraordinary exercise of that right in unusual circumstances. In general, the United States has sought specific treaty bases for maritime counterterrorism actions.

A number of instruments have been agreed at both national and international level to reduce and respond to the terrorist threat. Notable in the maritime context are two International Maritime Organisation instruments:

- the US-led 2004 amendments to the 1974 International Convention for the Safety of Life at Sea,[109] adding a new Part XI-2 on maritime security, which – among other things – made the new International Code for the Security of Ships and Port Facilities mandatory for companies operating ships of more than 500 tons on international voyages, as well as for seaports. Both must now prepare security plans and take various other security measures. Three different security levels are set. Failure to comply with the highest renders an energy cargo ship risky and subject to additional security controls before loading is permitted (or even to surcharges) at ports or terminals, so that it is undesirable for energy cargo ships ever to interface with ships or ports apparently less secure against terrorist threats than themselves; and
- US-led amendments to the US-inspired 1988 Convention for the Suppression of Unlawful Acts against the Safety of Maritime Navigation,[110] a 'prosecute or extradite' agreement requiring states either to submit a matter to the prosecution authorities in the normal way or to respond to an extradition request in respect of persons found on, or coming onto, their territory who are suspected of any of a list of terrorist offences committed against, or on board, ships.[111] The amendments in question, adopted in a 2005 protocol, turned the convention into a vehicle for deemed consent boardings of foreign flag vessels, including energy cargo ships, by (US or other) naval

---

108   Maritime Operations, a conference organised by the Defence Institution of International Legal Studies and the Russian Academy of Liberal Arts and Information Technologies Education, St Petersburg (June 2003) (www.diils.org).
109   London (November 1 1974), 1184 *UN Treaty Series* 277, as amended.
110   Rome (March 10 1988), 1678 *UNTS*.
111   Plant, "The Convention for the Suppression of Unlawful Acts against the Safety of Maritime Navigation", 39 *ICLQ* (1990), 27.

forces in search of terrorists guilty of a list of offences now expressly extended to cater for the post-9/11 terrorist threat and the related threat posed by proliferation of weapons of mass destruction).[112] Other prosecute or extradite conventions might also apply in the event of terrorist attacks against ships, notably the 1979 International Convention against the Taking of Hostages.[113]

None of the above-mentioned instruments attempts a definition of 'terrorism' or 'terrorist', the accepted wisdom until recently being that this was a political impossibility since (putting it crudely) one man's terrorist is another's freedom fighter. It has in the past also been generally considered that no definitions exist in customary international law. The extent of international law's coverage of the terrorist threat at sea has therefore been coincident only with the precision *vel non* with which the acts and persons addressed by those instruments have been described.

However, the Appeals Chamber to the first international court with jurisdiction over the crime of terrorism, the Special Tribunal for Lebanon, recently decided that a customary international law definition of 'terrorism' has gradually emerged. This definition, according to the chamber, comprises the following key cumulative elements:

- the perpetration of a criminal act (eg, murder, kidnapping, hostage-taking and arson) or threat to commit such an act;
- with the intent to spread fear among the population (which would generally entail the creation of public danger), or directly or indirectly coerce a national or international authority to take some action, or to refrain from taking it;
- when the act involves a transnational element.[114]

This general definition is likely to help states to direct their legislation towards effective prosecution of apprehended terrorist suspects. The chamber stated, indeed, that the customary rule can be held to impose a duty on states to prosecute those who commit acts of terrorism as defined under customary international law.[115] This is of particular utility in permitting prosecutions for acts with a transnational element falling within the gaps left by the maritime counterterrorism conventions. The 'prosecute or extradite' definition is also likely to be helpful in the implementation of UN Security Council resolutions on combating terrorism, including at sea.

Terrorism is treated as a war risk for insurance purposes.

---

112 See Klein, Natalie, "The Right of Visit and the 2005 Protocol on the Suppression of Unlawful Acts against the Safety of Maritime Navigation", 35 *Denv J Int'l L & Pol'y* (2007) 287.
113 New York (December 17 1979), 1316 *UNTS* 205.
114 *Interlocutory Decision on the Applicable Law: Terrorism, Conspiracy, Homicide, Perpetration, Cumulative Charging*, Case No STL-11-01/I (February 16 2011) (www.stl-tsl.org/x/file/TheRegistry/Library/CaseFiles/chambers/20110216_STL-11-01_R176bis_F0010_AC_Interlocutory_Decision_Filed_EN.pdf), Paragraphs 83 and 85.
115 *Ibid*, Paragraph 102.

3.2 Risks of piracy, armed robbery and ship hijacking

Ships' risk of suffering piratical attack, although temporarily suppressed by the Royal Navy during the course of the 19th century, is an ancient one. Accordingly, customary international law, now set out in the UN Convention on the Law of the Sea, has long permitted states universal jurisdiction to take action against piracy contrary to international law (piracy *iure gentium*). Together with armed robbery in territorial waters, such piracy has again become a growing problem since the 1980s. In recent years it has increasingly been characterised by ship and crew hijacking for ransom.

The risk of piracy arises in a number of sea areas worldwide,[116] but became particularly grave in the 1990s and early 2000s in the South China Sea and an adjacent chokepoint, the Straits of Malacca. Its decline there is largely attributable to the cooperation between the 16 states party to the Regional Cooperation Agreement on Combating Piracy and Armed Robbery against Ships in Asia.[117] Unfortunately, piracy reappeared in 2005 and again since 2007 off the 'failed state', Somalia, where it is now a serious problem.

Whatever the poverty-based or other reasons for piracy's revival, it has been facilitated by ease of access to sophisticated electronic means of communication and identification of target ships, and by easy access to arms and fast small craft capable of catching and boarding most ships.

In the Somali case, it has developed into a lucrative and self-sustaining business for at least 1,500 pirates. They have made hundreds of attacks, with a success rate above 25%, using, in addition to fast skiffs operating from shore, captured ships as mother ships from which such skiffs may be launched far out to sea and which derive a degree of protection from attack by keeping the original crew imprisoned on board. They have growing logistical capacities to finance operations, refuel at sea and hold for increasingly longer periods an ever-growing number of ships and crew members.[118] Higher and higher ransoms have been extracted for the hundreds of vessels hijacked and thousands of crew members held hostage. In 2009 alone the average ransom grew from around $1.5 million to $3.5 million per hijacking.[119] The annual cost to the world economy is estimated at between $5 billion and $7 billion,[120] of which a substantial part falls on the energy industry and its insurers. A recent alarming development is an increase in the number of crew members being mistreated or even murdered.

Tankers and bulkers are vulnerable despite their size. Somali pirates have even captured four laden ultra large crude carriers hundreds of miles out to sea, and held them and their crews for (according to the press) considerable ransoms:

- the Sirius Star (held between November 2008 and January 2009 for a reported

---

116 See www.icc-ccs.org/home/piracy-reporting-centre/prone-areas-and-warnings.
117 Tokyo (November 11 2004), 2398 *UNTS*. Concerning this cooperation, see www.recaap.org/index_home.html.
118 "Report of the Special Adviser to the Secretary-General on Legal Issues Related to Piracy off the Coast of Somalia", UN Document S/2011/30 (January 24 2011) (hereafter the "Lang Report"), especially Paragraphs 14-15 and 29-31.
119 "No stopping them", *The Economist* (February 5 2011), 60, at 61.
120 UN Department of Public Information press release SC/10164 (January 25 2011).

$3 million ransom);
- the Maran Centaurus (held from November 2009 to January 2010 for a ransom reported at between $5.5 million and $7 million);
- the Samho Dream (held between April and November 2010 for a reported $9.5 million ransom); and
- the Irene SL (held between February and April 2011 for a reported $13.5 million ransom).

Indeed, 2011 has seen clear evidence that Somali pirates are specifically targeting oil tankers exiting the Strait of Hormuz and are even prepared to use them as mother ships from which to launch attacks far out to sea (the Irene SL was captured off Oman as it exited that chokepoint on February 9 2011 and was later used as a mother ship).[121]

First at risk from Somali pirates were ships passing through the Gulf of Aden to and from the Red Sea and Suez Canal via the Strait of Bab el-Mandeb. Piratical attacks in the Persian Gulf having grown to an almost daily occurrence by late 2008,[122] several major energy cargo ship operators rerouted their tankers away from the Red Sea and Suez Canal route and around the Cape of Good Hope, despite the additional voyage time and fuel costs.[123] These operators have, however, remained the exception.[124] Attacks in the area have diminished; only 33 successful hijackings occurred in the two years following the establishment, early in 2009, of the naval-patrolled Internationally Recommended Transit Corridor for Ships Transiting the Gulf of Aden and the Maritime Security Centre-Horn of Africa[125] – an EU information exchange service with which vessels wishing to transit the affected waters under naval protection may register.[126]

Unfortunately, the major threat has merely been displaced further and further into the Indian Ocean, and now extends, for example, to energy cargo ships passing along the East Coast of Africa to and from the Persian Gulf to western markets via the Cape of Good Hope. Attacks now occur hundreds of miles offshore over an area much larger than Europe.[127] It follows both that the interests of regional states (including their energy cargo trading interests) are affected as far east as India and the Maldives, south as Madagascar and north as Oman, and that naval patrols have difficulty in patrolling the entire area. Attacks now occur more than once daily.

Numerous political and practical countermeasures have been taken, notably:
- the deployment of naval patrols by NATO, the European Union, a US-led

---

121 Osler, D, "Pirates targeting tankers at Strait of Hormuz exit points: Experts say industry may be forced to avoid the area after *Irene SL* seizure becomes eighth attack on a tanker in the last four weeks", *Lloyd's List* (February 10 2011); "Maritime Security Forum Thread, 13.5 Million ransom paid 'IRENE SL' who is no longer a PAG mothership threat in Close Protection World Area" (www.closeprotectionworld.co.uk/maritime-security-forum/46720-13-5-million-ransom-paid-irene-sl-who-no-longer-pag-mothership-threat.html).
122 UN Security Council Resolution 1950 (November 23 2010), Paragraph 3, which attributed this to states' failure to enforce the UN arms embargo on Somalia and private parties paying ransoms.
123 "Suez fears fall in revenues as fleets steer clear of pirates", *Financial Times* (November 21 2008), p8.
124 "Lang Report", *supra* footnote 118, Paragraph 28.
125 *Ibid*.
126 See www.mschoa.org/Pages/default.aspx.
127 *The Economist*, *supra* footnote 119, at 61.

coalition of states and China, India, Iran, Japan, Malaysia, Russia, Saudi Arabia, South Korea and Yemen to deter and, consistent with their various rules of engagement, sometimes to capture and detain or take forceful measures against the pirates and their boats and vessels. Military coordination takes place via shared awareness and deconfliction group meetings;
- the deployment of information systems both to assist merchant shipping (eg, the above-mentioned Maritime Security Centre-Horn of Africa) and to coordinate naval patrols (eg, the secure internet-based Mercury system, which facilitates low-level communications between all warships on anti-piracy patrol, whatever their nationality); and
- the US-inspired establishment[128] of a Contact Group on Piracy off the Coast of Somalia, which:
  - facilitates naval cooperation;
  - supports the development and implementation of avoidance, evasion and defensive best practices and advisories for ships that are sailing in the affected waters and threatened with, or come under, attack; and
  - administers a trust fund, focusing primarily on supporting local states' prosecution of pirate suspects.

The contact group builds on work conducted since the 1980s by the International Maritime Organisation and other UN agencies, in cooperation with industry and other stakeholders, on:
- data collection, research and dissemination;
- the preparation of practical guidance for governments and shipowners or operators and seamen on practical preventative measures;[129] and
- the promotion of cooperation among regional states, most pertinently through the 2009 Djibouti Code of Conduct Concerning the Repression of Piracy and Armed Robbery against Ships in the Western Indian Ocean[130] and the UN Anti-Piracy Action Plan of February 3 2011.[131]

As to legal means of controlling the piracy risk, Article 100 of the UN Convention on the Law of the Sea requires states to cooperate *to* the fullest possible extent on the high seas in the repression of 'piracy *iure gentium*', which is defined by Article 101 to include (for relevant purposes) any illegal acts (including secondary acts) of violence or detention, or any act of depredation, committed for private ends by a private

---

128   Pursuant to UN Security Council Resolution 1851 (December 16 2008).
129   See especially International Maritime Organisation documents: Guidance to Shipowners and Ship Operators, Shipmasters and Crews on Preventing and Suppressing Acts of Piracy and Armed Robbery against Ships, MSC.1/Circ 1334 (June 23 2009); Best Management Practices to Deter Piracy in the Gulf of Aden and off the Coast of Somalia Developed by the Industry, MSC.1/Circ. 1335 (September 29 2009) and its 3rd edition (June 2010); and Information on [the Internationally Recommended Transit Corridor for Ships Transiting the Gulf of Aden], SN.1/Circ 281 (August 3 2009).
130   Adopted on January 29 2009. See International Maritime Organisation Document C102/14 (April 3 2009), Attachment 1, Annex.
131   See www.imo.org/About/Events/WorldMaritimeDay/Documents/2011%20WMD%20theme%20Action %20 Plan%20handout.doc.

ship's personnel and directed against another ship (or persons or property on board it) outside the jurisdiction (ie, seaward of territorial waters) of any state. Article 105 permits any state, regardless of the normal principle of flag state jurisdiction, to seize a 'pirate ship' (defined by Article 103 as a ship intended to be used or being used by the persons in dominant control for committing an Article 101 offence) or a ship taken by pirates, and the courts of that state to determine the action to be taken with regard to the ship and property seized. All prosecutions must take place in national courts. No international court has jurisdiction to try pirates.

The first point to make arises from the emphasised wording: although these articles provide for exceptionally wide peacetime jurisdiction, they are essentially facultative. States are not specifically required to seize pirate ships or to detain captured pirate ships or pirates for judicial action. Indeed, while states – including parties to the UN Convention on the Law of the Sea – are obliged as a matter of international law to bring their internal laws into line with their international obligations, there is no express requirement in the convention to criminalise piracy, to provide for universal jurisdiction over it or to enact the precise piracy *iure gentium* provisions of the convention into domestic law. Many states have no domestic criminal piracy laws at all, and those laws that do exist vary greatly in scope and quality.[132] However, the picture is improving, apparently in view of the growing seriousness of Somali piracy.[133] The duty to cooperate in piracy's repression is, moreover, qualified by the words 'to the fullest possible extent', and a state unwilling to take action for reasons of policy or convenience might claim that cooperation is impossible for it given its particular circumstances.

Even assuming that a state is willing and able to act against pirates, it encounters four significant restrictions in the scope of piracy *iure gentium* – it enables warships (or other authorised public vessels) to intervene to protect ships by seizing a pirate vessel and arresting the pirates only when they do so:

- outside territorial waters;
- in relation to acts carried out from one ship against another (thus excluding acts by those already aboard the target ship);
- as concerns acts committed for private ends (thus arguably excluding politically motivated terrorist acts);[134] and
- in respect of acts occurring on the high seas.

These four restrictions, and in particular the first and last, represent serious constraints on the actions that warships are entitled to take to protect energy cargo ships. Fortunately, there are two possible legal means by which the effect of these constraints might be reduced (though not eliminated), although their use requires states to show political will.

---

132 The United Nations displays details of only 31 states' piracy laws at www.un.org/Depts/los/piracy/piracy_national_legislation.htm.
133 The Lang Report lists Belgium, France, Japan, Kenya, Maldives, the Netherlands, Seychelles, Spain and Tanzania as examples of those now establishing universal jurisdiction over piracy and prosecuting pirates. See *supra* footnote 118, Paragraphs 47 and 50-51.
134 *Sed contra* section 1, *supra*.

First, these restrictions might conceivably be modified or dispensed with in a particular situation where the UN Security Council comes to view as a threat to international peace and security (justifying action under Chapter VII of the UN Charter) persistent or particularly violent acts of piracy and armed robbery at sea (or at least, as in Somalia's case, the poor general security situation resulting in the piracy).[135] In June 2008 the council, acting under Chapter VII, decided to lift for an initial six-month period (renewed for the whole of 2009, 2010 and 2011)[136] the restriction precluding operations within Somali territorial waters for states acting with the consent of the Transitional Federal Government of Somalia (notwithstanding that its writ runs in effect to only part of the capital city).[137] Such states' warships could, after notification to that government, enter Somali territorial waters and use all necessary means, consistent with third state ships' right of innocent passage, for the purpose of repressing acts of piracy and armed robbery at sea. Somali territorial waters are treated in practice as the 12 nautical-mile zone permitted by the UN Convention on the Law of the Sea, rather than the old excessive claimed zone of 200 nautical miles. At least one resolution even authorised counterpiracy operations on land.[138] Remarkably, the Security Council goes further in some respects than it normally would when imposing sanctions under Chapter VII against a state. The Somali piracy resolutions nevertheless preserve all of the other restrictions on warship action against piracy contained in the UN Convention on the Law of the Sea.

Second, certain of the four restrictions might be obviated where a state party takes action, pursuant either to the 1988 Convention for the Suppression of Unlawful Acts against the Safety of Maritime Navigation[139] or (as appropriate) to that convention as amended by the 2005 protocol, against pirates, armed robbers or hijackers. In particular:

- such a state is obliged, and not merely facilitated, to submit a case to its prosecution authorities or to commence requested extradition proceedings against an alleged offender within the meaning of the convention and coming within its jurisdiction (including aboard its vessels); and
- port or coastal states have a (qualified) duty to take delivery of an alleged offender from a ship requesting this.[140]

Other 'prosecute or extradite' conventions, such as the Hostages Convention,[141] are also of potential utility. Unfortunately, states have not always been diligent in acting under these conventions and it is unclear to what extent they are being applied, if at all, to Somali piracy.

---

135 All of the below-cited UN Security Council resolutions state this as the basis of Chapter VII action.
136 UN Security Council Resolutions 1846 (November 2 2008), Paragraphs 10-11, 1897 (November 30 2009), Paragraphs 7-8, and 1950 (November 23 2010), Paragraphs 7-8.
137 UN Security Council Resolution 1816 (June 2 2008), Paragraphs 7 and 8.
138 Paragraph 6 of UN Security Council Resolution 1851 (December 16 2008) uses the phrase 'in Somalia'.
139 *Supra* footnote 110.
140 Articles 10 and 8 respectively, both of which are amended by the protocol. See further Plant, *supra* footnote 111, at 48-50, and Lang Report, *supra* footnote 118, Paragraph 45.
141 *Supra* footnote 113.

The international law of piracy *iure gentium* often differs from national piracy laws, which are of great practical importance since they govern, among other things:
- the terms under which captured offenders may be detained on board a warship and perhaps submitted to prosecuting authorities;
- criminal proceedings for offences of piracy brought against alleged offenders coming within the jurisdiction;[142] and
- piracy clauses in maritime commercial instruments and marine insurance policy clauses on piracy, barratry, violent theft and kidnap and ransom.[143]

As regards the first and second bullet points, it is significant that 90% of captured Somali pirates are released without charge.[144] This occurs for a number of reasons:
- States might have no domestic law criminalising piracy or other wrongs committed at sea, so that no charges may be brought or pre-trial detention effected.
- States might have made criminal legal provision against piracy, but none extending universal (or other extraterritorial) jurisdiction over piracy offences on the high seas (or in foreign, including Somali, territorial waters).
- Some states' domestic procedural safeguards might in practical terms restrict a warship to capture and release. German, Russian and Spanish warships, for example, have no right to detain a captured pirate who is no longer posing a threat for longer than one or two days, unless he or she is brought before a judge. Bringing such individuals before a judge is a virtual impossibility where the nearest courts are in a foreign port, often distant, and the procedures for taking custody and commencing process in that port are neither automatic nor swift.[145]
- States might be deterred from detaining pirates on board their warships by the potential applicability of human rights norms. In particular, Article 5 of the European Convention on Human Rights and Fundamental Freedoms,[146] concerning the right to liberty and security of the person and imposing restrictions on arrest and detention (including the right to judicial supervision), appears in principle to apply to Somali pirates detained by European warships.[147] A case that is likely to determine whether it does was filed with the European Court of Human Rights on March 16 2010.[148]
- States might simply be unwilling to submit for prosecution pirates captured by their warships off Somalia because of the lack of on-board detention

---

142 See UN Documents S/2010/394 (July 26 2010) and Lang Report, *supra* footnote 118, Paragraphs 45 *et seq*.
143 As to English law, see Clift, R, of Hill Dickinson, Solicitors, "Piracy: a brief overview" (www.iumi.com/index.cfm?id=7295).
144 Lang Report, *supra* footnote 118, Paragraph 14.
145 *Ibid*, Paragraph 53 and Note 34. The report made recommendations concerning this and other aspects of prosecution and transfer of suspects for trial.
146 Rome (November 4 1950), 213 *UNTS* 221.
147 See *Medvedyev v France* (Case 3394/03, European Court of Human Rights, Grand Chamber judgments of July 10 2008 and March 29 2009).
148 *Samatar v France* (Case 17301/10), which concerns the detention of Somali citizens pending France's opening of an investigation into hostage taking aboard a French merchant ship.

facilities or the cost or inconvenience of removing them to home territory for trial.

In these circumstances, naval states patrolling against Somali pirates have sought the cooperation of local states in accepting detainees into their custody and commencing prosecution proceedings. This has had some success, but is highly dependent upon aid to build capacity in respect of the anti-piracy laws and court infrastructure of those countries. Kenyan law has been of particular importance in recent years, as memoranda of understanding between Kenya and the naval powers operating anti-piracy patrols off Somalia formerly provided for the prosecution of Somali pirates in Kenya's courts – this being conditional on aid to build its judicial and legal capacity, largely supplied by the European Union. This practice was discontinued in late 2009, after 50 trials resulting in conviction – ironically, just as Kenyan law was amended to better reflect the provisions of the UN Convention on the Law of the Sea on piracy and those of the Convention for the Suppression of Unlawful Acts against the Safety of Maritime Navigation.[149] This has resulted in the sending of some pirates to US or other 'home' courts. It remains unclear to what extent, if at all, Kenya will resume anti-piracy trials.[150]

In December 2008, moreover, the UN Security Council authorised the carriage of ship riders, so that energy cargo ships may now carry on board a law enforcement official of a country willing to take custody of pirates with a view to prosecution.[151] It is unclear how far this has been acted upon.

In conclusion, although some 738 individuals suspected or found guilty of piracy had been transferred to judicial authorities by the end of 2010 and were being detained in 13 different countries, the appointed UN legal expert, Jack Lang, reported his fear in January 2011 that this was exceptional, and that 'capture and release' had become the rule among all patrolling navies.[152] He pressed for immediate action to improve the criminal law-based approach to a solution. On April 11 2011 the UN Security Council obliged with a far-ranging resolution, (1976), largely devoted to criminal law approaches.

As regards piracy clauses in maritime commercial instruments and marine insurance policies, there is no clear definition of 'piracy' under English marine insurance law, although guidance can be obtained from the 1906 act and from case law. The threat of violence and its occurrence at sea appear to be essential elements.[153] In contrast to the law of piracy *iure gentium*, 'sea' under English law includes territorial waters, and 'pirates' include passengers on board who mutiny and rioters who attack a ship from shore.

---

149 Albeit defectively, so as to require amendment before further trials may be commenced. See Lang Report, *supra* footnote 118, Paragraph 50 and Note 33.
150 See 104 *Am J Int Law* (2010), 397, especially Roach, J Ashley, "Countering Piracy off Somalia: International Law and International Institutions", 397, 414-15; and Thuo Gathii, J, "Kenya's Piracy Prosecutions", 416-36. See also Roach, "Suppressing Somali Piracy – Next Steps", 14(39) *ASIL Insights* (December 1 2010), especially Note 12 and accompanying text.
151 UN Security Council Resolution 1851, Paragraph 6.
152 Lang Report, *supra* footnote 118, Paragraphs 42-43.
153 The *Andreas Lemos* [1982] 2 Lloyd's Law Reports 483 (QB).

Piracy and armed robbery are traditionally and normally treated as perils of the sea covered under major hull clauses, such as the International Time Clauses (Hulls). Similarly, P&I liabilities incurred by P&I club members and arising as a consequence of, or caused during, a pirate attack or hijacking, or an incident of armed robbery, will be covered under the club's rules in the normal way, subject to the normal requirement of innocence and prudence. Piracy is not an excluded cause of liability, like war (broadly defined) or terrorism. Accordingly, it does not fall within the terms of war risks clauses, unless the piratical conduct amounts to the perils covered by those clauses, such as civil strife or terrorism. It follows that the determination of where risk falls might depend on the identification of incidents as, for example, piracy rather than terrorism, or armed robbery rather than civil strife or riot. The distinctions might be fine ones.[154]

When, however, piracy becomes unusually serious in particular sea areas, such as the Malacca Straits during the 1990s and early 2000s and more recently off Somalia, many marine insurance policies treat it as a war risk (in the latter case, with a quadrupling of typical premiums).

Insurers have, in addition, responded to the new risks posed by the Somali pirates' practice of ship and crew kidnap and ransom, which are inadequately covered by standard hull, war and P&I insurance, by developing new 'kidnap and ransom' policies. These extend cover to:

- loss of earnings incurred by owners or charterers of detained ships and cargoes;
- loss of life or injury during kidnap;
- ransom payments; and
- emergency team expenses.

The premiums for this cover are very high,[155] and likely to lead to attempts to obtain contributions from other parties and to possible disputes over this. The situation is complicated by the secrecy surrounding ransom negotiations and payments and by the potential illegality of such payments under relevant national laws. Under English law, for example, a payment to Somali pirates to release captives is not contrary to public policy and so is lawful.[156] Such a payment would be unlawful, however, if it were demanded for political purposes or to fund terrorism, as this would breach the relevant provisions of the Proceeds of Crime Act 2002 (and associated money laundering regulations) and the Terrorism Acts 2000 and 2006. There is little evidence of any link between the Somali pirates and terrorism,[157] although working arrangements have been made since 2010 to launch pirate attacks

---

154 Clift asserts that 11 participants in an act might count as pirates but 13 in the same act as rioters. *Supra* footnote 143.
155 Kidnap and ransom policies were first developed in 2008, with premiums around $3,000 for cover up to the then maximum ransoms, $3 million; by April 2009 premiums had increased tenfold (*Insurance Daily* (April 27 2009) (www.insurancedaily.co.uk/2009/04/14/marine-kr-premiums-rise-tenfold-on-somali-pirate-threat)) before falling back again (Neligan, M and Turner, Lorraine, "Piracy premiums take a breather but menace remains", Reuters (April 1 2010) (feraljundi.com/2010/04/07/kidnap-and-ransom-piracy-premiums-take-a-breather-but-menace-remains).
156 *Masefield AG v Amlin Corporate Member Ltd* ([2010] EWHC 280 (Comm)).

from areas under the control of the Islamic separatist movements Al-Shabaab and Hizbul Islam, which have links with Al Qaeda.[158]

The total increase in insurance premiums attributable to Somali piracy has been estimated at $600 million per year.[159]

A recent alternative approach is to arm personnel for protection of self or ship, or to hire private security companies to supply armed on-board guards or escort vessels.

A number of maritime escort private security companies are in operation, the UK-owned Protection Vessels International being generally regarded as the market leader. Unfortunately, their use to reduce the risks of piracy itself introduces new risks, including those of breaching domestic firearms legislation, contradicting International Maritime Organisation guidance to merchant ships and loss of governmental control over the human rights aspects of apprehension and detention of pirates. In view of these risks, the UK government for one has strongly discouraged the use of private security companies.[160] By contrast, the United States and others have not.[161]

These new risks have been reduced by a Swiss initiative, conducted in consultation with the United Kingdom and the United States, which led to the 2010 International Code of Conduct for Private Security Service Providers.[162] The code applies to private security services generally, and not just maritime services. In signing this, Protection Vessels International and several other maritime private security companies have affirmed their willingness to operate their maritime escort services with respect for applicable human rights (and international humanitarian law) standards, and for an independent code governance and oversight mechanism.

Several provisions of the code are particularly pertinent to anti-piracy escort operations. With regard to weapons, signatory companies are to obtain weapons licences required by the flag state's or any other applicable law, arrange suitable weapons training for their personnel and ensure that those personnel do not possess weapons which are illegal under any applicable law (Paragraphs 56-59). With regard to the detention of pirates, "Signatory Companies will only, and will require their Personnel will only, guard, transport, or question detainees if: (a) the Company has been specifically contracted to do so by a state; and (b) its Personnel are trained in the applicable national and international law" (Paragraph 33). Lastly, with regard to apprehending pirates, "Signatory Companies will, and will require their Personnel to, not take or hold any persons except when apprehending persons to defend themselves or others against an imminent threat of violence, or following an attack or crime committed by such persons against Company Personnel, or against clients

---

157　Chalk, P, "The Maritime Dimension of International Security: Terrorism, Piracy, and Challenges for the United States", Rand Corporation,(June 2008) (www.rand.org/pubs/monographs/MG697.html).
158　Lang Report, *supra* footnote 118, Paragraphs 19 and 24.
159　*The Economist*, *supra* footnote 119, at 62.
160　See Clift, *supra* footnote 143, slide 24.
161　*Ibid*, slide 25.
162　Swiss Confederation, Nyon (November 9 2010) (www.news.admin.ch/NSBSubscriber/message/attachments/21143.pdf. The first 59 company signatories are listed at www.icoc-psp.org/uploads/INTERNATIONAL_CODE_OF_CONDUCT_Final_with_Company_Names.pdf).

or property under their protection, pending the handover of such detained persons to the Competent Authority at the earliest opportunity. Any such apprehension must be consistent with applicable national or international law and be reported to the Client without delay" (Paragraph 34). In all cases, "Signatory Companies will, and will require that their Personnel to, treat all [apprehended or detained] persons humanely and consistent with their status and protections under applicable human rights law".

Where, notwithstanding this new code, a flag state continues to prohibit the carriage or use of firearms on board its ships or to discourage the use of armed escorts, the ship's voyage might become non-innocent or imprudent so as to deprive it of its marine insurance cover. This problem seems likely to be at least partly resolved, from the second half of 2011, by a new scheme marking the entry into this sensitive business area of a large, reputable London insurance broker, Jardine Lloyd Thompson.[163] The detail, including possible participation by public and other (including insurance) interests, needs to be fleshed out, but the idea is to offer, on a not-for-profit basis, combined insurance cover and armed escort through the Gulf of Aden on a flat-fee premium basis (projected to average $21,500 per vessel transit) aimed to be cost neutral or better. Public contributions are sought and contribution by the scheme to regional anti-pirate capacity building intended. The Convoy Escort Programme will involve 18 ship escort vessels flagged in a single state, each with a crew of five and an armed security team of eight, who will intercept and deter pirates prior to hijack attempts. It seems likely that the company intended to provide the escort service will sign the 2010 code.

To the degree that the scheme might appear to involve any flag state commission (by way of letters of marque and reprisal or otherwise) to any ship to attack and capture foreign ships as a privateer, it is stressed that the Convoy Escort Programme would be a flag naval company operating under the maritime and criminal laws of the flag state. Privateering was a 16th to 19th century practice under which private vessels fitted out for war were commissioned by belligerent governments to attack enemies and enemy objects. In 1856 the European powers (though not the United States)[164] agreed to abolish the practice (doing so by Article 1 of the Declaration of Paris), essentially because they came to recognise that private armed ships, maintained at private cost for private gain could not be kept under proper control.[165] The Convoy Escort Programme cannot be said to involve a revival of privateering for at least two reasons. First, no state of war exists between Somalia and any state involved with the programme. Second, the escort vessels will use force only to protect the programme's clients' assets (and, if necessary, themselves) on a non-profit basis, and will not seek to take pirates' assets for personal gain.

Still less will Convoy Escort Programme personnel be mercenaries.[166] Mercenaries are to be understood as foreigners who, although not members of the armed forces

---

163 Osler, D, "Dobson Fleet management to run private anti-piracy scheme", *Lloyd's List* (January 25 2011) and "Broker outlines anti-piracy plan", *Lloyd's List* (February 3 2011).
164 Article 1(8) of the US Constitution still gives Congress the power to grant letters of marque and reprisal.
165 Bederman, D, "Privateering", *Max Planck Encyclopaedia*.
166 As suggested by Macintyre, B, in "Echoes of corsairs sailing the high seas", *The Times* (February 18 2011).

of a belligerent, are specially recruited to fight in an armed conflict, do fight and are motivated essentially by the desire for private gain.[167] It seems no more likely that Convoy Escort Programme on-board personnel will fit within this definition than within that of 'privateers', and for the same reasons.

The Bahamas and the Marshall Islands, supported by other leading flag states, have now proposed that the International Maritime Organisation support the establishment of an association of independent maritime security providers that would set standards and be subject to the code of conduct and to vetting procedures.[168]

An even more recent development has been the Dutch government's paying its military to place armed soldiers aboard key merchant vessels, including one carrying a renewable energy cargo – a wind turbine.[169]

3.3 Safety, environmental protection law and chokepoint blockage risks

Given that energy cargo ships of different flags are likely generally to be subjected to the same safety and environmental protection standards, on a non-discriminatory basis,[170] with relatively few unilateral or regional exceptions, little need be added here to the treatment of those subjects in the introduction. This is not, of course, to say that even properly maintained and managed energy cargo ships might not encounter safety risks or risk of causing a major pollution incident in certain waters or circumstances. Complete elimination of these risks is impossible.

The greatest risk to be feared is that of a tanker incident resulting in major spillage of a persistent oil cargo in a state's territorial waters or exclusive economic zone, causing serious pollution damage and necessitating costly clean-up. This might result in civil liability towards the coastal state and injured private claimants and/or criminal fines and administrative penalties imposed by the coastal state.

As to civil liability, it is generally possible to limit liability for oil pollution damage to amounts set under the international oil pollution compensation scheme. The United States, it has been pointed out, represents an exception.

For a state party to the 1969 International Convention on Civil Liability for Oil Pollution Damage and its 1992 protocol (the Civil Liability Convention),[171] liability for oil pollution damage in its waters resulting from a tanker incident is channelled to the tanker owner alone. That owner can limit its liability by establishing a fund in that state to satisfy claims. Limitation is according to the tanker's tonnage (and not to the size of spill or amount of pollution damage). These limits are set with

---

167   See Article 47 of the 1977 Additional Protocol I to the Geneva Conventions (1949), 1125 *UNTS* 3; and, for a definition of 'mercenarism', the International Convention against the Recruitment, Use, Financing and Training of Mercenaries, New York (December 4 1989), 2163 *UNTS* 75.
168   Proposal for consideration at the 89th meeting of the Maritime Safety Committee (May 11-20 2011).
169   Wiese Bockmann, Michelle, "Dutch government pays $1m to put soldiers on two vessels: Dutch military to board Vroon and Heermaa vessels as they transit piracy-prone waters", *Lloyd's List* (March 23 2011).
170   The main International Maritime Organisation conventions, for example, provide for no more favourable treatment of ships of non-parties. See, for example, Article 5 of the 1973/78 International Convention for the Prevention of Pollution from Ships and Article II.3 of the 1978 Protocol to the 1974 International Convention for the Safety of Life at Sea.
171   Brussels (November 29 1969), 973 *UNTS* 3, 9 ILM 45 (1970); London (November 27 1992) 1956 UNTS 255. Some 123 states, representing 96.7% of world tonnage, are party to the 1992 protocol. Only 37, representing a mere 2.95% of world tonnage, remain party to the original convention, with its more restricted liability regime and much lower limitation levels.

insurance and reinsurance market capacities (and ultimately, the continued viability of energy cargo shipping) in mind. For a tanker over 140,000 gross tonnes, the owner's liability is now limited to 89.77 million special drawing rights ($138.55 million). For a tanker of between 5,000 and 140,000 gross tonnes, it is limited to 4.5 million special drawing rights ($6.95 million), plus 631 special drawing rights ($972) for each additional gross tonne over 5,000; and for a tanker less than 5,000 gross tonnes, to 4.5 million special drawing rights ($6.95 million).

If the pollution damage exceeds the compensation available under that convention and protocol, recovery up to 230 million special drawing rights ($354.2 million), including the shipowner's share, is possible from the fund established by the 1971 International Convention on the Establishment of an International Fund for Compensation for Oil Pollution Damage, as superseded by a 1992 protocol, amended in 2000.[172] This fund is made up of contributions from major oil importers.

This international liability regime ensures that the burden of compensation is spread between shipowner and cargo interests. The 2000 amendments greatly increased the shipowner's share of the total maximum amounts payable (from 47% to 68% in a typical case).

To ease the burden on oil importers even further, a voluntary agreement has been reached among owners of small tankers indemnified through members of the International Group of P&I Clubs. Under the terms of the Small Tanker Oil Pollution Indemnification Agreement 2006, liability in respect of incidents involving participating tankers up to 29,548 gross tonnes in all member states of the 1992 International Oil Pollution Compensation Fund is increased to 20 million special drawing rights ($30.8 million).

Under the optional 2003 Protocol establishing an International Oil Pollution Compensation Supplementary Fund,[173] higher levels of recovery are possible: up to 750 million special drawing rights ($1.155 billion) – a level designed to cover the very worst pollution incidents. Recovery is from a supplementary fund financed by the oil importers of states parties to the protocol (to date, 27 mainly developed states).

To reduce this major new (though as yet untested) burden on oil importers, a second voluntary tanker owners agreement – the Tanker Oil Pollution Indemnification Agreement 2006 – provides for indemnification of the Supplementary Fund for 50% of the amounts paid in compensation by that fund in respect of incidents involving tankers entered in one of the P&I clubs' members of the International Group.

In respect of an oil tanker spill in US waters, liability is channelled by the Oil Pollution Act 1990 not to the owner alone, but to any "Responsible Party", including the owner, a demise charterer or the operator. This party is entitled to limit its liability to the lower limits set under that act: $23.5 million for a single-hull and $17.1 million for a double-hull tanker over 3,000 gross registered tonnes. Liability for higher amounts might, however, arise concurrently under other federal legislation,

---

172 Brussels (December 18 1971), 1110 *UNTS* 57; London (November 27 1992); London (October 18 2000) 1953 *UNTS* 330.
173 London (May 16 2003), [2005] *ATNIF* 21.

such as the Comprehensive Environmental Response, Compensation and Liability Act.[174] In addition, the 1990 act does not pre-empt US states' rights (often exercised) to set higher limits, or even impose unlimited liability. Significant increases to the limits set by the 1990 act have, moreover, been debated in Congress since the Deepwater Horizon spill of April 2010.

Lastly, claims up to $1 billion above the responsible party's limits may be made to, and paid out of, the Oil Spill Liability Trust Fund, currently funded by a $0.08 per-barrel tax on domestic and imported oil.

Given the possibilities for limitation, civil liability amounts might well be exceeded by coastal state criminal fines and administrative penalties, which need not be restricted to the shipowner (or, in the case of a spill in US waters, the responsible party). While, for example, damages against the shipowner and the International Oil Pollution Compensation Fund in respect of the 1999 Erika oil spill off France were limited to €192 million ($251.9 million), the French courts imposed fines totalling €900 million ($1.18 billion), made up of the maximum fines under French law on the classification society (€375 million), the charterer (€375 million), the head of the management company (€75 million) and the head of the owner corporation (€75 million).[175]

P&I cover for an energy cargo ship extends in principle to oil pollution liabilities and related fines. There are, however, limits to the amount of pollution cover that can be offered. This is normally $500 million – except in the United States, where it may be more than double that figure.

There is always the risk, moreover, that the liability limits will be broken, giving rise to unlimited liability. The Civil Liability Convention limits are broken "[i]f the incident occurred as a result of the actual fault or privity of the owner".[176] The US Oil Pollution Act 1990 limits are more easily broken – that is, if a responsible party is found guilty of gross negligence, wilful misconduct or violation of a federal safety, construction or operating regulation. The last is a potentially wide ground, not seen in the Civil Liability Convention, that might be interpreted to refer to minor or even non-causative violations. The limits set by the Comprehensive Environmental Response, Compensation and Liability Act may also be broken relatively easily.[177]

P&I cover does not, moreover, extend to the financial consequences of non-compliance with statutory requirements, including the relevant main International Maritime Organisation requirements, set out in the 1974 International Convention for the Safety of Life at Sea[178] and the 1973/78 International Convention for the Prevention of Pollution from Ships.[179]

A less feared, but important, risk is that of blockage of a chokepoint by the wreck of a tanker or tankers following an accident, or a belligerent or terrorist attack. A

---

174 42 USC § 9601, 9607.
175 "Erika oil spill in Brittany could set legal precedent for responsibility in maritime pollution" (www.celticcountries.com/magazine/environment/erika-oil-spill-in-brittany-could-set-legal-precedent-for-responsibility-in-maritime-pollution).
176 Article V(2) of the Civil Liability Convention.
177 See 42 USC § 9607(c).
178 Footnote 109 *supra*.
179 London (November 1973 and June 1 1978), 1340 *UNTS* 61.

recent warning of potential consequences might be taken from a tanker collision resulting in blockage for six days of the Mississippi River below New Orleans, in July 2008. Over 200 ships suffered delay and oil spill clean-up and business interruption costs exceeded $125 million.[180] Given the narrowness of the fairway in the twisting Bosphorus, the Turkish Straits present the greatest peacetime[181] risk in this regard,[182] although Western fears since the 1970s of a deliberate blockage of the Strait of Hormuz have also been notorious. As the result of international concern about Iran's nuclear programme, for example, "oil-exporting Gulf states adopted a contingency plan on June 14 [2006] in case of a blockage of shipping through the mouths of the Gulf and the Red Sea".[183]

3.4 Direct action protest risks

Global environmental non-governmental organisation Greenpeace supports abandonment of fossil and nuclear fuel use in favour of renewable energy. It has conducted protests, for example, against oil company-sponsored scepticism about climate change and the use of single-hull oil tankers (which present a greater environmental risk than double hulls in many circumstances). Since the 1970s, moreover, Greenpeace has conducted numerous ship-based direct action (rather than merely symbolic) protests in which its protesters have actively sought to prevent, or at least to delay, the movement by sea of fossil (and nuclear) fuels.[184] In conducting these, Greenpeace relies on a right of peaceful protest implicit in the closely related freedoms of expression and of assembly recognised in the main international instruments on human rights, including:

- Articles 19 and 20 of the Universal Declaration of Human Rights;[185]
- Articles 19(2) and 21 of the International Covenant on Civil and Political Rights;[186]
- Articles 10(1) and 11(1) of the European Convention on Human Rights and Fundamental Freedoms;
- Articles IV, XXI and XXIV of the American Declaration of the Rights and Duties of Man;[187]
- Articles 13 and 15 of the American Convention on Human Rights;[188] and
- Articles 9(2) and 11 of the African Charter on Human and Peoples' Rights.[189]

---

180 The Tintomara-Tug Mel Oliver/ACL tank barge collision of July 23 2008. See US Coastguard House Testimony (September 16 2008), at transportation.house.gov/Media/File/Coast%20Guard/20080916/CG%20testimony.pdf.
181 During the 1956-1957 armed conflict, Egypt deliberately blocked the Suez Canal by sinking over 50 ships, causing its closure for months and costly UN salvage operations.
182 See text at footnotes 98 and 99 *supra*.
183 "Gulf States Prepare for Blockage of Hormuz Strait", DefenseNews, Reuters, Abu Dhabi, UAE (June 15 2006).
184 See Plant, "International Law and Direct Action Protests at Sea: Twenty Years On", XXXIII *Neth Y'book Int L* (2002), 75, Appendix 2.
185 New York (December 10 1948), UN Document A/811.
186 New York (December 16 1966), 999 *UNTS* 171.
187 OAS Resolution XXX 1948.
188 San José, Costa Rica (November 22 1969), 114 *UNTS* 123.
189 Banjul (June 27 1981), 21 ILM 58 (1982).

The protests take place against energy cargo ships either in port or at a terminal, in which case the national law of the port state is likely to govern,[190] or at sea, in which case international navigation rights become relevant; these both allow the protest vessel (and fast boats launched from it) to approach the energy cargo ship and the energy cargo ships to seek to avoid, outrun or outmanoeuvre the protesters.

In conducting sea-borne protests, Greenpeace tries wherever possible to reduce the scope for exercise of jurisdiction over it – for example, by using ships of different nationalities from those of the target energy cargo ships, by using multinational crews and/or by acting beyond coastal state jurisdictional limits.

Direct actions on the high seas (and in exclusive economic zones) give rise to requirements for protest vessels or their boats to sail close to a target ship and perhaps harass her navigation or obstruct her activities, including, on occasions, by occupation; when boarding and occupying a ship, protesters will typically chain themselves to equipment or enter a survival capsule from which it is difficult to remove them. Any action taken to expel (or detain) them will be governed by the national law of the energy cargo ship's flag state. Target energy cargo ships often have no lawful options but to resort to reasonable evasive or physical preventative measures and/or legal restraining orders in order to prevent or control such direct actions, even those involving risk of collision.[191] Coastal states have no jurisdiction, and at-sea enforcement measures against protesters by the flag state of a targeted energy cargo ship are surrounded by jurisdictional difficulties. They are in any event undesirable because they might further increase the risk of collision.

One protest is of particular importance in illustrating direct action protest risks to energy cargo ships, because it created a clear risk of collision or grounding for a laden oil tanker near a chokepoint and a possible risk of its blocking the chokepoint. On July 5 2002 the Turkish coastguard boarded the Crude Dio, a fully laden, 160,000 dead weight-tonne Greek-registered tanker, and arrested Greenpeace protesters. These had sailed their Dutch-registered protest ship, Esperanza, north from Istanbul the previous day in order to protest against the tanker's passage through the Turkish Straits while she was drifting just outside Turkish territorial waters awaiting clearance to enter the Bosphorus. The protest ship and her boats hindered the tanker from getting under way and heading for Turkish waters. The coastguard delayed intervening for several hours until it had jurisdiction to do so when the tanker finally drifted into territorial waters. At-sea enforcement appeared to be justified by:

- the high risk that the drifting tanker, which might have proved unable to recover power, posed of a major pollution (or blockage) incident;[192] and
- Greenpeace's allegedly falsely obtaining port clearance from Istanbul to go south, not north to the protest site.

The international community's increased concern with terrorist threats since the 9/11 attacks, and especially since the adoption of the International Code for the

---

190 Perhaps concurrently with the law of the flag state, when the energy cargo ship is occupied.
191 Plant, *supra* footnote 184, deals with the complex legal issues, especially at pp82-110.
192 The *Lloyd's List* editorial of July 6 2002 called the protest action "lunatic".

Security of Ships and Port Facilities, has put in question the compatibility of maritime direct action protests with:
- the security of each ship subjected to such a protest; and
- the security of the next port of call of a target ship that has been boarded and occupied at sea by protesters from non-governmental organisations, whether or not they are still aboard her when she approaches that port.

In 2002 Cyprus and Malta proposed that the International Maritime Organisation's impending Maritime Security Conference adopt a resolution calling upon the organisation to develop requirements for protesters to desist from demonstrations that might pose a threat to maritime safety in general and in particular, in the light of the events of 9/11, to the security of a vessel or port facility.[193] They were, however, unsuccessful.

4. Conclusion

The above-described major risks to energy cargo ships (war, terrorism, piracy, armed robbery and ship hijacking at sea, safety, environmental protection law and blockage risks, and direct action protest risks) taken together, or even separately, might influence an international energy project away from use of a maritime route or routes.

While it is difficult to predict when and how often individual major threats will arise, it is possible to suggest that some will be rare, such as truly risky direct action protests. In addition, it is possible to locate with a reasonable degree of certainty the geographical areas where major risks are more likely to arise. The greatest oil pollution liability risks are likely to arise for energy cargo ships trading to the United States, especially post-Deepwater Horizon. Blockage risks can arise only where the channel or fairway is very narrow, as in the Panama and Suez Canals or in the Bosphorus, or where a local state might be sufficiently politically unstable,[194] or desperate and determined enough either to close the waterway to traffic or even deliberately to scuttle a number of vessels. War risks are more likely to arise in unstable regions off Africa and Asia, terrorism off failed states or those affected by radicalism or fundamentalism, and piracy, hijacking and armed robbery off failed states or very poor regions. The big questions are:
- whether these areas overlap the main energy routes – in effect, from the Middle East to respectively Europe, the United States and the Far East – or can be avoided; and
- how significant are the major risks, in comparison with other factors, in determining the inclusion *vel non* of a maritime segment in a transboundary energy project.

---

193   International Maritime Organisation Document MSC 75/17/45. Such a resolution would have been binding, as a matter of internal institutional obligation, on non-governmental organisations with consultative status with the organisation, including Greenpeace International. See further Plant, *supra* footnote 184, at 111-12.
194   There was much concern about closure of the Suez Canal during the anti-Mubarak movement in Egypt early in 2011.

It is salutary to note that, alarming as was the eight-year Iran-Iraq tanker war, less than 1% of energy cargo tonnage was attacked.[195] It is also important to note that, disturbing as has been the spread of Somali anti-tanker piracy to new areas far offshore, the loss of life among seamen has until recently been kept very low, and the financial costs and losses to the industry – despite growing ransom payments – have been a fraction of the value of the cargoes kept safe or recovered from the pirates.

All websites referred to in the footnotes were most recently accessed on April 24 2011 and exchange rates calculated on February 9 2011.

---

195   Walker, *supra* footnote 60, at 75.

Cross-border regulatory risk: the EU exemption regime

Leigh Hancher
Allen & Overy LLP

1. Introduction

European and regional energy market integration requires sufficient cross-border infrastructure capacity, a common approach towards new interconnector and pipeline investment and the efficient use of electricity and gas networks. The investment challenge is a considerable one. The European Commission has estimated that €1 trillion will have to be invested into the EU energy system by 2020, about half of which will be required for networks, gas storage and smart grids. About €200 billion in investment will be needed for energy transmission networks alone. It is predicted that only about 50% of this investment will be taken up by the market by 2020, leaving a substantial gap caused by delays in planning and environmental permitting, delays in obtaining access to finance and a lack of suitable risk mitigating instruments.[1]

The European Union is keen to set its ambitions at a high level, even if these same ambitions are not always easy to realise. In 2002 the Barcelona European Council set a fairly modest target of 10% for electricity generation capacity coming from interconnector capacity. A decade later, as the European Commission confirmed in a document on energy infrastructure priorities, few EU countries have reached that target. Yet at the same time, interconnection and cross-border infrastructures are increasingly seen as an important means to enhance security of supply, develop liquidity, respond to the growth of renewables and deal efficiently with surplus power. In the case of direct current electricity interconnectors, they also provide for balancing and ancillary services. As this chapter seeks to explain, the legal framework for realising major infrastructural projects is by no means clear and predictable.

There are three possible approaches to new infrastructure investment. A first approach sees transmission system operators develop new cross-border infrastructure as part of their price-controlled transmission business, where the costs and revenues are underwritten by consumers. A second approach, termed the 'merchant approach', sees interconnectors developed by private investment as standalone projects outside the price-controlled transmission business. The developer is exposed

---

[1] EU Commission, "Energy Infrastructure Priorities for 2002 and Beyond" (Com (2010) 677/4, November 2010).

to the market demand for capacity and price (usually determined by auctions), both of which are uncertain. In order for this model to function, the investor usually seeks some revenue protection, often provided through exemptions from certain aspects of the EU regulatory framework. Lastly, government intervention or government-sponsored projects are conceivable with the aim of supporting security of supply where there would otherwise be no economic rationale for a project.[2] As explained below, the so-called 'Third Package' of energy legislation adopted in August 2009 includes, in EU Regulation 714/2009[3], provisions for national regulatory authorities to cooperate at regional level to enable an adequate level of interconnector capacity, including through new interconnection within and between regions, in order to allow for the development of effective competition and improved security of supply.[4]

The next section briefly describes the main principles of the current regime and evaluates the experiences relating to the application of the provisions relevant for direct current interconnectors and major gas infrastructure. The chapter goes on to consider the impact of the Third Package, before offering some conclusions on the impact of the current and future EU exemption regime.

2. The EU legislative framework

The EU legal framework for realising cross-border infrastructure has developed in a piecemeal fashion; it does not always sit easily with the EU institution's efforts to create an internal electricity market – efforts that date back to the adoption of the EU First Electricity Directive (96/92/EC)[5] and of the EU First Gas Directive (98/30).[6] In 2003 the EU Second Electricity Directive (2003/54/EC)[7] replaced the EU First Electricity Directive and the EU Second Gas Directive (20003/55/EC)[8] replaced the EU First Gas Directive.

These directives laid down a framework of common rules for the creation of an internal electricity and gas market. In particular, these directives stated that non-discriminatory, transparent and fairly priced network access, including cross-border flows of electricity and gas between member states (so-called 'regulated third-party access'), was an essential precondition for effective competition.[9] In order to ensure a competitive environment, EU Regulation 1228/2003 was adopted.[10] EU Regulation 1775/2005 was adopted in 2005 to regulate certain aspects of gas transmission. The EU Second Electricity and Gas Directives required member states to ensure non-discriminatory third-party access by enforcing fair tariffs and terms and conditions for the transportation of electricity and gas.

---

2   European Regulators' Group for Electricity and Gas, "Cross Border Framework for Transmission Network Infrastructure" (October 2006).
3   OJ [2003] L 176/1.
4   In addition, the Third Package comprises, with respect to the electricity market, the EU Third Electricity Directive (2009/72/EC), which repeals the EU Second Electricity Directive (2003/54/EC) (OJ [2009] L 211/55, p55).
5   OJ [1997] L 27/20.
6   OJ [1998] L 204/1.
7   OJ [2003] L 176/37.
8   OJ [2003] L 176/57.
9   See Recitals 5 et seq of the EU Second Electricity Directive.
10 OJ [2003] L 176/1.

On March 3 2011 the EU Third Electricity and Gas Directives (2009/72 and 2009/73, respectively)[11] superseded the EU Second Electricity and Gas Directives. On the same date EU Regulation 714/2009[12] replaced EU Regulation 1228/2003, EU Regulation 715/2009[13] replaced EU Regulation 1775/2005 and the EU Energy Agency Regulation (713/2009) came into force.[14] These measures form the Third Package, which entered into force as of March 3 2011 at national level.

In order to ensure the long-term adequacy of the provision of electricity in Europe, under the EU Second Electricity Directive the transmission system operators are, to a certain extent, obliged to invest into new infrastructure. In contrast, Article 13 of the EU Third Gas Directive makes this obligation explicit for gas infrastructure. In return, compensation for the ensuing costs should be included in the regulated tariffs. The directives require that tariffs allow for the necessary network investments in order to create a well-connected and well-functioning internal electricity market.[15] Hence, in accordance with this legislation, the default approach to electricity interconnectors, as well as major infrastructural investment in the gas sector, is a regulated investment.

However, in addition to investments by transmission system operators, the EU institutions have also encouraged, at least to a certain extent, private (or at least voluntary) investments in order to accelerate the establishment of interconnections between national grids and pipeline systems. An essential tool in order to encourage private investments in new electricity interconnector capacity is the case-by-case assessment envisaged by Article 7 of EU Regulation 1228/2003 – now Article 17 of EU Regulation 714/2009. This provision allows for the relevant national authorities (ie, the regulators or ministries in the areas that are to be interconnected, and the European Commission) to relax, to a certain degree and where appropriate, some of the regulatory requirements imposed by the EU Second Electricity Directive and EU Regulation 1228/2003 (especially with regard to third-party access). Such relaxation aims to encourage investments by third parties in direct current interconnectors, where these parties are separate from the existing transmission system operators – at least in terms of their legal form.

Article 22 of the EU Second Gas Directive (now Article 36 of the EU Third Gas Directive) envisages a similar derogation from certain provisions of that directive for major investments in new gas infrastructure, as well as major upgrades to existing infrastructure. Article 22 can therefore apply to facilities developed in a single EU member state, such as liquefied natural gas facilities.

These procedures should ultimately lead to an exemption decision specifying those regulatory provisions that are not applicable for a new interconnector or infrastructure project for a specified duration – normally the period corresponding to the expected pay-back time of the project. As a result, private investment, usually based on project financing methods (as opposed to investment financed by grid

---

11   OJ [2009] L211/55- 135.
12   OJ [2009] L211/15.
13   OJ [2009] L211/36.
14   OJ [2009] L211/1.
15   Article 23(2)(a) of the EU Second Electricity Directive.

tariffs), is to be encouraged. These exempted projects are also referred to as 'merchant projects' insofar as they are financed by private investment or do not form part of the regulated asset base of the national transmission system operators.

Nevertheless, exemptions for merchant projects, although permitted in certain circumstances, are the exception to the rule. Furthermore, and as explained in this chapter, the exemption procedures have proved difficult for regulators to apply in practice, with the European Commission often requiring additional conditions. This mixed experience, together with the challenges of developing a sufficient new cross-border infrastructure to meet the needs of a low-carbon economy by 2020, has led several national regulatory authorities, including the United Kingdom's Office of the Gas and Electricity Markets (OFGEM)[16] and France's Energy Regulation Commission,[17] to consider a more systematic approach to interconnector investment.

So far, in the electricity sector three merchant projects have been awarded such an exemption under EU Regulation 1228/2003 – that is, interconnectors between Estonia and Finland (Estlink), between England and the Netherlands (BritNed), and between Ireland and Wales (East-West Cables).[18] In addition, in November 2010 a limited exemption from third-party access was granted for a part of the capacity in a small alternating current interconnector between Austria and Italy for a period of 16 years.

In the gas sector, the exemption procedure has been applied to a number of liquefied natural gas facilities (Rovigo, South Hook, Dragon and Isle of Grain), and to cross-border gas pipelines (IGI Poseidon, Ostsee Pipeline Anbindungs-Leitung (OPAL) and Nabucco). By contrast, few gas storage facilities have been subject to third-party access under the EU Second Gas Directive, unless these facilities are "technically and economically necessary for providing efficient access to the system" (Article 19). Consequently, recourse to the exemption regime has rarely been necessary.

Obviously, the scope and nature of these various projects, as well as the exemptions granted, differ greatly. This remains true even within the electricity sector, where the scope of the exemption is limited to direct current cross-border interconnectors.

3. The scope of the exemptions: substantive conditions

3.1 Electricity

Article 7 of EU Regulation 1228/2003 provides that an interconnector may be exempted by the national authorities from the provisions of Article 6(6) of the same regulation and Articles 20 and 23(2), (3) and (4) of the EU Second Electricity Directive. According to Article 6(6) of the regulation, revenues derived from the use of interconnector capacity may be used for the following purposes only:

---

16 See OFGEM's consultation document, "Electricity Interconnector Policy 12/10" (January 26 2010).
17 See the commission's consultation document on the application of Article 7 of EU Regulation 1128/2003 (www.cre.fr/fr/content/download/9642/165309/file/100504CP_NIE.pdf).
18 Exemption decisions are available at ec.europa.eu/energy/infrastructure/infrastructure/electricity/electricity_exemptions_en.htm.

- to guaranty the actual availability of the allocated capacity;
- to maintain network investments or increase interconnection capacities; and
- to be taken into account by the regulatory authorities when approving the methodology for calculating network tariffs and/or assessing whether tariffs should be modified.

This provision is usually referred to as the 'revenue cap' and is intended to ensure that the transmission system operators concerned reinvest the profits derived from congested interconnectors.

Article 7 of EU Regulation 1228/2003 provides that an exemption may be granted under the following conditions:

(a) *the investment must enhance competition in electricity supply;*
(b) *the level of risk attached to the investment is such that the investment would not take place unless an exemption is granted;*
(c) *the interconnector must be owned by a natural or legal person which is separate at least in terms of its legal form from the system operators in whose systems that interconnector will be built;*
(d) *charges are levied on users of that interconnector;*
(e) *since the partial market opening referred to in Article 19 of Directive 96/92/EC,[19] no part of the capital or operating costs of the interconnector has been recovered from any component of charges made for the use of transmission or distribution systems linked by the interconnector;*
(f) *the exemption is not to the detriment of competition or the effective functioning of the internal electricity market, or the efficient functioning of the regulated system to which the interconnector is linked.*

The relevant authorities (ie, ministries or national regulatory authorities) of the member states in which the interconnector is to be built decide on the exemption on a case-by-case basis. Even though interconnectors can be exempted from certain provisions of the directive, pursuant to Article 7(4) the scope of the exemption may well vary as the national authorities may limit that scope to certain regulatory provisions or to a certain part of the overall capacity. As discussed below, the individual exemption is usually granted subject to conditions. It follows that 'exempted' does not mean 'unregulated'.

3.2 Gas

Article 22 of the EU Second Electricity Directive is essentially similar in scope to Article 7 of EU Regulation 1228/2003, but a number of important differences exist.

First, Article 22 may apply to new as well as existing infrastructure that is modified or in which the capacity is significantly increased. Second, the investment must enhance security of supply. Third, as there is no equivalent to a revenue cap in the gas sector, there is no need for the authorities to check that the funding of the gas infrastructure will be covered by customers of the transmission system operators.

---

19   The EU First Electricity Directive, see footnote 5 above.

Investors in gas infrastructure are primarily interested in reserving capacity for their own shareholders or other investors. Accordingly, they are keen to obtain an exemption from the rules on regulated third-party access in order to ensure that capacity can be allocated in such a way that future revenue streams are secured. If those investors are also dominant gas supply companies, however, the grant of an exemption could enhance their market power.

4. Procedural conditions

Under the current rules, it is for each national authority concerned to reach a decision on an exemption application and to publish its duly reasoned decision. Each authority must then notify its decision to the European Commission, including all relevant details regarding the reasoning of the decision, as well as all relevant information necessary in order to assess the decision. The commission has two months to review the decision (with a possible one-month extension). If the commission does not agree with the national decision(s), it may request the national authority to amend or withdraw the exemption decision. So far, the referring authorities have complied with the commission's decisions.[20] Should the regulators not comply with such a request of the commission, the final exemption decision is subject to the decision of a committee of member states (see Articles 7(5) and 13 of EU Regulation 1228/2003 and Article 22(4)(e) of the EU Second Gas Directive). So far, this latter procedure has not been invoked.

5. Commission policy under the Second Package

In May 2009 the European Commission published a working paper on its assessment of exemption decisions containing a detailed analysis of the reasoning of Article 7 of EU Regulation 1228/2003 and Article 22 of the EU Second Gas Directive, its preconditions and the conclusions that may be drawn for the assessment of a given project.[21] The document is intended to give guidance to national authorities and potential applicants on the interpretation of the different conditions set out in Article 7 of EU Regulation 1228/2003 and Article 22 of the EU Second Gas Directive in light of the practical experience gained since publication of the commission's first interpretive note on the matter in 2004. The commission hopes that this updated guidance will enable a more coordinated and coherent approach on the part of the national authorities. Regrettably, however, the May 2009 note does not anticipate the entry into force of the Third Package. Hence, any applications made after December 2010 will not have the benefit of full guidance on how they are likely to be assessed under the new regime.

The working paper takes as its starting point that only the default regime can

---

20 Recent European Commission decisions requesting amendments to the national decisions include the Nabucco pipeline from Turkey to Austria via Hungary, Romania and Bulgaria, and the Ostsee Pipeline Anbindungs-Leitung (known as OPAL) from northern Germany to the Czech Republic (see below). All exemption decisions of the commission with regard to the gas sector are available at ec.europa.eu/energy/infrastructure/infrastructure/gas/gas_exemptions_en.htm.

21 "Commission staff working document on Article 22 of Directive 2003/55/EC concerning common rules for the internal market in natural gas and Article 7 of Regulation (EC) No 1228/2003 on conditions for access to the network for cross-border exchanges in electricity" (May 6 2009, SEC (2009) 642 final).

guarantee a real internal market; it follows that any derogation will be exceptional and can be granted only if it is unlikely significantly to frustrate the internal market objective. In particular, an exemption request by dominant undertakings will be treated with exceptional care and the availability of alternative facilities for third parties will be an important factor. The document also places considerable emphasis on the issue of risk and how that aspect of the project is to be assessed. The document insists on the need for the investors to test market demand – usually by way of an open season – as a crucial element of their risk assessment. Under the Third Package, market testing is compulsory.

6. Decisions to date

6.1 Electricity

(a) **BritNed**

BritNed is a joint venture of TenneT (the Dutch transmission system operator) and National Grid (the UK transmission system operator) to construct and operate the first electricity link between the Dutch and UK electricity markets. The link became operational as of April 2011. The capacity available for trading is 1,000 megawatts (MW) and the nominal cable capacity (peak capacity[22]) is 1,320MW. The project costs are estimated at €600 million.[23] As transmission system operators, neither TenneT nor National Grid has an interest in supply activities. Furthermore, according to its licensing regime, National Grid is not allowed to include interconnectors in its regulated asset base or to cover the costs of interconnectors with general tariff revenue. Project costs must be recovered by marketing the cable capacity. BritNed will apply a blend of implicit and explicit auctions[24] on the basis of capacity contracts of different durations, but with a maximum duration of one year.[25] The revenues are expected to be mainly derived from price differences between the Dutch and UK electricity markets. Since the prices vary significantly from year to year, revenue forecasts remain uncertain.

Although this business model does not depend on long-term contracts, in order to obtain legal certainty BritNed applied for an exemption from third-party access. The main reason for requesting an exemption was the exposure to the risks entailed by a possible revenue cap. BritNed claimed that this would result in a risk asymmetry: if revenues were less than expected, investors would have to bear any loss without being able to fall back on a mechanism for socialising or compensating them. The national regulators could rely upon *ex post* regulation if the project was

---

22 CAB D(2007)/1258, exemption decision on the BritNed interconnector by the European Commission (October 18 2007).
23 See www.britned.com.
24 'Explicit auction' means that transmission capacity is auctioned separately and independently from the market place where electricity is auctioned. This method offers a simple means to handle the capacity on an interconnector. 'Implicit auctions' include the day-ahead transmission capacity in the auctioning of electricity. The resulting prices in each market reflect the actual costs of energy and transmission, so that electricity flows from low price markets to high price markets fostering price convergence.
25 BritNed application for EU exemption (June 12 2006), pp12 *et seq* (www.ofgem.gov.uk).

commercially successful and cap returns or demand capacity expansions (Article 23(4) of the EU Second Electricity Directive).

BritNed argued that these uncertainties – in addition to the fact that high revenues would be capped, whereas high losses would not – led to a degree of risk associated with the investment that was commercially prohibitive.[26] BritNed therefore applied for an exemption from Article 6(6) of EU Regulation 1228/2003 for a period of 25 years in order to be able to amortise its investment.

Both OFGEM and the Dutch regulator, the DTE, had taken a favourable view on the application. On referral of the two national decisions, the European Commission deemed the project to be pro-competitive, since by establishing the first interconnection between the Netherlands and the United Kingdom, it expands market coupling in the northwestern European market, contributes to the convergence of marginal electricity prices, leads to less volatile and lower average prices and reduces market concentration.[27] Although the commission accepted that BritNed was exposed to risk asymmetry, it required the imposition of certain conditions in the national exemption decisions.

First, the commission maintained that the application had highlighted the possibility of bad years in respect of the prospective revenues, but ignored exceptionally good years, which are equally likely and could enable BritNed to amortise the investment very quickly.[28]

Second, the commission expressed concern as to the effects on competition. An exemption is to be granted only if it is not to the detriment of competition. Otherwise, customers do not benefit. This could be the case where a project is the first of its kind and therefore establishes a monopolistic market position. In such circumstances an investor might be tempted to under-dimension the project in order to maximise revenues – for example, by keeping capacity artificially scarce and revenues high by installing a suboptimal capacity. The commission was not convinced that the proposed size of the interconnection cable offers the optimal balance between rewarding BritNed for undertaking the investment and benefiting consumers on both sides.[29]

The commission held that the 25-year exemption was in principle justified, but requested the national regulators to amend their decisions to include a condition that obliges BritNed to report on the total costs and revenues of the project after 10 years. If this report reveals an actual rate of return on the investment of more than one point above the estimated rate, BritNed will have to increase transmission capacity or the rate of return will be capped.[30]

(b) **Estlink**

The Estlink cable is a direct current interconnector between Harku in Estonia and Espoo in Finland. It has been fully operational since January 2007 and provides a

---

26 BritNed application, p5; BritNed exemption decision, p8 *et seq.*
27 BritNed exemption decision, p 11(a).
28 BritNed exemption decision, p 11(b).
29 BritNed exemption decision, p11(f).
30 BritNed exemption decision, p 13.

transmission capacity of 350MW. The project costs amounted to €110 million. Estlink was constructed and commissioned by Nordic Energy Link, a subsidiary of Eesti Energia AS, the main producer, supplier and distributor of electricity in Estonia.[31] In order to develop the Estlink project, Easti Energia was joined under an elaborate corporate construction scheme by:
- Latvenergo, the main producer, supplier and distributor of electricity in Latvia;
- Lietuvos Energija, the transmission system operator in Lithuania; and
- Finestlink Oy, a subsidiary of Pohjolan Voima and Helsingin Energia, a municipal power and heat producer and supply company from Finland.[32]

In contrast to BritNed, investors in the Estlink cable could book all capacity for a certain period in order to redeem their investment. Subsequently, depending on certain developments but no later than December 31 2013, the Estlink cable will be transferred to the transmission system operators of Finland, Estonia, Latvia and Lithuania at a predetermined price, and will be operated as a full transmission system operator cable.[33] An exemption from third-party access pursuant to Article 7(1) of EU Regulation 1228/2003 was requested for the initial period. The Finnish and Estonian authorities took a positive decision. The European Commission also concluded that the Estlink cable is clearly pro-competitive as it opens the Nordic electricity market for cheap electricity from the Baltic states, which have substantial overcapacities. In addition, even if the cable is more likely to transport electricity from the Baltic to the Nordic market, the possibility of a reverse flow would already serve as a price signal to the Baltic market.[34] Further, the commission acknowledged that the investors essentially rely on the possibility to sell a considerable amount of their electricity via the Estlink cable, and that the risk attached to the investment would be prohibitive unless a sufficient transmission capacity were ensured to the investors by means of an exemption from third-party access.[35] However, the commission had a number of concerns. Since all capacity towards the Baltic states is reserved for the Baltic investors, the commission considered that this reservation could prevent entry into the Baltic market by Finnish suppliers via the Estlink cable; it also assessed potential incentives for the Baltic investors to hoard their capacity.

These concerns were ultimately dispelled by the details of the agreements between the investors and the intended business model. The business rationale for the Estlink project is to exploit the price differential between the Baltic and the Nordic electricity markets. As long as this price difference exists, it is unlikely that any competitor from Finland will be prevented from selling on the Baltic market. Once this price difference disappears, as is predicted to occur by 2013, revenues will

---

31 See www.nordicenergylink.com; Exemption Decision E/2005/001, Estlink Project, by the European Commission (April 27 2005), pp5 *et seq.*
32 According to the agreement, Eesti Energia is the main shareholder with 39.9% of the shares, Latvenergo and Lietuvos Energija have 25% each and the remaining 10.1% are divided between Pohjolan Voima and Helsingin Energia.
33 Estlink exemption decision, pp2 and 6.
34 Estlink exemption decision, pp2 and 7.
35 Estlink exemption decision, p7.

decline significantly and the Estlink cable will then be sold to the transmission system operators.[36]

As the investors will have to pay a fee for the reserved capacity, regardless of whether any of this capacity is used, there is a significant disincentive for capacity hoarding. However, in order to reduce these risks further, the commission recommended the imposition of transparency requirements on investors by means of *ex ante* forecasts as well as *ex post* evaluation of the cross-border trade via the Estlink interconnector.[37]

(c)   The East-West Cables
In the East-West Cables project, Imera Ltd is an Irish project development company set up to establish electricity interconnection between Ireland and Great Britain. In 2008 Imera applied for an exemption under Article 7 of EU Regulation 1228/2003 on behalf of its two subsidiaries, East West Cable One Ltd and East West Cable Two Ltd – both special purpose companies set up to develop and construct two direct current interconnection cables between Ireland and Great Britain. In contrast to the BritNed and Estlink projects, Imera has no links with any existing player in the electricity market.[38]

The first cable connects Ireland and Wales and was planned to be operational in 2010. The second cable, expected to become operational in 2011, also connects Ireland and Wales. Each cable has a capacity of 350MW. As the two cables formed one project in terms of financing and tendering, they were treated as one project – the East-West Cables project.[39]

Imera applied for a full exemption from regulated third-party access as well as from the limitation of revenues under Article 6(6) of EU Regulation 1228/2003 for a period of 25 years for the first cable and 20 years for the second cable.[40] Imera, like BritNed, claimed risk asymmetry. In order to minimise the risks that arise from short-term price fluctuations, the business model for the marketing of the cable capacity is mainly based on long-term contracts of at least 10 years.[41] Imera also applied for an exemption from regulated third-party access as protection against possible *ex post* regulation in case the project was more successful than forecast.[42] In order to prevent foreclosure effects, Imera suggested the facilitation of a secondary market on which capacities would be auctioned on a short-term basis, as well as a capacity cap for single customers.[43]

The European Commission found the project to be pro-competitive. Despite an existing interconnection between Northern Ireland and Scotland via the Moyle interconnector in Northern Ireland, the commission acknowledged the importance

---

36   Estlink exemption decision, p2.
37   Estlink exemption decision, pp3 *et seq.*
38   Imera application for EU exemption (April 2008), p5 (www.ofgem.gov.uk).
39   The regulators treated the two cables as single project. See European Commission exemption decision on the East-West Cables project (December 19 2008), p.5 (ec.europa.eu/energy/infrastructure/infrastructure/electricity/electricity_exemptions_en.htm).
40   Imera application, p4.
41   East-West Cables exemption decision, p 21.
42   Imera application, p4.
43   Imera application, p4.

of interconnection capacities, especially for the Irish market. Ireland is dependent on wind energy for future electricity supply. Wind energy may be exported via the interconnector and when there is a shortage of wind, conventional energy may be imported.[44] However, the commission found that, in principle, there was no reason why a well-designed regulatory scheme could not deal with the rules raised by Imera. A parallel, fully regulated project by Irish transmission system operator Eirgrid for interconnector between Ireland and Great Britain was also planned. The commission concluded that a second interconnector added a significant amount of uncertainty to the East-West Cables project since it was less certain that there is sufficient additional market demand to make both interconnector projects profitable. Thus, the commission agreed that the significantly higher risk of the East-West Cables project in the presence of the Eirgrid interconnector justified an exemption, provided that the Eirgrid interconnector proceeded as planned.[45]

Finally, the commission considered whether the exemption was not to the detriment of competition or the effective functioning of the internal electricity market; it accepted that a capacity cap for dominant undertakings and the facilitation of a secondary market with use-it-or-lose-it schemes would remove this risk.[46] It requested the national regulators to amend their exemption decisions accordingly.[47] Imera was also obliged to implement effective intra-day trading.

6.2 Gas[48]

(a) Nabucco

The Nabucco pipeline should make possible the transportation of gas into Europe from the Middle East and Caspian region. It is over 3,300 kilometres long and has a planned maximum capacity of 31 billion cubic metres per year. It will cross Turkey, Bulgaria, Romania and Hungary and end at the Austrian Baumgarten gas hub. Potentially, Nabucco could supply up to 10% of the European gas demand, and primarily countries that are dependent on one supply route. It is designated a 'trans-European energy network priority project' and will receive substantial financing from the European Economic Recovery Plan. A consortium of five gas transmission system operators is responsible for the construction of the pipeline, together with German energy company RWE. Four separate European regulatory authorities were involved in the exemption procedure and an intergovernmental agreement was concluded with Turkey on July 13 2009 to secure transit. The Austrian exemption decision was the first to be notified to the European Commission and served as a blueprint for the remaining national decisions. The commission issued two decisions requesting the Austrian regulator to modify its exemption decision. In 2009 the

---

44 East-West Cables exemption decision, p 9.
45 East-West Cables exemption decision, p 23.
46 'Use it or lose it' schemes provide for any allocated capacity to be returned to the market in case it is not used (eg, Article 6 (4) of EU Regulation 1228/2003 and Article 16(4) of EU Regulation 714/2009), in order to avoid strategic hoarding or reservation of capacity.
47 East-West Cables exemption decision, pp39 and 56.
48 All exemption decisions of the European Commission with regard to the gas sector are available at ec.europa.eu/energy/infrastructure/infrastructure/gas/gas_exemptions_en.htm.

exemption for the other three countries was finalised. In all cases the commission requested a strengthening of the conditions imposed on the developers of the pipeline to ensure harmonisation of the different regulatory regimes and compliance with Article 22 of the EU Second Gas Directive.

The conditions imposed in the national exemption decisions and in the intergovernmental agreement include the following. Nabucco is exempted for 25 years, commencing from the date that the first part of the pipeline becomes operational, from regulated third-party access and tariff regulation. Third-party access applies, however, to at least 50% of the capacity, so that any supplier can ship gas throughout the entire route. The remaining 50% is given by a first option to the pipeline owners or their affiliates; if this option is not taken up, the capacity is offered on the open market under the 'use it or lose it' principle. The total capacity will be marketed by Nabucco International, a joint venture between the six members of the consortium. Uniform tariffs and uniform terms and conditions will be applied to standard contracts in all five countries. The consortium must implement an open season at regular intervals. However, if following the open season the firm requests for capacity reservation exceed the maximum available capacity of 25.5 billion cubic metres per year, then Nabucco International must extend the capacity of the pipeline to make available to third parties additional transportation capacity to meet effective demand – provided that this extension is technically and economically feasible.

Furthermore, an undertaking holding a dominant position in one or more of the relevant upstream or downstream gas markets should not be allowed to book more than 50% of the total capacity at the exit points in the relevant market. This means that a dominant undertaking in one EU member state cannot book more than 50% capacity at the exit point in the Austrian section of the Nabucco pipeline. If there is not sufficient market interest to book the remaining exit capacity, the dominant company may book more capacity, but must offer the gas volume exceeding the 50% capacity cap to the market. This was a crucial condition for the commission in approving the national exemption decisions.

(b) *IGI Poseidon*

The commission approved the 25-year exemption to this pipeline linking Italy and Greece and developed by Edison and DEPSA. The Italian authorities had adopted a decision granting an exemption on behalf of the Greek authorities, given that at the time of the application the Greek authorities had not implemented the EU Second Gas Directive into national law. The commission imposed a number of conditions on the national decision, including a requirement of market testing to establish whether additional capacity should be constructed to meet the requirements of third parties.

(c) *OPAL*

OPAL is a gas pipeline linking the planned Nord Stream pipeline in Northern Germany to the German-Czech border and executed by a joint venture between Wingas and E.ON Ruhrgas. The German regulator, BNetzA, granted an exemption for the flow of gas into the Czech Republic for 22 years, but not for flows within Germany or for reverse flows. BNetzA had also imposed conditions concerning hoarding and congestion management. It also required that the company's articles of association be submitted to it prior to their adoption to ensure that the pipeline was operated independently from the interests of its shareholders. Lastly, a deadline for construction was set with a further limitation that failure to meet the approved commissioning date would lead to expiry of the exemption.

The European Commission amended the national decision on June 12 2009 and imposed additional conditions. The commission was concerned that the pipeline would not enhance completion on the Czech market. Consequently, the commission imposed a capacity cap of 50% on any dominant undertaking acquiring capacity at the Czech border. Notably, the commission stated that Gazprom and its subsidiary would constitute a single undertaking for the purposes of calculating this capacity cap. However, the cap could be exceeded if a gas release of 3 billion cubic metres a year were implemented. The gas release was to be accompanied by a corresponding capacity release and BNetzA had to approve both release programmes. BNetzA modified the national exemption in line with these conditions. In the meantime, the investors have lodged a challenge to the commission decision to treat Gazprom and its subsidiary as a single entity. The applicants have also claimed that they were denied the opportunity of a fair hearing by the commission prior to the adoption of its decision.[49]

(d) **BBL**

This case involved an interconnector between the United Kingdom and the Netherlands, undertaken by Dutch gas transmission system operator Gasunie, with the capacity being sold to three companies – Gasunie Trade and Supply (now known as GasTerra), E.On Ruhrgas and Wingas. GasTerra had already signed an agreement to sell a corresponding volume of gas to Centrica and additional complications arose as a result of the fact that Shell and ExxonMobil are major shareholders in GasTerra. An exemption was granted by both national authorities for a period of 10 years only (despite the fact that long-term agreements had been signed for 15 years). As a result of an intervention by the European Commission, the exemption excludes reverse flow into the United Kingdom as such reverse flow is subject to full regulated third-party access.

(e) ***Liquefied natural gas facilities***

The exemption has also been applied to several liquefied natural gas facilities, including the Isle of Grain (developed by National Grid Transco as a merchant project) and South Hook, both in the United Kingdom. The latter is a joint venture between Qatar Gas and ExxonMobil. A third facility, the Dragon liquefied natural gas

---

[49] Registered as Case T-318/09, September 25 2009. In support of its action, the applicant first claims that the defendant infringed its right to a fair hearing by denying it any opportunity to comment on the condition imposing a burden on it before it was adopted. Second, the applicant contends that the defendant infringed the applicant's right of access to documents by denying it any access to the file of the proceedings.
Finally, the applicant complains that the defendant misapplied Article 22(4) of Directive 2003/55/EC and the principle of proportionality, the principle of equal treatment and the duty to state reasons (Article 253 of the EC Treaty).

terminal, was also dealt with by the UK regulator, OFGEM.

Two further facilities have been exempted in Italy: the Brindisi liquefied natural gas terminal and the Rovigo terminal.

In all five cases, only a single national regulator was involved. The two Italian cases were dealt with under an Italian law that stipulates that if only 80% of capacity is to be reserved for a period of 20 years, the projects will automatically qualify for exemption. Both cases concerned new entrants to the Italian gas market. Consequently, neither case raised any competition concerns. However, the Brindisi project had been commenced well before the exemption was granted. It was still considered as 'new' within the meaning of the EU Second Gas Directive, as it was not due to be completed before the cut-off date of August 2003.

7. The Third Package

A major criticism of the regime as discussed above is that there is little assurance for investors that the various national authorities concerned will evaluate their applications for exemption in a coherent and coordinated way. The Third Package has introduced important changes in this respect. In particular, the new Agency for the Cooperation of Energy Regulators may play an important role in the decision-making process, as explained further below. However, these regulations and directives also include new obligations that will apply equally to exempted and non-exempted interconnectors.[50] Recital 23 of EU Regulation 714/2009 specifies that all exemptions granted under EU Regulation 1228/2003 continue to apply until the scheduled expiry date, as detailed in the granted exemption decision. A similar provision is included in the EU Third Gas Directive.

In particular, Article 9 of both the EU Third Electricity and Gas Directives requires ownership unbundling of all transmission system operators. EU member states may choose between several options that are less far reaching. In accordance with Article 17 of EU Regulation 714/2009 and Article 36 of the EU Third Gas Directive, an exemption can also be requested from Article 9, in addition to third-party access and regulatory supervision. Just as under the predecessors of these provisions, Article 7 of EU Regulation 1228/2003 and Article 22 of the EU Second Gas Directive, any exemption is given only for a limited duration. On the expiry of the exemption, the relevant national unbundling regime(s) will apply in full. Furthermore, Article 10 of the two new directives requires that all transmission system operators be certified in order to ensure compliance with the unbundling rules. It is unclear whether exempted projects would also have to obtain certification, given that Article 10 is not listed as one of the provisions to which the exemptions provided in Article 17 of EU Regulation 714/2009 apply (or, for that matter, Article 36 of the EU Third Gas Directive, which corresponds to Article 22 of the EU Second Gas Directive).

The new directives and regulations also include new rules on congestion management and capacity allocation that apply equally to exempted and non-exempted facilities. The national regulatory authorities must ensure that these rules are respected prior to granting the exemption. Furthermore, the new package also

---

50   OJ [2009] L211.

introduces requirements on transmission system operators to develop a 10-year development plan coordinated through their European networks ENTSO–E and G. These organisations are also responsible for drawing up network codes. These codes are developed by ENTSO-E and ENTSO-G in cooperation with the Agency for the Cooperation of Energy Regulators and are approved as binding rules by the European Commission.[51]

It is by no means clear how exempted projects, which are not necessarily operated by transmission system operators, can or should fit into this new regulatory framework, given that the exemption will, as explained above, be partial only.

The Third Package gives new competences to the new Agency for the Cooperation of Energy Regulators, as of March 2011, in relation to cross-border infrastructure. The new procedures are as follows:

- The agency may submit an advisory opinion on the requests for exemption to the national regulatory authorities. The opinion may be used as a basis for the joint decision within two months of the date on which the request for exemption was received by the last of the national regulatory authorities. If all national regulatory authorities agree on the exemption request within six months of the date of the last filed application, they shall inform the agency of their decision accordingly.[52]
- Where all national regulatory authorities concerned have been unable to reach an agreement within six months or, alternatively, upon the joint request of all national regulatory authorities concerned, the agency can decide upon the exemption request in place of the national regulatory authorities. If this occurs, the agency must comply with the procedures set out in either EU Regulation 714/2009 or the EU Third Gas Directive (as if it were itself the national regulatory authority). Before taking a decision, the agency must consult the national regulatory authorities and the applicants.[53]

The competences under the bullet points above are also spelled out in Article 8 of the EU Energy Agency Regulation.

The agency/national regulatory authority must transmit its final decision to the European Commission. The commission has the same powers as under the Second Package – that is, to review and either veto or require amendments to the agency/authority decision. Even if, under these new procedures, the agency takes the place of the national regulatory authorities in the EU member states and the commission approves the agency's decision, that decision is not necessarily binding on the member states; they can opt to treat it as an opinion if the national regulatory authority in their own system provides only opinions on exemption applications.[54]

In addition, under the new rules the commission's exemption decision will lose its effect two years from its issue date if construction of the project has not started and five years from its issue date if the project has not become operational. The EU

---

51 See Articles 6 to 8 of Regulation 714/2009 (electricity) and Articles 6 to 8 of Regulation 715/2009 (gas).
52 See Article 36(4) of the EU Third Gas Directive and Article 17(4) of Regulation 714/2009.
53 See Article 36(4) of the EU Third Gas Directive and Article 17(5) of Regulation 714/2009.
54 See Article 36 (7) and Article 17(6).

Energy Agency Regulation also allows the commission to adopt guidelines on the conditions of application.

8. Conclusion

The one-size-fits-all approach to regulated third-party access and regulated tariffs in the Second and Third Packages is perhaps not entirely satisfactory. Thus, the exemption procedures reviewed in this chapter are often seen as useful safety valves to accommodate different economic circumstances, especially where the facilities in question are not essential or monopolistic. However, as discussed in this chapter, the exemption procedures are in themselves based on a one-size-fits-all approach. The same conditions for assessment and the same procedures apply, irrespective of whether the exemption is sought for a major pipeline, a direct current interconnector or a liquefied natural gas terminal.

Furthermore, the exemption, if awarded, is only partial. This may make sense for an interconnector or pipeline, but less so for a liquefied natural gas terminal where even light-handed regulation may be unnecessary. The entry into force of the new Third Package will not substantially improve this situation. Indeed, it is likely that there will be more regulation of exempted facilities in order to align their operations with the default regime.

Obviously, in cross-border situations the experience with applying the exemption regime has been made more complex where more than one national regulatory authority is involved. The new procedures introduced by the Third Package should alleviate some of the burden on investors seeking to secure a coherent response from two or more countries, as long as these countries are members of the European Union, the European Economic Area or, to a certain extent, the Energy Community Treaty.

Nevertheless, given the scale of infrastructure investment that is officially recognised as needed to meet the European Commission's ambitious 20-20-20 goals, it is perhaps surprising that despite the adoption of the Third Package reforms, the EU exemption regime as discussed here remains at best partial and at worst unpredictable in its scope.

Host government agreements: the investor perspective

Charles Lindsay
Allen & Overy LLP

1. Introduction

A major energy infrastructure investment will very often involve a degree of limited recourse debt financing. One way or the other, the recourse of lenders of that financing for repayment of their loans will be limited to the revenues produced by the project itself. The lenders will not have general recourse to the other assets of the sponsors of the project. For the purposes of this chapter, it is assumed that this limitation of recourse is achieved by the incorporation by the sponsors of a special purpose vehicle company (a project company) which will be the borrower of the debt financing for the purposes of developing the project. The size of financing required will normally be large enough to warrant an international syndicate of banks or group of bondholders. The risks relating to the project in question will therefore need to be limited to risks that are typically borne by international lenders of limited recourse financing or, as it is commonly called, project financing. This chapter is not intended to be a discussion of the general principles of project finance bankability,[1] but the approach to the risks identified in this chapter is driven by those principles.

The perceived risks presented in the host state will, of course, depend on the state itself, its history of hosting similar projects and its general record as a friendly environment for foreign investment. That said, even states which have a very substantial history of project financings can present risks for a large energy infrastructure project that can nevertheless have a potentially devastating economic effect on the stakeholders in the project. The United Kingdom is a good example of a jurisdiction in which hundreds of energy and public-private partnership projects have been project financed, but where some projects, particularly in the North Sea, have suffered from intervening changes in law or standards.

This chapter assumes that the project in question entails the construction and operation of a transnational oil or gas pipeline. The project will have been sponsored by a group of international companies which will have ownership of hydrocarbons that they will want to ship to end markets through the pipeline. The sponsors will require long-term rights to capacity in the pipeline to make these shipments.[2]

---

[1] For a discussion of the basic principles of bankability see Vinter, Graham and Price, Gareth, *Project Finance*, Sweet & Maxwell (Third Edition) at Chapter 6.
[2] In the case of gas pipelines in the European Union or in other jurisdictions that have signed up to the various EU packages of gas legislation, long-term reservation of this capacity may be constrained by regulation requiring third-party access rights to the pipeline infrastructure. This is dealt with in the chapter of this book on cross-border infrastructure projects.

2. The investors

To analyse the risks associated with a project, it is necessary to consider the interests of the investors in the project. The various project investors and their ultimate risks are detailed below.

2.1 Equity investors

Equity investors are shareholders in the project company that develops, owns and operates the project. They will have subscribed and paid for shares in the project company, and are likely also to have advanced subordinated debt to the project company. They may also have an ongoing commitment to the project company's lenders to provide further equity or subordinated debt to the project company in the event of the occurrence of certain contingencies. They are likely also to have to provide some form of guarantee or debt service undertaking to the project company's lenders that will remain in place at least until completion of the development of the project. The equity investors' risks are therefore that:

- their investment will not produce the internal rate of return that was required for them to make their investment decision in respect of the project;
- circumstances might occur in which they become obliged to the project company's lenders to provide further equity or subordinated debt funding to the project company; and
- their guarantee or debt service undertaking in respect of the project company's debt will be called upon.

2.2 Project financiers

Project financiers are lenders of debt to the project company, and providers of guarantees of debt. Lenders may be commercial banks, bondholders or multilateral lending agencies such as the European Bank for Reconstruction and Development, the International Finance Corporation, the European Investment Bank or the Asian Development Bank. Guarantors of these loans are typically export credit agencies, providers of political risk insurance and, as noted above, the shareholders of the project company – at least until completion of the project. The lenders' risk is that the repayment instalments of their debt and the interest payments on it are not paid as scheduled. The guarantors' risk is that their guarantees will be called upon.

2.3 Shippers

Shippers ship hydrocarbons through the project. Their risks are that:

- their ability to ship hydrocarbons through the pipeline may be interrupted temporarily or permanently;
- the cost (whether as tariffs, transit fees or taxes) of shipping their hydrocarbons through the pipeline will become so high as to render the pipeline an uneconomic export route for their hydrocarbons; and
- where the shippers are selling hydrocarbons to offtakers in the host state, the offtakers will fail to take or pay for the hydrocarbons to be sold to them or that the price that they pay for these hydrocarbons becomes so low as to be uneconomic for the shippers.

At least some of the shippers are likely to be equity investors in the project too, increasing their overall exposure to risks associated with the project.

3. Tools for mitigation of risks for investors

3.1 Host government agreements

An agreement with the host state can effectively address a number of these risks. The parties to such an agreement would be the host state itself (or some organ of the state that represents it), the project company and, possibly, the equity investors. The shippers would have an interest in being third-party beneficiaries under the host government agreement, even though they are unlikely to be party to the agreement itself.

In broad terms the purpose of a host government agreement is to facilitate the development of the project and to provide a degree of stability to the project with a view to preserving its economic equilibrium. Host government agreements come in many different forms and have many different names (eg, host state agreements or concession agreements). Some of the detailed terms of a host government agreement are discussed in section 4 below.

A host government agreement will often work alongside a project-specific intergovernmental agreement. (See section 3.2 below and the chapter of this book on intergovernmental agreements for more.)

The Energy Charter Secretariat established under the Energy Charter Treaty has published a model form of host government agreement recommended for use in pipeline projects.[3] It is fair to say that if the host state, the project company and the equity investors were to enter into an agreement containing substantially the same terms as the model agreement, most host state risks would be very satisfactorily dealt with, on the face of the document. So why not simply present the model agreement to the host state, request a document based on it and expect this to solve all the problems posed by the host state risks? This question is answered by raising some further questions:

- Would the host state accept the model agreement? While the model terms, especially in their first edition, were based loosely on existing agreements from projects in the Caspian region, these agreements are very comprehensive and far-reaching, and place a number of major risks firmly with the host state. As pointed out in the chapter of this book on managing risks to transboundary energy infrastructure, host states have recently tended towards resisting risks being allocated to them in this manner. Consequently, it would be reasonable to expect that the host state would fiercely resist much of the risk apportionment proposed by the model agreement. The model agreement is, after all, intended only as a starting point for negotiations.
- Is the host government agreement worth the paper it is written on? The host government agreement is an agreement with a sovereign state and therefore,

---

3   Energy Charter Secretariat, Model Intergovernmental and Host Government Agreements for Cross-Border Pipelines (Second Edition) (www.encharter.org/fileadmin/user_upload/document/ma-en.pdf).

depending on how important issues of applicable law and enforcement are addressed, it may not necessarily afford the same degree of predictability as an agreement with a private party. After all, the host state is both counterparty and legislator. While the host government agreement may contain a waiver of the host state's sovereign immunity,[4] there remains the risk that the host state will be able to evade its obligations under the agreement by legislating its way out of them. Host government agreements are commonly ratified by the legislature of the host state, which would have the effect of making these agreements domestic law and enforceable under general law, as well as in contract. However, there may well be no constitutional obstacle to the legislature amending its contractual obligations or domestic law by subsequent specific or general legislation. These concerns can be significantly dealt with in the following manners:

- by annexing the host government agreement to the project's related intergovernmental agreement and making the terms of it an integral part of the treaty-level intergovernmental agreement. This would allow the other host states party to the intergovernmental agreement (and any investor which might also have standing) to enforce these terms.[5] But, in addition, being part of the intergovernmental agreement would, in theory, give the terms of the host government agreement the status of treaty law; in some jurisdictions this would have the effect of giving those terms precedence over domestic law – even subsequently enacted domestic law. Even if, as a legal matter, this precedence is not established for certain in the host state,[6] this technique may be of value before any international court or tribunal (if provided for in the intergovernmental agreement), since a state may not invoke its domestic law to avoid international law obligations. This technique appears to provide a perception of entrenchment of the terms of the host government agreement and has tended to give investors a satisfactory level of comfort as to the durability and enforceability of the terms of the host government agreement;

- by providing that the law applicable to the host government agreement is not, at least exclusively, the law of the host state, coupled with a neutral forum for the resolution of disputes (eg, international arbitration or a foreign court). Options for the applicable law include choice of a neutral governing law (often English law is acceptable), or a hybrid choice of the law of the host state together with either the laws of another domestic legal system, public international law, or, exceptionally, transnational rules of law (eg, the so-called '*lex mercatoria*'), together with a stipulation that the latter should prevail in the event of inconsistency with the host state law. These techniques lack the certainty that investors would ideally prefer and unsurprisingly they have in the past been the

---

4 See the chapter of this book on investor treaty protection for the significance of the waiver of sovereign immunity
5 See the chapter of this book on intergovernmental agreements.
6 Investors are unlikely to get lawyers to issue a clean legal opinion on the point.

subject of international arbitration proceedings; however, they can achieve a degree of stabilisation where other options may not exist; and
- by structuring the key terms of the host government agreement so that the investors are able to obtain bilateral or multilateral investment treaty protection from any breaches and any legislative or other attempts the host state might make to evade its obligations. This matter is dealt with in the chapter of this book on bilateral and multilateral investment treaties. In the context of host government agreements, it is important to note certain limitations to the effectiveness of this protection – in particular, that the awards based on these treaties are often financial in nature (as opposed to granting restitution or preventive relief), since even though other forms of relief are usually available, international tribunals appear reluctant to direct a state to take specific measures within its jurisdiction. The end result may be that the claimant merely has a large financial claim against a sovereign state that has no inclination to pay it. Enforcing such financial awards against some sovereign states is notoriously difficult.

- Would international standards be respected? Certain terms of a host government agreement may well offend principles adopted by multilateral lending agencies such as the International Finance Corporation and the European Bank for Reconstruction and Development. These principles would certainly need to be complied with in a project seeking financing from either of these institutions. These principles may also be used as a benchmark for standards to which commercial international lenders will comply. In particular, the sensitive areas here are provisions of host government agreements that constrain (or through which the imposition of an obligation on the host state to pay compensation have the effect of constraining) the sovereign right of the host state to change laws relating to the environment, labour, and health and safety standards. Furthermore, a host government agreement might contain covenants on the host state to carry out acts that may affect the human rights of the citizens of that state – for example, in relation to land acquisition for the project and security for the project. Even if there are no multilateral lending agencies involved in the financing, the influence of non-governmental organisations should not be discounted in this context. The Baku-Tbilisi-Ceyhan project faced serious opposition in relation to the host government agreement with Turkey from Amnesty International and the World Wildlife Fund; this resulted in the need to renegotiate that agreement, which caused considerable delay and additional expense to the project.[7] Some major international oil companies also have framework agreements with Amnesty International under which they undertake to comply with some of that organisation's principles.

---

7   The details of the problems faced in connection with the host government agreements in the Baku-Tbilisi-Ceyhan project are discussed in Boyd-Carpenter, Harry and Labadi, Walid, "Striking a balance: intergovernmental and host government agreements in the context of the Baku-Tbilisi-Ceyhan pipeline project", *LiT online*, Law in Transition, European Bank for Reconstruction and Development.

A host government agreement can be a very useful and powerful tool for managing risk, especially when used in conjunction with a related intergovernmental agreement; however, host government agreements should not be thought of as a sure way of dealing with the host state risks. Section 4 below sets out in detail some of the risks that can be mitigated through host government agreements and the extent to which residual risks are likely to remain.

3.2 Intergovernmental agreements
Many of the risks likely to arise in relation to the host state will be expected to be addressed, at least in general terms, in the intergovernmental agreement. For a discussion of this, see the chapter on intergovernmental agreements.

It should be remembered, of course, that the investors will not be parties to the intergovernmental agreement related to their project and will have no direct rights under it. Therefore, while the intergovernmental agreement may provide a high level of political comfort and, possibly, go a long way towards entrenching the terms of the host government agreement if the text of the host government agreement is attached to it (see section 3.1 above), it generally will involve the limitation that intergovernmental agreements – unlike bilateral investment treaties, the investment chapters of free trade agreements and multilateral treaties like the Energy Charter Treaty – typically do not provide any direct rights of investors to bring arbitration proceedings against any of the parties.

Intergovernmental agreements tend also to provide obligations and statements in very general terms. So, for example, an intergovernmental agreement will often provide that the host states should create the most favourable regulatory environment for the project to which it relates. The actual implementation of this obligation, which is what is meaningful for the project investors, will be found in the domestic legislation of the host state. This domestic legislation might be enacted through general legislation or project-specific legislation by ratification of a host government agreement. In short, intergovernmental agreements by themselves will not always go far enough to deal with many of the issues raised above, where these matters are not already satisfactorily dealt with in existing domestic law of the host state.

3.3 Treaty protection
Bilateral investment treaties and multilateral investment treaties (eg, the Energy Charter Treaty) provide investors with direct rights against a host state for sovereign interference with investments – specifically against:
- unreasonable, arbitrary or discriminatory treatment;
- failure to afford investments fair and equitable treatment, full protection, security, and treatment that is not less favourable than that afforded to local competitors or competitors from third states; and
- failure to observe obligations entered into with investments, freedom of transfers, compensation in the event of loss or damage due to civil unrest, and compensation in the event of expropriation of project assets or measures having equivalent effect to expropriation.

There are limits to such protections, however: as is noted in the chapter on intergovernmental agreements, for instance, these treaties do not generally prohibit expropriation; instead, they provide that where expropriation occurs, there should be compensation paid according to internationally accepted standards.

Other limitations on the usefulness of treaty protection are, first, that it may be difficult to establish that the investor in the project is also treated as a protected investor for the purposes of the relevant treaty and, second, that the investor actually has an investment in the host state which will be protected under the treaty. All this is assuming that the host state and the state in which the investors are established[8] are parties to a relevant investment treaty; with an international project such as a cross-border pipeline, several jurisdictions are likely to be involved. Well-advised investors ensure that such structures are in place to guarantee the protection of an investment treaty should a dispute arise.

3.4 Aligning interests with host state

One possible way of smoothing the progress of a project and reaching a degree of stabilisation of the project in the host state is for the project to be carried out by a joint venture between the equity investors and either the host state or some form of emanation of the host state, such as a state-owned company. This could be achieved by giving the host state a shareholding in the project company, for example, which would give it both some economic value in the project (in addition to its rights to tax the project's profits) and an impression that it has a degree of direct control over a project that is likely to be a key piece of the state's national infrastructure. If necessary, the equity contributions of the host state can be structured as contributions in kind – whether as provision of services or assets that may be integrated into the project – so that the host state is not required to contribute any hard cash for its equity participation.

4. Specific risks for the investors and ways to mitigate them

This section examines some of the risks that the investors may encounter in the project which can be characterised as risks that are specific to the host state itself (as opposed to technical, contractual or commercial risks). The matters dealt with in this section are all capable of having the adverse effects on the project investors set out in section 2 above.

4.1 Lack of political support

Where the host state has a developed regulatory and permitting system in place providing clear procedures for permitting and land acquisition, as well as objective non-discriminatory methods for evaluation of relevant applications in respect of these matters, it is perhaps naïve in many cases to expect the project to proceed through these parts of the development phase in a timely manner without high-level political support for the project. Where the regulatory framework is less complete, such political support is even more vital. Lack of political support or, worse, political

---

8   The question of establishment is not simply a question of the jurisdiction of incorporation of the relevant investor – see the chapter of this book on intergovernmental agreements.

opposition at a high level can result in the imposition of insurmountable obstacles for the development of the project; the risk here is that the project will simply not get off the ground. Political support for the project may be evidenced by the existence of an intergovernmental agreement, but one should bear in mind that many countries have disparate political power bases and the political support required to produce an intergovernmental agreement may not involve the individuals or government departments whose support is required for the project to succeed in-country. A host government agreement will typically contain statements of political support and undertakings to facilitate the project. Of course, a degree of political support will be required in order for a host government agreement to be entered into in the first place, but intergovernmental agreements often contain an undertaking to enter into a host government agreement; as such, the host government agreement can represent a joining up of disparate power bases in the host country and produce a more coordinated political approach to the project.

4.2 **Permitting framework of the host state**

The laws that form the basis for the permitting and transit rights for the project may be inadequate in some way. They might provide no clear rules governing the rights to develop or operate the project, or the rules may be excessively complex, onerous, expensive to comply with or even obstructive. The risk here is, again, that the project will simply not get off the ground because of this.

This can be mitigated by including in a host government agreement provisions that grant all necessary permits for the project which the national legislature is capable of granting, thus effectively making the host government agreement a single permit for the entire project. Alternatively, it may be sufficient to use the host government agreement to provide procedures and rules to fill holes in the legislative framework where there is a perceived defect in the permitting regime. In either case, this will work only if the host government agreement is enacted into the domestic law of the host state. The single permit approach goes only so far because many host states may lack the ability to offer a single authorisation that will cover the entire project as certain permitting powers may be constitutionally devolved on the local authorities. In such cases the host state can give undertakings in the host government agreement that it will provide all assistance possible to ensure that the local permits are granted in a timely manner; however, by its nature such an obligation could never amount to a high degree of comfort that this will happen.

This is just one of the many reasons why the importance of local, as well as national, political support for a project of this nature is important. A project may need to be re-routed or may be seriously delayed due to slow processing, or outright rejections, of permit applications, and this can have very serious financial and timing consequences. Such was the experience, for example, of the developers of the Corrib onshore gas pipeline, which had to be re-routed because of the rejection of a planning approval for part of its route near its landfall on the west coast of Ireland.

In some host states a gas transmission system operator may benefit from special statutory rights that allow it to obtain pipeline permits on an accelerated basis. Some thought should therefore be given to the possibility of forming a joint venture with the transmission system operator if this would allow the project to benefit from these rights too.

4.3 Land rights

A cross-border pipeline project will require a potentially huge package of land rights. This will involve easements[9] and access rights for the pipeline route itself, the extent of which will change during the surveying, engineering, construction and operation phases of the project.[10] Given the possible vast number of landowners and occupiers of land in the pipeline corridor, many of whom may not even be easily identifiable, obtaining the necessary land rights in a timely manner will present a challenge for any transit pipeline project. The risk that the cost of acquisition of the land, whether by private treaty or by expropriation of rights, will increase beyond what it economically feasible for the investors is great, as is the potential for associated delays.

The host government agreement may include a provision that sets out all of the necessary land rights for the project with the host state undertaking to deliver all of these rights to the project in a timely manner; this may itself be subject to constitutional constraints – the power of expropriation of certain types of land may be vested in a local government and not the national government. In any event, the ability of the project company to require the host state to deliver all the rights will not by itself remove all the investors' burdens in relation to the acquisition of land rights: the issues described in section 3.1 above are particularly acute in the context of acquisition of land rights and the project company will probably have to comply with standards of international finance institutions, which may affect the timing and cost of the acquisition of these land rights (eg, see Performance Requirement 5(5) of the European Bank for Reconstruction and Development's Environmental and Social Policy, which requires the developers of a project to attempt to acquire land rights by private arrangement before doing so by way of expropriation).[11]

4.4 Transit rights

Securing transit rights is of course essential to establish freedom of transit for the hydrocarbons to be shipped. Where the host state is a member of the Energy Charter Treaty, it will be under an international treaty obligation to permit freedom of transit, although these transit provisions are not enforceable by private parties such as the project investors. Even in cases where the host state is a signatory of the Energy Charter Treaty, the question of transit rights is likely to be dealt with in the related intergovernmental agreement. What is provided in that agreement should also be repeated in the host government agreement to ensure that the project investors have the ability to enforce the host state's obligation to provide freedom of transit directly.

---

9   Otherwise known as rights of way or servitudes.
10 For an idea of the extent of land rights required, see Appendix I, Part II of the Energy Charter Secretariat's Model Host Government Agreement.
11 See www.ebrd.com/downloads/about/sustainability/2008policy.pdf, p34.

4.5 Use of foreign labour and materials

The host state may impose restrictions on immigration and imports. While a degree of local content in the construction and operation of the project may not be objectionable, the project company will probably wish to utilise skilled (or cheap) foreign labour. Accordingly, it will need to know that local laws permit this. Typically, where the host state is under no international treaty obligation in this respect, a host government agreement will contain provisions governing this issue.

4.6 Technical standards

The host state may be a jurisdiction in which it is unclear which technical standards will apply to the pipeline. Another risk is that the technical standards of the host state are not consistent with those of the countries to which the pipeline will be connected upstream and downstream. As mentioned in the chapter on intergovernmental agreements, it is important to have uniformity of technical standards so that the entire cross-border pipeline operates as one integral system instead of several pipelines patched together. While this key issue will be addressed in general terms in the related intergovernmental agreement, the detail will need to be provided for somewhere; a host government agreement that forms part of the domestic law of the host state is a possible repository for this detail.

4.7 Changes in the law

If the law of the host country changes from that in place at the outset of the project, there is considerable scope for delays, interruptions and capital and operational expenditure increases as a consequence. It may also constrain the capability of the pipeline infrastructure itself in a manner that affects the overall performance of the project company under its contracts with the shippers. Examples include changes in laws relating to labour, health and safety, the environment and taxation. In the past, project investors have sought to limit the risk of changes in law by including covenants in their host government agreements that require the host state either:

- to repeal, or exempt the project from, any change of law that has the adverse effects referred to above; or
- to pay compensation to the project company for the financial consequences of the change of law in order to restore the economic equilibrium of the project.

On the face of it, this proposition sounds attractive to a project investor. In reality, however, it is undeliverable. This is because, quite apart from the natural reluctance of the host state to have its legislative ability (or, put another way, its sovereignty) constrained in this manner, project investors are likely to face opposition to this from international financial institutions and non-governmental organisations. A provision like the one requiring the host government to repeal, or exempt the project from, any change of law is likely to be viewed as a freezing provision; as such it is sometimes considered imperialist or, at the very least, undesirable. Non-governmental organisations have for some time campaigned against freezing provisions in host government agreements and there is some indication that the opinion of international financial institutions is coming into line with this approach.[12] Although not all financings will involve international financial institutions, the positions of these bodies on this issue are increasingly treated as benchmarks of acceptable international practice by commercial financiers and project sponsors. In any event, any provision that would in any way place legal or financial obstacles in the way of a host state's attempts to bring its labour, environmental or health and safety laws into line with accepted international standards (eg, EU standards) would likely be considered unacceptable for these purposes. Conversely, provisions that seek to preserve only the fiscal position of the project company are considered more acceptable.[13] In short, project investors should not expect a high degree of protection against change of law in host government agreements, neither should they assume that they will obtain protection from anything more than discriminatory changes in law. Even then, they should expect some considerable negotiation of the question of what 'discriminatory' means for these purposes.

4.8 Taxation

Not only will the project need to involve certainty as to the cross-border treatment of taxation (which is primarily a subject for the intergovernmental agreement), but it will also require comfort as to taxation matters within the host state. The taxation system in the host state may not be sufficiently sophisticated to deliver the predictability and required treatment in order for the equity investors to model the project company's net revenues to generate the required cover ratios for its project financing. There may also be a view that the taxation system in the host state is unstable, particularly if the host state is going through a period of political and economic transition. A project company may seek to use a host government agreement to obtain specific tax treatment on particular matters. A more extreme approach would be for the project company to enter into a project-specific profit tax agreement with the host state under which the general tax laws of the host state are disapplied and the tax regime is tailor-made for the project. It is to be assumed that this tailor-made regime would be entrenched to the same extent as the host government agreement.

The host government agreement may also deal with issues relating to value added tax (VAT). Many jurisdictions are notoriously slow and unreliable in paying VAT refunds that are owed to project developers during the construction phase of their project at the time when, because these developers have no project revenues, they do not have any output VAT liability to offset against their input VAT paid to their construction contractors. This can amount to a substantial funding requirement for the project. Severe delays in VAT refunds have left some projects

---

12 Stabilisation Clauses and Human Rights – A research project conducted for IFC and the United Nations Special Representative to the Secretary General on Business and Human Rights, March 11 2008, Shemberg, Andrea and Aizawa, Motoko .
13 *Ibid.*
14 For example, the toll road projects in Greece following the onset of the Greek sovereign debt crisis in 2009-2010 that were in their construction phases all suffered severe delays in receipt of their VAT refunds having this effect.

with insufficient funds to meet their costs to complete.[14] The project company may seek to remove this risk by obtaining a project-specific VAT exemption; this may be effected by a ruling under an existing law or through a provision in the host government agreement. If this is unavailable, a fall-back position would be for the project company to obtain the right to sell its VAT refund receivables to a third party which has output VAT liability at the time, and for that third party to have the right to offset these receivables against its output VAT liability.[15]

4.9 Expropriation of the project

The most catastrophic development for the project investors would be the expropriation by the host state of the project itself. This risk and its mitigation in the context of the intergovernmental agreement and non-project specific international treaties are discussed elsewhere in this book. However, host government agreements have a valuable role to play in the investor protection scheme.

Experience from investor-state arbitrations has shown that it is very desirable for investors to have more than one legal basis on which to make their claims. Thus, multiple claims might be made against a host state under a bilateral investment treaty, a multilateral investment treaty and a contract such as a host government agreement. It is therefore useful to give the project investor a contractual right under the host government agreement on which to base its claim.

A host government agreement may be used to enhance the project investor's position with regard to expropriation in general. In particular, the host government agreement may provide for a more sophisticated methodology for calculating compensation payable for expropriation than is provided by the constitution of the host state or, indeed, by investment treaties. Host government agreements generally provide for the compensation to be prompt, adequate and effective, which typically means the market value of the investment immediately prior to the expropriation, payable without delay in a freely convertible currency. The valuation of a long-term investment such as a pipeline is a complex matter. Arguably, it should be based on the net present value of the net revenues of the project company projected over the life of the project. A host government agreement could expressly stipulate such a calculation methodology, which would guide the arbitrators in an international investor-state arbitration brought under an investment treaty.

A host government agreement may be used to seek to confirm the scope of the type of investor that will be protected under an investment treaty. Typically, under such a treaty, in order to qualify as a protected investor an investor would need to show that it has the nationality of one of the states that is party to the treaty, and that it holds an investment in the host state. When structuring the financing of a cross-border pipeline project, it is desirable for the project infrastructure to be owned in each host state by a single company which would raise finance and enter into a one-stop-shop shipping contract with the shippers. Many treaties provide that a company incorporated in the host state may be treated as a foreign investor entitled to bring a claim in its own name if the parties have agreed to treat it as such, because

---

15   See Section 26(18) of the Energy Charter Treaty's Model Host Government Agreement.

it is foreign-controlled. The host government agreement can specify such arrangement if the project company is locally incorporated. As mentioned above, shippers may also be the equity investors in the project; as such, they will have an investment in the host state country by reason of being shareholders in the project company that holds assets in that state, even if the project company is incorporated abroad. However, the host state is unlikely to treat as investments the assets embodied in these companies' shipping contracts with a project company incorporated outside the host state. This can be resolved if the project company is itself incorporated in the host state. If that is impossible or undesirable, the shippers could nevertheless benefit from provisions in the host government agreement under which the host state agrees to compensate them for loss of the benefit of their contracts with the project company. This might be done by making the shippers third-party beneficiaries under the host government agreement; as such, they would have a contract with the host state that would likely qualify as an investment in the host state for the purposes of investment treaty protection.

Tax risks

Stuart F Schaffer
Baker Botts LLP

A goal of any foreign company in sponsoring a large energy project in a particular host country will usually be to minimise the amount of taxes and other government take imposed on the sponsors over the life of the project. The decision of whether to sponsor a large energy project in a particular host country thus needs to take into account the tax costs and risks associated with the development and operation of the project.

1. **Applicable taxes**

   Initially the potential project sponsors will need to identify the taxes that will or could apply to the project, and the sponsors in the host country under the law as currently in effect. This usually will require that the sponsors engage local tax advisers for this purpose. The process of engaging local tax advisers should be initiated very early in the predevelopment process. Many developing countries have a limited number of independent and competent tax advisers. In the case of a competitive process, the potential project sponsors would not wish to be in a position where they are required to engage a less desirable local tax adviser because the best tax firms in the country have already been engaged by potential sponsors of competing projects. Similarly, it is less desirable to utilise a local tax adviser that at the same time is representing one or more competitors (and perhaps also the government).

2. **Taxes at multiple governmental levels**

   Taxes applicable to a large energy project in the host country often are imposed by multiple governmental authorities. In many countries, taxes are imposed by both the national government and one or more levels of regional and local government. For example, property taxes are frequently imposed at local government level. Moreover, there may be some taxes for which the government revenues are divided among the national and regional/local governmental bodies. In some such cases, the tax rates might be set separately by each applicable government body. In other cases, a single tax rate may be imposed, but the revenues are divided between the government levels based on specified percentages, which might change from year to year.

   In many countries, different agencies or authorities at a particular level of the government might be responsible for imposing or administering taxes and other government take. For example, income or profit taxes are usually administered by the Ministry of Finance, while taxes or tariffs on imports and exports are usually administered by the state customs authority.

Lastly, in some places there may be uncertainty as to which government has the authority to impose and collect taxes or certain types of taxes. For example, the taxes applicable to energy projects in the Kurdistan region of Iraq might ultimately include some combination of taxes imposed by the Iraqi federal government and the Kurdistan regional government. Another possibility is that the project might be located in an area that is subject to the disputed jurisdiction of two or more states.

In the case of a host country where taxes are imposed by multiple governmental levels or authorities, project sponsors that intend to negotiate the taxes applicable to the project as part of a host government agreement negotiation might need to engage in discussions with those various governmental bodies. This could greatly increase the time and effort needed to conduct the tax negotiations. Whether such multiple tax negotiations will be necessary will depend on both the laws of the host country and political considerations.

3. Tax benefits

A potential project sponsor should decide very early in the project development process whether it will be necessary to enter into negotiations with the host government to reduce the amount of taxes otherwise applicable to the project. In many jurisdictions benefits or incentives are available under the generally imposed tax laws to certain investors in the jurisdiction. These incentives might consist of tax exemptions or reductions for particular industries or regions of the country, or types of project. The incentives might be available for the life of the project or only for a specified maximum period of time. They often are available to foreign investors only. The tax laws of the host country will generally describe how such tax incentives might be obtained by project sponsors; however, in many jurisdictions the usual workings of tax incentives might be gleaned only from local tax advisers.

Another set of generally applicable tax benefits to foreign investors in a particular country might be available under a tax treaty. Many countries have entered into bilateral or multilateral tax treaties that provide tax breaks for host country investors which are resident in another country. These tax breaks often include:

- the reduction or elimination of withholding taxes on certain income of a foreign person that is sourced in the host country;
- exemption from income tax for foreign persons for certain types of income (eg, shipping income from international operations or certain capital gains) or from income that is not attributable to a permanent establishment in the host country; and
- assurance by a taxpayer's country of residence that income taxes payable in the host country will be creditable against home country taxes, or that income earned in the host country will be exempt from home country taxation.

Project sponsors will want to determine whether the generally available tax incentives in the host country will be sufficient to produce an acceptable level of host country tax obligations or whether negotiation of additional tax incentives will be necessary.

4. Prioritising

In determining whether there will be a need to negotiate with the host government for tax breaks, the project sponsors will want to compute the net present value of the expected liability for the various applicable taxes. In addition, the sponsors will often prefer relief from those taxes that will be incurred in the development and construction stages of the project – as opposed to those taxes that will be largest during the operational phase of the project, when the sponsors will have substantial revenues from which to pay any taxes. For example, many sponsors will have the greatest concern for liability for taxes, such as value added tax (VAT) or other transfer taxes, that are imposed on fees payable to engineering, procurement and construction contractors. Another source of substantial tax cost in the development or construction phase of a project could be taxes or duties imposed on the import of construction materials and supplies (eg, pipe for a pipeline).

Of course while the sponsors might be focused particularly on minimising their burden for those taxes that may be imposed during the development and construction phases of the project, the host government might be focused on maximising its revenues during the early stages of the project so that it will be able to demonstrate to the populace early benefits from the project. That said, the benefits of local employment during construction of a large project might satisfy this goal.

5. Amending generally applicable tax laws

If the sponsors of a large energy project decide that it will be necessary to negotiate a reduction in their host country tax burden, an important initial step will be to determine how the host country tax laws can be amended as applicable to the sponsors' project in a manner that will be binding and enforceable. Some countries' laws specifically contemplate the ability to amend the generally applicable tax laws in a host government agreement. For example, the petroleum law or the tax laws in many countries provide that the taxes applicable to a project may be revised in a production sharing contract or other agreement entered into between the project sponsors and the host government or relevant state-owned company. In many cases, those laws contain limitations on the amount or types of tax break that can be provided in the host government agreement.

Where the law of the host country does not permit modifications of the tax laws to be contained in a host government agreement, the tax laws of the host country will need to be analysed to see how modifications to such tax laws can be achieved. The tax laws might permit tax breaks to be granted to individual taxpayers only through an amendment to the generally imposed tax law itself.

Sometimes the tax laws of the host country will not specify how benefits can be granted to a particular taxpayer, in which case general constitutional or legal principles in the country will need to be applied. For example, the host government agreement or the tax provisions in the host government agreement might have to be independently given the force of law by ruler's decree or by parliamentary action. In some countries, the tax laws may not be overridden, except by intergovernmental treaty. In such a case, the tax provisions for a transnational project would need to be contained in, or at least approved by, a treaty.

In case a new law or treaty is needed to enable the tax benefits that have been negotiated with the host government to be given the force of law, the new law or treaty will usually need to be enacted or entered into based on the procedures mandated in the host country for a law (including the specific procedures for a tax law) or a treaty.

6. **Fiscal stability**

The fiscal stability clause is a crucial provision in a typical host government agreement for a large energy project. The host government agreement should include a fiscal stability clause that provides protection to the project sponsors from tax changes that will adversely impact the economic results for the sponsors.

A fiscal stability clause will typically refer specifically to changes in taxes. The clause should refer to changes in the amount of taxes, not just to changes in applicable tax rates. The amount of a sponsor's host country tax liability can be affected as much by changes in the determination of the taxable base (eg, the computation of income or deductions in the case of a profit tax) as by changes in tax rates. Additionally, the fiscal stability clause should be triggered not only by changes in tax laws or regulations, but also by changes in administrative practice or judicial or administrative interpretations of the laws and regulations. Lastly, the fiscal stability clause should be broad enough to protect a sponsor against increases in home country taxes resulting from changes in host country laws (eg, amendments to the host country profit tax law could affect to what extent those profit taxes qualify for foreign tax credits in the sponsor's home country).

7. **Negotiation strategy**

The potential project sponsors will want to determine in the early stages of development of the project what their strategy will be for negotiating the tax-related provisions of the host government agreement. It is important to be able to know who will be negotiating the tax provisions on behalf of the host government. Often the authority that is responsible for leading the negotiations on the host government agreement on behalf of the government (eg, the Ministry of Energy or Natural Resources) will not be the body that will negotiate the tax provisions (eg, the Ministry of Finance). This can sometimes make it more difficult to obtain the desired tax benefits because the Ministry of Finance is rarely the governmental body that will be enjoying the benefits of, or will be championing, the project from the host country's perspective. In such a case, it will often be desirable to include the Ministry of Energy or Natural Resources in the tax negotiations, or at least to keep the Ministry of Energy or Natural Resources fully apprised of the status of the tax negotiations.

The project sponsors should also determine whether the negotiations of the tax-related provisions of the host government agreement should be conducted on behalf of the sponsors by the team that will be negotiating the main body of the host government agreement or, instead, by a specialised tax team. The advantage of using a specialised tax team is to put the particular knowledge and expertise of those individuals to work in real time during the negotiations. A potential disadvantage of using a specialised tax team is that in such cases the tax issues are often left open until late in the negotiations, which can reduce the sponsors' leverage.

8. **Tax provisions of host government agreement**

The most basic decision to be made in structuring the tax clause of a host government agreement is how to phrase the general default rule regarding the sponsors' liability for taxes in the host country. A typical tax clause in a host government agreement will state as a general rule either that the sponsors will be subject to all taxes applicable under host country law except as otherwise provided in the host government agreement, or that the sponsors will be exempt from all taxes applicable under host country law, except as otherwise provided in the host government agreement. This determination will usually be based on the overall business deal of the parties regarding taxes and on whether one or the other approach is mandated by the petroleum law or other applicable law that authorises host government agreements.

Another decision to be made in structuring the tax clause of a host government agreement is how detailed the clause will be. Often this determination is based on how acceptable the project sponsors believe the generally applicable tax laws of the host country to be. For example, if the project sponsors have determined that the generally applicable profit tax of the host country will not qualify for foreign tax credits in their home countries, it may be considered necessary to draft a replacement profit tax law as part of the host government agreement.

The length and level of detail of the tax clause in the host government agreement also will be affected by whether the tax clause will cover procedural as well as substantive tax rules. Procedural rules can relate to such matters as tax return filings, payment of estimated taxes, the currency in which taxes are computed and paid, the language in which tax returns and records are prepared and maintained, the statutory period of limitations for the taxes, the applicability and rates of interest and penalties for non-payment or late payment of taxes, and examination and appeals procedures in the event of a tax dispute. Whether such procedural provisions are needed will depend on the tax procedural and compliance rules generally applicable in the host country, both under the law as written and in practice.

Another structural issue for the tax provisions in a host government agreement is whether the provisions should be included in the body of the agreement or as a schedule or exhibit appended to the agreement. A potential benefit of the attachment approach is that the tax provisions could be more easily reproduced and circulated to the persons responsible for administering and implementing the tax provisions in the host country.

Another basic decision to be made concerns the selection of the beneficiaries of the tax benefits contained in the host government tax clause. Will the tax benefits apply only to the sponsors themselves? Or will tax benefits also be granted to persons such as suppliers and subcontractors, the operating company, employees of sponsors, subcontractors and the operating company, the shareholders or owners of the sponsors, lenders to the project and customers (eg, shippers of oil or gas through a pipeline)? In this regard, it is important to note that taxes incurred by subcontractors, employees and lenders, among others, are often borne by the project sponsors, either through gross-up or reimbursement provisions of a contract or as a practical matter in setting contract terms.

Another consideration is whether the tax clause of the host government agreement should merely speak as a law in stating the taxes to which the relevant persons will or will not be subject, or whether there should also be indemnification of the project sponsors from any such taxes that are actually imposed. If such an indemnification is to be included, who will be the indemnifying party, the government itself, the relevant state-owned company or both? Another means of ensuring the benefit of the tax clause is to provide for the right to offset any taxes that are imposed in contravention of the tax clause against other amounts payable under the host government agreement by the sponsors to the government or the state-owned company.

In order for the sponsors to derive the full benefit of the tax provisions of the host government agreement, those provisions may need to apply to periods before or after the term of the host government agreement. Tax benefits may be sought for predevelopment activities that take place before the host government agreement becomes effective (eg, for fees paid for initial studies and assessments). The tax benefits should also continue to apply to taxes imposed on a sponsor that are due after the host government agreement has expired or terminated, or after the sponsor has assigned its interest in the host government agreement to another person.

Lastly, in the case of a host government agreement that is applicable to, and executed by, multiple sponsors of a project that will be acting in the form of an unincorporated joint venture, the tax provisions of the host government agreement will usually state that the sponsors' liability for taxes in the host country is to be several and not joint or joint and several.

9. **Tax issues raised by non-tax provisions**

Sometimes the provisions of a host government agreement outside the tax clause will have tax implications for the project sponsors that should be addressed in the tax clause. For example, the host government agreement might require the sponsors to operate the project in a particular legal form, such as through a jointly owned legal entity formed in the host country or through an unincorporated joint venture among subsidiaries of the sponsors formed in the host country. The requirement that certain types of entity carry on the project could have tax implications for the sponsors in either the host country or the sponsors' home countries. These implications could relate to the availability of incentives under the generally applicable tax laws of the host country, or the availability of foreign tax credits or applicability of anti-deferral rules under home country tax laws.

The host government agreement might require that specified payments be made by the project sponsors, such as bonus payments to the host country or state-owned company and contributions to development or other funds in the host country. A tax-related issue could be whether such mandatory contributions or other payments are deductible for income tax purposes in the host or home country.

Many host government agreements allow the project sponsors to recover the costs that they incur in developing or operating the project before large portions of revenues from the project are required to be shared with the host country (ie, cost recovery). The host country might insist that the rules for determining which costs are recoverable by the sponsors also apply for determining which costs are deductible for income tax purposes in the host country. This may raise foreign tax credit issues for the sponsors in their home countries. Another issue is whether host country taxes (or host country taxes other than income taxes) will be cost recoverable.

Many host government agreements require the ownership of the project itself or the assets making up the project to be transferred from the sponsors to the host country government or the relevant state-owned company at some time (whether upon expiration or termination of the host government agreement or at the time of cost recovery or project payout). Such a transfer of ownership could have tax implications in the host country or the sponsors' home countries. For example, will the transfer be subject to VAT?

Lastly, many host government agreements require (or allow) the project to be operated on a day-to-day basis by an operating company, which may be an affiliate of a sponsor of the project or jointly owned by all the sponsors. A tax issue is whether the activities of the operator, or whether transfers of funds between the sponsors and the operator, will be subject to host country taxes.

10. Host country respect of agreement tax provisions

Sometimes successfully negotiating a tax clause in an agreement with a host government is only half the battle towards achieving the desired tax results in the host country. An issue in some developing countries is how likely it will be that the tax benefits granted in the host government agreement will be received in practice by the project sponsors. One requirement to achieve the negotiated tax results is that the results of the negotiation be communicated to the persons who administer the tax laws in the host country. Something that can improve the communication and information dissemination process is to provide in the host government agreement that the tax benefits set forth in the agreement will be published in the form of rules or regulations of the relevant tax agency of the host country. In some countries the issue will not be whether the local officials will be made aware of the tax benefits under the host government agreement, but whether they will make those benefits available in practice to the sponsors. This issue relates to the level of respect for the rule of law in the host country. The risks may be affected by such issues as the budget situation in the host country and the amount of corruption inherent to the system.

11. Issues for multiple country projects

For those projects that will be carried out in multiple host countries, the sponsors will often prefer consistency of the tax rules as applicable in each country. This requires comparable and consistent host government agreement tax provisions. Whether this is achievable will depend in part on how similar the host countries' applicable tax laws are.

The strategy of the sponsors for negotiating the host government agreements in the various host countries can affect the tax clauses in the different host government

agreements. The choices available to the sponsors are usually to negotiate:
- the tax clauses with the host countries in parallel, which sometimes will require that different tax negotiating teams be used; or
- the tax clauses one at a time using the same tax negotiating team.

An advantage of the latter approach is that it will make it somewhat easier to achieve consistency in results among the different host countries because the same persons are conducting the negotiations in each country on behalf of the sponsors. A potential disadvantage is that the starting point for the second negotiation may be the end result of the first negotiation, which will often result in a less favourable tax clause for the sponsors.

Another important tax issue raised by a multiple country project is the need to allocate tax items between the host countries in a manner that does not lead to double taxation of the sponsors. For income tax purposes, where there are items of income and deduction applicable to more than one of the host countries, the items will need to be allocated between the countries in a manner that fairly and properly reflect the profit allocable to each country. This requires that the host countries allocate such items based on consistent methods.

One way that can be effective in attaining consistency in the application of taxes in two host countries is for the tax matters that are particularly affected by multiple country operations to be set forth in a document to which both host countries are bound. This could be either a bilateral treaty or a host government agreement signed by representatives of both host countries. An advantage of the latter is that the sponsors, as parties to the agreement, will usually have more ability to influence the content of the tax clause. A bilateral treaty will generally be negotiated only by the host countries themselves.

12. Tax structuring issues

How the project sponsors structure their investment in the project will sometimes have an impact on their tax liabilities. In the case of a multisponsor project, whether the sponsors organise themselves into a jointly owned separate legal entity or an unincorporated joint venture could affect both host and home country tax liabilities. It could at least affect the determination of who is the taxpayer for host country tax purposes.

Another structuring decision is whether the sponsor or, in the case of multiple sponsors, the joint venturers or owners of a separate legal entity that operates the project are entities that are incorporated in the host country or in a foreign country. This could also have an effect on the liability for host country and home country taxes. Sometimes the decision of where to organise the sponsor entities will be affected by the desire to benefit from a particular bilateral investment treaty with the host country.

In the case of a project located in multiple host countries, another tax structuring issue is whether the sponsors should conduct the project through separate entities for each separate host country. Use of a single project entity operating in multiple countries could raise risks as to how tax items are to be allocated between the countries.

Another common tax structuring goal is to reduce host country profit taxes through 'earnings stripping' techniques, which generally involve deductible payments being made by a project sponsor to an affiliated entity located outside the host country. Such earnings stripping payments can take the form of shareholder loans, royalties for licensed intangible property rights and fees for consulting or other services. It may be desirable to address the ability to enter into such intercompany arrangements in the host government agreement.

13. Forms of host country taxes and other government take

In computing its total expected liability for government take in the host country and in determining whether and how to negotiate reductions in the amount of taxes that are generally applicable under host country law, the sponsor of a large energy project will need to consider the following forms of tax and government take.

13.1 Income or profit taxes

The sponsors generally will be subject to tax in the host country on their net income or profit from the project. Many countries impose a higher tax rate on profits from energy projects than would apply under the generally imposed income tax law. Sometimes a host government will prefer to negotiate a tax on gross revenues from a large foreign-owned energy project, instead of a tax on profits, in order to derive a more predictable revenue stream and to avoid disputes over the deductibility of certain expenses, such as overhead and amounts paid to related foreign persons.

The sponsors might want to ensure that the gain on an assignment of a sponsor's interest in the project or in the host government agreement will be exempt from income tax in the host country. If such an exemption applies, an issue is whether the assignee will succeed to the assignor's tax attributes, including the tax basis in the assets of the project.

A frequent tax concern for foreign sponsors is whether the host country income tax qualifies as a creditable tax for the purposes of home country foreign tax credit. For this issue to be relevant to a foreign sponsor, the sponsor's home country income tax law would have to provide that double taxation of a taxpayer's foreign source income is to be avoided through a foreign tax credit regime rather than through a regime under which foreign source income is exempt from taxation. In a foreign tax credit regime, a taxpayer's home country income tax liability on foreign source income would generally be reduced by its liability for foreign income taxes on such income (ie, a tax credit). In addition, the sponsor would need to have the practical ability to benefit from foreign tax credits in the home country. For example, foreign tax credits are not as valuable to a sponsor that has a large tax loss carryforward in its home country.

In many countries, whether a foreign income tax is a creditable tax is an all-or-nothing proposition – the tax either qualifies for foreign tax credits or it does not. A common requirement for home country creditability is that the foreign tax has been enacted and is administered as a tax in the host country. In most countries, contractual modifications of the generally imposed income tax law will not preclude creditability.

Another common requirement for foreign tax creditability is that the foreign tax qualify as a tax on net income for which taxpayers are generally allowed deductions for all reasonable costs. Types of expense that may be non-deductible for host country profit tax purposes, and that therefore may raise creditability concerns, are interest, overhead and intercompany expenses, bonuses, development fund contributions and loss carryforwards.

Foreign tax credit issues also may be raised in a sponsor's home country if the sponsor is required under either host country law or the host government agreement to sell production from the project in the host country at prices that are below fair market value. Creditability of the host country income tax also could be affected by its separate imposition with respect to the net income from each separate project conducted by a particular sponsor (including a requirement that the sponsor conduct each project in a host country through separate entities). Lastly, some countries might not allow foreign tax credits for any taxes that are required to be paid in kind rather than in cash.

A common structure for imposing profit tax in host countries is for the tax to be the liability of the project sponsors but for the tax to be paid by the foreign government or the relevant state-owned company on the sponsor's behalf (often referred to as a 'tax-paid regime'). If such a tax is considered to be creditable in the sponsor's home country, then the sponsor's total effective income tax liability will be less than if the sponsor had been exempt from host country income taxes. This phenomenon is illustrated in the table below.

	Tax-paid regime	Exemption regime
Net income	$100	$100
Host country profit tax rate	25%	0%
Host country profit tax liability	$25	$0
Home country profit tax rate	30%	30%
Home country profit tax liability	$30-$25=$5	$30
Total home country and host country profit tax liability	$5	$30

The above example has been simplified because, in order for the host country profit tax under a tax-paid regime to be creditable, most home countries will require that the host country impose tax not only on the $100 in net income, but also on the additional income resulting from the assumption of the sponsor's tax liability by the foreign government or state-owned company under the tax-paid regime (known as 'tax on tax'). In the above example, the taxable income would actually be equal to $100 \times \dfrac{1}{1-25\%} = \$133.33$.

An issue raised by a tax-paid host country tax regime in cases where the tax is payable by the relevant state-owned company is that most home countries will not allow foreign tax credits for such host country taxes unless the income tax is actually paid by the state-owned company. The project sponsor will need to consider in such a case whether the state-owned company has the wherewithal to pay the tax and whether it will actually pay the tax.

In a tax-paid regime, the sponsor will want to ensure that host country income tax returns are properly filed (often the host government agreement will provide that the sponsor itself is to prepare and file such returns) and that the sponsor receives receipts for payment of the income tax. An official host country tax receipt is often required as a condition to obtaining home county foreign tax credits.

The other forms of host country taxes and government take are out-of-pocket costs to the sponsor of the project because they generally will not qualify for home country tax credits.

13.2 VAT

VAT will be imposed in many host countries as a percentage of the price paid for certain goods and services and a percentage of the value of imported goods and services. VAT can be a large upfront cost of a project, as it may be imposed on amounts payable to the engineering, procurement and construction contractor and the cost of materials used in the construction of the project. A unique aspect of VAT is that the cost of input VAT to the recipient or importer of the goods or services may be recoverable out of output VAT, which is collected by the taxpayer from persons to which the recipient or importer sells its production from the project. Many countries provide that any VAT that is not so recovered will be refunded to the taxpayer after some period of time.

Unfortunately, certain sales of production, such as export sales, often are exempt from VAT and thus do not allow for recovery of input VAT paid by the sponsor. In addition, many developing countries will not refund VAT (or will take a long time to pay refunds and will usually do so without interest).

In countries where there is a likelihood that input VAT will not be recovered or refunded, the sponsors will want to negotiate for a VAT exemption with credit for the project. This means that VAT will not be charged to the sponsors by their subcontractors and suppliers. In many host countries, it will be difficult to negotiate such a complete VAT exemption with credit for a project because this can cost the host government a great amount of potential revenue, and such an exemption can easily be abused by lower-tier suppliers and subcontractors. Where a host government is willing to agree to a broad VAT exemption, it will often insist on strict compliance requirements, such as the requirement that qualifying subcontractors and suppliers obtain exemption certificates from the taxing authority on a regular basis.

13.3 Customs duties

Customs duties are often imposed by host countries on the import of equipment, materials and supplies used for a project. These are based on the percentage of value of the imported goods. Customs duties are not recoverable, although there may be exemptions for goods that are later re-exported from the host country (or re-exported within some specified period of time). Customs duties sometimes are also imposed on the export of the production from a project, again based on the value of the exported goods. Lastly, many countries impose other charges, such as registration, documentation or other fees, for customs functions in the case of imports and exports of goods.

13.4 Property or asset taxes

These taxes may be imposed annually on the value of real property or personal property situated in the host country on a certain date. These taxes are often imposed at the local government level.

13.5 Other taxes

There may be other smaller taxes in the host country that add to the cost of operating the project, such as stamp duties on the execution or transfer of certain contracts and excise taxes on certain goods such as vehicles.

13.6 Taxes on foreign subcontractors

Foreign subcontractors will often be subject to a tax burden in the host country in connection with a project. These can include all or some of the taxes described above. At times withholding taxes may be imposed on a foreign subcontractor's revenues from the project in lieu of taxes on profits or net income. These withholding taxes often are borne by the sponsors of the project, perhaps through contractual reimbursement or gross-up provisions in the subcontract. Many host countries will find it politically difficult to grant exemptions or benefits in regard to taxes on foreign subcontractors, as such benefits will appear to encourage the use of foreign subcontractors instead of local subcontractors. The project sponsors may want at least to include in the host government agreement a provision stating that the sponsors are not liable for the failure of subcontractors to satisfy their host country tax liabilities.

13.7 Taxes on employees

Employees of sponsors and subcontractors working on the project in the host country may be subject to income tax and social security or social insurance contributions in the host country. The sponsors may wish to negotiate exemptions or reductions in these liabilities for their expatriate employees. At the very least, they may seek to limit host country income taxes on expatriate employees to their compensation for work performed on the project itself, in order to avoid the treatment of expatriate employees as host country tax residents who are subject to income tax on their worldwide income, including investment income. Another area for potential relief is host country taxes or duties imposed on the import of personal belongings by expatriate employees and their families.

13.8 **Withholding taxes on foreign persons**

A host country may impose withholding taxes on certain amounts payable by the sponsors in connection with the project to persons that are not tax residents of the host country. These withholding taxes might apply to dividends (or deemed dividends) paid by the project entity to its foreign shareholders or interest paid to foreign lenders. In the case of loans from foreign financial institutions, the sponsors may often have the contractual obligation to gross up the lender for any such interest withholding taxes in the host country.

13.9 **Other forms of government take**

The host government agreement or host country law may provide for forms of government take, other than taxes, to be borne by project sponsors. Government take might include:
- the provision of either a carried interest in the project or a share of revenues from the project in excess of revenues for the sponsors' cost recovery (eg, 'profit oil');
- the payment of royalties based on production, bonuses based on the signing or effectiveness of the host government agreement, or annual fees or rentals;
- the achievement of certain project milestones;
- payments towards a development fund or other contributions; and
- the requirement to fund training or other programmes in the host country.

A host state perspective on risk

Judith H Kim
Geoffrey Picton-Turbervill
Ashurst LLP

1. Introduction

This chapter focuses on the issues that a host state will wish to address in the development of any transboundary energy infrastructure. In order to give some colour and context to the issues, the author has built them into an illustrative case study for the import of pipeline gas. While the detail of the issues is obviously specific to this case, the same principles apply whether the project involves natural gas, liquefied natural gas, electricity or any other energy source. The author has sought to identify some of the key issues from a host state's perspective, and to look at how the way these are addressed and may be changing as markets develop.

2. Background

Bordavia is a country with a young and growing population, a stable and predictable government and a bright future. It is enjoying an unprecedented period of economic growth – business and industry are booming and all signs indicate that this growth will continue for the long term. But Bordavia is facing a crisis: it needs more energy to fuel this growth and it needs it quickly in order to satisfy the rising demand for energy and to continue feeding the growth of the nation.

With a good base of natural resources, Bordavia has in the past been self-sufficient from an energy perspective. However, it no longer has sufficient natural gas of its own to fuel its power plants and provide feedstock for industrial and domestic consumers. Sitting within an energy-rich region, it has neighbours with sufficient excess natural gas. One of these neighbours is willing to sell natural gas to Bordavia and some international companies and banks are willing to invest in Bordavia and the region in which it is situated.

A consortium of companies sees an investment opportunity. Accordingly, it proposes to develop a project to bring gas to Bordavia by investing in an upstream gas development in a neighbouring country, and building a pipeline that will transport natural gas from that neighbouring country to Bordavia.[1] A project of this nature is complex. It will involve at least two governments (and many other parties in addition to the consortium members), cross international borders and require commercial, economic, financial, technical, operational, environmental and other

---

1 Transboundary electricity supply from a power plant or a liquefied natural gas import terminal project could also have been used as an example of a transboundary energy project. However, the author chose a two-country pipeline project for the sake of simplicity. Nevertheless, the concerns of the host state set out in this chapter are analogous to the concerns applicable to any transboundary energy project.

expertise. In addition, each participant in the project will have different objectives (commercial and otherwise) and concerns that they will want addressed before they proceed with the project. Different participants may also have more than one role in the project and, hence, different perspectives on the issues involved.

Without wanting to detract from the objectives and concerns of the many other participants to a project of this nature (on which numerous articles and books have been written, in particular from the perspective of the foreign investors and the international banks that finance such projects),[2] this chapter focuses on the objectives and concerns specific to Bordavia, in its capacity as the host state importing natural gas.

3. Objectives of the host state

A host state's requirement to fulfil its energy needs is a matter of national interest; at times it will involve issues of international cooperation and interdependence. Accordingly, the concerns and objectives of any host state in securing adequate energy supplies will often override the usual commercial concerns that drive purely commercial entities. So it is not surprising that the issues that Bordavia will consider in relation to the transboundary pipeline project proposed by the consortium are not primarily focused on the rate of return of the investment or legal risk mitigation, even though financial analysis and risk mitigation will of course form a key part of its overall assessment of project liability. A transboundary pipeline project will be appealing to Bordavia only if it fulfils its one overriding objective – that is, to put in place a secure supply of energy for Bordavian consumers, which is a matter of national interest. The project will thus need to be structured with specific attributes that, taken together, will satisfy this objective.

'Security of supply' has been defined in numerous different ways – for example, as "a flow of energy supply to meet demand in a manner and at a price level that does not disrupt the course of the economy, in an environmentally sustainable manner". More simply, the International Energy Agency says that energy is secure if it is "adequate, affordable and reliable".

In the context of a transboundary pipeline project, natural gas supply and delivery will typically need to satisfy the following basic conditions in order to ensure that a secure supply has been put in place which will be of long-term benefit to Bordavia and its people:[3]

The gas must be delivered on schedule to a specified delivery point – this is required to ensure that downstream demand expectations can be managed and coordinated, and that there is a clear point where the parties have agreed that risk, title in and liability for the natural gas passes from the seller to the buyer.

The gas must meet pre-agreed quality specifications to ensure that downstream

---

2   For a general and informative discussion on this topic, see Browning, William E and Dimitroff, Thomas J, "Transboundary pipeline development and risk mitigation", in Picton-Turbervill, Geoffrey (ed), *Oil and Gas: A Practical Handbook* (2009).
3   These basic conditions would be required for any transboundary energy project, albeit with appropriate adjustments to take into account the specifics of the supply arrangement, such as the supply of electricity or liquefied natural gas.

consumers will receive the expected energy value from the natural gas (neither too much nor too little), and that the natural gas supplied will be suitable for its intended use by downstream consumers, whether in power plants or petrochemical facilities, or for domestic use.

The gas will need to meet pre-agreed delivery conditions (eg, pressure) to ensure that the physical plant, machinery and equipment receiving the natural gas from the seller is compatible with, and will not be damaged by, natural gas delivered by the seller, and so that the buyer can control the rate of flow of natural gas within its own networks – again to minimise adverse effects on downstream consumers.

The gas will need to be available in the quantities required, with flexibility as far as possible to increase or decrease volumes to take into account demand variations to ensure that fluctuations in downstream demand can be managed efficiently and without causing the buyer to have to seek consent from the seller in each instance.

The gas will need to be competitively priced – meaning that the price of natural gas meets consumer expectations and market conditions, and is not subject to unexpected or unpredictable changes. Ideally, the price of natural gas would be able to be determined objectively either by way of an established gas market or by reference to published indexes to ensure that it remains competitive with the price of alternative fuels. In addition, the host state will want the right to review the price from time to time to ensure that the price being paid is in line with the market.

The factors outlined above will be fundamental issues for Bordavia in its negotiation of the supply terms if it is to ensure, as far as possible, that its downstream customers are provided with the quantities of gas that they require, at a suitable quality, at a price that meets their expectations and on a sustainable basis. Aside from the physical requirement for the gas to drive the country's economic growth, the government will be investing significant political capital in assuring its electorate of a secure supply; there are plenty of examples around the world of the political fall-out that follows energy shortages. Thus, the government's negotiation will focus on achieving these objectives; anything that dilutes them will render the project less attractive.

Ideally, of course, the government will look to have more than one source of supply – however tightly the deal is negotiated, there is always a risk of supply interruption through unforeseen circumstances; the only mitigant to that is to have alternatives on hand, whether that be indigenous supplies, gas in storage or parallel third-party services. The availability of competing suppliers will also provide competition and greater leverage for the government in negotiations.

In addition to the fundamental issues outlined above, Bordavia will want to address other considerations. Although these are secondary to the basic conditions required to ensure a secure supply of energy, they are nevertheless significant and warrant attention and action. They include:
- protection of the environment – the impact of the transboundary infrastructure project on the environment will need to be minimised;
- sustainability of the project in the context of any affected communities – negative impacts on affected communities will need to be minimised, and investors and the host state will need to determine how they can develop

projects for, or otherwise make contributions to, affected communities;
- health and safety of workers – this will be a priority for investors and the host state during the construction and operating life of the project;
- reputation of the project in the global sphere – the host state will want the project developed and operated in compliance with the highest global standards for similar major projects, not only from a health, safety and environmental management perspective, but also to maintain a good reputation for the project and for the investing environment in the host state; and
- physical security of the infrastructure – the investors and the government will need to ensure that the physical assets are properly protected against sabotage, terrorism and other acts, so as to ensure that there is no interruption of natural gas supply.

4. Achieving the host state's objectives

In order to fulfil its overriding objective of putting in place a secure supply of energy, a host state will approach negotiations with investors as a 'zero-sum' game, where every issue discussed is looked at in the broader context of the state's security of supply, insofar as every gain the investors make in the negotiation may result in a loss of an element of that security to the host state, and vice versa. The host state will therefore try to balance every term and condition that ensures, or has an impact on, security of natural gas supply in its favour. It will also, as part of an internal assessment, seek to put a monetary value on concessions being considered; if the negotiations form part of a diversity of supply strategy – and there are other, different sources offering competing supplies – the comparative value of each supply source will be critical in establishing which provides the greater security. A supplier which is, on the face of it, lower cost may turn out to be significantly higher cost if that lower cost is accompanied by greater risks of interruption to supply, whether through equipment failure, upstream technical problems or government interference. Thus, placing an actual value on contractual concessions may be a key element in the host state's analysis that will significantly affect its attitude and approach.

In order to guarantee the necessary security of supply, the host state will seek to negotiate contracts for the transboundary pipeline project that contain the following:
- firm commitments on start dates of significant project milestones, such as commencement of construction of the required infrastructure, dates for testing and commissioning of the infrastructure, and ultimately the start date of the supply and delivery of natural gas. Firm commitments on start dates provide certainty and allow the expectations and demand of downstream customers to be managed, while failure to achieve agreed start dates would trigger contractual remedies such as liquidated damages or rights to terminate in favour of the host state;
- as few conditions precedent as possible to the commencement of the project that are not in the host state's direct control, so as to ensure that the counterparties are firmly committed to the project. Probably the most significant conditions precedent that are not in control of the host state

relate to government approvals from the nation supplying natural gas, and confirmation that financing from international banks has been secured on reasonable terms;
- strict controls over the specification of natural gas, including legal and practical ways of minimising the risk of receiving natural gas that does not meet required specifications (eg, contractual requirements for testing of natural gas well before the delivery point) and remedies in favour of the host state for failure to supply on-specification natural gas, such as payment of liquidated damages (sometimes as a discount against the cost of future supplies), or in some cases, provision of alternative natural gas;
- clear pricing provisions with appropriate price review mechanisms to ensure that the price for natural gas stays competitive with market prices and prices for alternative fuels. The contract should also minimise the risk of price increases – for example, via cost pass-through mechanisms or upward price review provisions;
- firm supply obligations with specific volumes to be supplied. However, for the host state the contracts should reflect the necessary flexibility in volume and scheduling supply terms to meet likely demand variations downstream (eg, daily and monthly swing volumes and no daily 'take or pay' obligation), while minimising its liability for failure to take volumes (for whatever reason). In addition, the contract should include a firm supply obligation to be backed up by remedies for failure to supply natural gas in favour of the host state (including liquidated damages and, in some cases, requiring investors to provide natural gas from alternative sources in order to guarantee uninterrupted supply);
- provisions to ensure availability of the necessary upstream natural gas reserves[4] and firm commitments on the part of the seller to supply from agreed reserves, with minimal rights to be excused for upstream technical problems. Given the critical importance of this aspect, there may be circumstances in which the host state would want the right to step in and take over the operation of the reserve or the related infrastructure if necessary to ensure uninterrupted supply;
- limited scope for investors to rely on *force majeure* claims, so that *force majeure* events include only events that are truly unavoidable and beyond the sellers' control, while at the same time seeking to ensure that the host state is suitably protected in circumstances where, for whatever reason, it cannot take nominated quantities by reason of *force majeure*; and
- limited, if any, change of law provisions. Investors will likely seek extensive protection against changes in law or regulation in the host state that may have an adverse impact on their returns, on the grounds that if the project is to be viable, it needs assurance of stability. To the extent that the host state agrees such protection, the associated costs will pass through into the ultimate energy costs for the downstream consumer and thus will have a

---

4   Or, in the case of a power plant project, provisions ensuring availability of adequate fuel supply.

direct impact on pricing. The host state will therefore be keen to limit the scope of these provisions to the greatest possible extent.

The host state may also seek to negotiate positions in the upstream and transportation projects, on the basis that the greater degree of involvement it can have in those essential elements in the chain, the more influence it will have over events or actions that may affect its in-country supply. A host state, as a party to a supply contract and with no other involvement, is inevitably less able to influence factors affecting that supply than if it is also involved in the other elements that feed into it. Thus, although a greater overall project involvement adds complexity, it also adds value for the host state.

If Bordavia is able to secure the best position on each of the terms and conditions that support and affect the supply of natural gas (which would indeed be a significant achievement), it will, from a contractual perspective, have put in place as close as it is possible to come to a secure supply of energy for Bordavian consumers.

Of course, firm contractual supply provisions backed up by the obligation to pay liquidated damages for failure to supply do not, of themselves, guarantee the supply of gas on the terms agreed. They simply provide a contractual remedy if those obligations are not honoured; that remedy is likely to be the right to receive monetary compensation, which is of limited practical value. This is because Bordavia needs natural gas delivered to its consumers, on schedule and on specification; receipt of liquidated damages or the right to pursue the sellers for breach of contract for failure to supply (probably through dispute resolution provisions that will take many years to implement and enforce) is not a solution if natural gas is not being delivered and the lights are going out. Bordavia does not want its remedy to be the ability to make claims against the sellers; it will be facing a huge problem if natural gas is not being supplied, regardless of whether there is a right to claim monetary damages. So Bordavia will seek to ensure, in whatever way possible, that alternative sources of natural gas supply will be available as a backup to its primary supply, and that the risk of supply interruption or failure is minimised. This may involve:

- commitments from the sellers either to make available alternative sources or to bring alternative supplies on stream (a major commitment on the part of the sellers); and/or
- the host state ensuring the availability of different sources of back-up supply, whether through storage, alternative third-party suppliers or locally available reserves.

Clearly, procuring alternative supply arrangements will carry cost and other implications that will need to be factored into the assessment of commercial viability of the project. However, this will be a key issue for the host state as practical solutions to a supply interruption will be far more important than the existence of a legal remedy for default, which is very much the remedy of last resort.

5. **Recent trends in host states' approaches to residual risk**
Bordavia's approach to the negotiations in connection with the development of a

proposed transboundary pipeline project will need to be consistent with attracting investors (particularly international banks) to undertake and invest in the project. This is where a tension exists: Bordavia needs to meet its objective on terms that will ensure a secure supply of energy, but its objectives will need to be compatible, at some point, with those of the investors if a viable project is to exist – otherwise there is a risk that the investors will take the view that the project is too difficult, and that it will not be possible to agree on acceptable terms. If this tension is not resolved, the investors may decide to invest elsewhere, where there is less perceived risk or where they think that they will obtain better terms and conditions. This is particularly true in the context of project financing, where international banks are especially sensitive to risk, and where host state resistance in negotiations leading to unresolved risk could render a project difficult to finance at a reasonable cost.

This tension will always exist. However, it seems that attitudes are changing and that host states are becoming less willing to provide investors with the sort of comprehensive protection that used to be readily available for major projects. This is consistent, no doubt, with the trend towards a toughening of terms on the upstream that has recently been evident in many countries around the world. In allocating risks in a transaction, the logical approach is to apportion specific risks to the parties best able to control, manage or bear those risks. However, when transboundary energy infrastructure projects were first being developed, many of the states hosting these projects were both unfamiliar to the investors proposing to invest in them, and regarded as high risk environments. Because the host states needed these early projects (and the investment that was tied to them), the investors were often able to use this leverage to extract terms and conditions from host states that were significantly favourable to investors. Thus, host states would frequently agree to assume project risks that were not fully within their (or any party's) control (we will refer to these as 'residual risks') – for example, by:

- providing investors with full protection for any change of law that might have a detrimental economic or other adverse impact on the project;
- allowing investors a complete pass-through of costs to the host state; and
- agreeing to extensive and broad *force majeure* protections, in addition to protections against host state actions.

The Dhabol power project in India was a good example of this, being one of the first projects developed as part of the liberalisation of India's power sector in the 1990s. The terms were such that the resulting cost of power was not sustainable in the Indian market and the project itself was not therefore sustainable long term. So a key objective for both the host state and the investors must be to create a balanced commercial structure that is sustainable over a long period and will stand the test of time; if that is not achieved and the project is too heavily weighted in one direction or another, it will inevitably fail.

In keeping with this, host states now appear less willing to take on residual risks and to provide comprehensive protections for investors. Host states are becoming more willing to challenge the assumption that they will assume extensive risk; they are also more likely to take the view that an infrastructure investor should not necessarily be treated any more favourably than any other investor in the country, and should accept normal investor risks insofar as they apply to other investors (eg, that the rate of tax will change, that regulations will change and so on). Host states tend to draw a distinction between changes applicable to any investors and those that are discriminatory; they are willing to accept responsibility for discriminatory changes (eg, in the nature of expropriation or creeping expropriation), but not for those changes that are generally applicable, which are regarded as a general risk if doing business to be assumed as part of any commercial transaction. This, of course, is only the basis on which investors have assumed risk in more developed countries for many years – for example, upstream oil and gas companies have been investing in the UK North Sea for decades without any protection against changes in law or regulation, and have found themselves at the wrong end of several tax changes over the past few years that have significantly affected their returns. So this is not a risk that is confined to less developed economies; ultimately, a judgement about the country of investment and its overall risk profile is required.

Whether a host state will be successful in resisting investor demands for residual risk protection will depend on the appetite of the investors for developing the project. The latter will be influenced by the assessment of how commercially viable the project is, which is to be determined on the basis of various factors including a risk/benefit analysis and the investors' view of the potential of the project to generate revenues. Different investors will take different approaches. In three similar projects over the last few years in the same country in Africa, one project did not proceed after years of negotiation because the host state and the investors could not bridge the gap between their respective requirements. By contrast, the two others went ahead and are operating successfully, because the respective investors were willing to accept a greater degree of country risk and to invest on the basis of less than complete protection. So, to a significant degree, it comes down to investor appetite for risk and, of course, the specific attributes of a host state – such as the stability of the government, the perception of certainty and predictability of host state's legal framework and whether there are any security concerns. What is clear is that a good host state reputation is required to attract investments; potential investors will view a reputation of certainty and predictability favourably, and such reputation will bolster a host state's position that it will not protect investors against residual risk. However, even host states without a strong reputation of certainty and predictability seem to be increasingly less willing to provide investors with a full protection against residual risk.

Bordavia wants to achieve its overriding objective of putting a secure energy supply in place. Accordingly, it will not want to agree to protecting investors against residual risks. Yet this approach will need to be tempered against what Bordavia thinks the investors will accept in order to avoid the situation where the investors decide to take their money and resources elsewhere.

6. Conclusion

In any negotiation there will be a tension between the parties; when one party gains an advantage, the other party will likely be subject to an equal loss. When a host state negotiates terms for a transboundary energy infrastructure project, every concession that it makes by undertaking residual risk to protect investors will cut directly into its ability to fulfil its overriding objective of putting a secure energy supply in place. This zero-sum dynamic explains why host states are less willing to assume residual risk in favour of investors. The willingness of host states to undertake risk not wholly within their control in the context of major energy project development is declining, notwithstanding that the demand for new projects, including transboundary energy infrastructure projects, remains high. Whether investors will participate in projects where the host state is not willing to assume residual risks will depend on the appetite of investors for the specific projects, how material the perceived risks really are, and whether there are other investors waiting in the wings.

Joint venture risks and responsibilities

Thomas J Dimitroff
Infrastructure Development Partnership LLP

1. **Introduction**

   The broad spectrum of risks addressed in the various chapters of this book share a common feature: they lie largely outside of the developer's sphere of control and may be described as risks that are external to energy infrastructure projects.[1] The authors of this book consistently counsel that the likelihood of external risks maturing into threats capable of effecting cost and schedule, and resulting in loss – whether legal or reputational – is greatly reduced when they have been identified and proactively managed early in a project's lifecycle. In other words, while external risks fall outside the developer's sphere of control, they may nevertheless be brought within the developer's sphere of influence.[2]

   On the face of it, the management of risks internal to energy infrastructure projects should be straightforward as they lie entirely within the developer's sphere of control and should remain so throughout the lifecycle of a project. However, internal risks, if not managed from the very outset of a project, can quickly migrate outside of a developer's sphere of control and form patterns that quickly set in concrete. At that point, internal risks may not only jeopardise delivery of the business objective, but also render the business more vulnerable to external risks. In addition, unmanaged internal risk can elevate the risk that business can pose to people, surrounding communities, the environment and potentially the macro-economy.

   A common feature of most businesses is that the business objective is delivered by groups of people acting together under a defined strategy, in accordance with an agreed plan and budget, and within a clear governance system. Internal risks arise when there is a breakdown or dysfunction in the agreed process of how groups of people making up a business work together.

   Energy infrastructure projects are often implemented through incorporated joint

---

1   For example, security and environmental risks may be considered external to a project. While a developer can and should take steps to identify and assess these risks, the developer must also determine the extent to which the mitigating measures deployed may pose risks to people, communities and the environment. Consideration should also be given to the risk of 'blow-back' that may in turn pose further risks to the delivery of the project's business objectives.
2   Risks may fall entirely outside of a developer's sphere of control and influence; they may be altogether unforeseen. While this chapter does not directly address the topic of unforeseen risks, the governance recommendations provided – in particular early planning, identification and alignment of shareholder interests, defined values and qualities of good leadership – remain among the prescription for mitigating both foreseen and unforeseen risks.

ventures between multiple energy companies.³ However, the combination of companies can magnify business complexity and present internal risks that complicate and stymie decision making. The attributes of individual shareholders within a joint venture can also present internal risks when they are not properly understood, and where adequate governance systems have not been put into place to recognise common values and deliver shared objectives. Indeed, when internal risks presented by joint ventures are not properly managed, they may present the single greatest risk to the successful delivery of the business, including energy infrastructure projects.⁴

This chapter examines the internal risks posed to energy infrastructure projects implemented through incorporated joint ventures. The chapter begins by exploring the nature of internal risks to corporations and the location of those risks in the domain of corporate governance. The responsibility of corporations to manage these risks on behalf of their shareholders and on behalf of affected external stakeholders is also considered, with particular emphasis on corporate responsibility. The attributes and business rationales of various types of shareholding company are next examined with a view towards assessing the internal risks that the shareholding companies may pose to joint venture arrangements.⁵ The chapter concludes by discussing emerging trends in corporate responsibility, the extraterritorial application of compliance requirements and the risks that these trends may pose to joint ventures. These trends point towards the increasing importance of internal risk management in strategic decision making for international joint ventures in the energy sector.

2. Corporations and internal risks

Corporations can be highly efficient vehicles for mobilising capital, providing returns to investors, and developing and selling products and services to consumers. The corporate vehicle enables shareholders to pool and leverage their capital while limiting individual liabilities to the amount of shareholder capital invested. Shareholders can deploy their capital free from the day-to-day management of the corporation. Directors should supervise management and set strategy in the best interests of the corporation and its shareholders. Accordingly, an essential feature of a corporation is that the ownership, direction and management functions are typically split and discharged by groups of different people.⁶

Like other forms of institutions, such as school boards or clubs, corporations can

---

3   Note that the terms 'corporation' and 'company' are used interchangeably throughout the chapter.
4   See McKinsey's study (1991) that assessed the performance of 49 joint ventures and determined that only 51% succeeded. A follow-on study conducted in 2001, which surveyed over 2,000 joint ventures and alliances, showed a success rate of only 53% (see *Harvard Business Review,* February 2004, pp1-15). While resource-driven energy infrastructure joint ventures may achieve their objectives despite inherent dysfunctionality, their ability to properly manage external risks is compromised, with a resulting impact on schedule, budget, efficiencies and strategic opportunities.
5   The intention of this chapter is not to undertake a comparative analysis of different systems of corporate law, but rather to promote understanding of different business drivers which inform the governance frameworks of energy infrastructure joint ventures.
6   As this chapter demonstrates, the understanding of a corporation may shift dramatically when we turn to examine non-US and non-UK corporations and certain state-owned corporations (see section 4.3 below).

and do create formal rules to govern how they act. These governance rules can provide for additional rules that render the actions of corporate personnel, or people acting on behalf of the corporation, as the actions of the corporation itself.[7] Governance rules are also internal to the corporation and do not ordinarily have normative implications outside of the corporation itself. For example, a governance rule cannot be used to create a legal or ethical obligation rendering a corporation's action legal or illegal, right or wrong, simply by saying or declaring it so.[8] However, corporate actions themselves can have legal and ethical implications external to the corporation.[9]

When governance rules are breached or found to be inadequate, and where corporate behaviour breaks down, the resulting risks can be said to be internal. As internal risks deteriorate into threats, the corporation becomes increasingly vulnerable. External risks that may otherwise be identified and managed early on are often missed, along with the responsibility that corporations have to avoid posing risks to others. As roles within the corporation are divided and subdivided among groups of people, risks are most effectively managed when a corporation's board of directors devolves sufficient management authority to strike the correct balance between the risks that come with creative business decisions and the assurance that comes with supervision. Too much of either will lead a corporation to failure.

3. Corporate responsibilities

Corporations share many attributes with people, such as the capacity to enjoy 'human' rights (eg, the right of free speech) and the capacity to be subject to legal obligations. However, corporations also have attributes that people do not have, including the capacity to survive their founders and exist simultaneously in multiple jurisdictions. While corporations exist separately and apart from people, it should be made clear that they can act only through people. This is not to suggest that a corporation lacks the capacity to act. To say that a corporation cannot act without people does not mean that a corporation does not have its own actions, or that the actions of people (or groups of people) who act for the corporation are the same as the corporation's own actions.[10]

The directors acting on behalf of corporations incorporated within common law jurisdictions owe fiduciary duties to both the corporation and its shareholders – for instance, the duty to act in such a way as to maximise shareholder returns. Directors and officers of corporations also have other duties, such as the duty to direct company affairs so that it does not steal from, kill or otherwise harm others. These obligations exist as a matter of law – that is, if they are not discharged, legal

---

7   See Gardner, John, "Reasons for Teamwork", *Legal Theory* 8, Cambridge University Press (2002), p495. Point further developed in discussions with John Gardner, University College fellow and Oxford chair of jurisprudence, University of Oxford (January 2006).
8   *Ibid*, footnote 7.
9   Accordingly, the tax, legal and regulatory risks addressed in this book are considered to be risks that lie outside of projects and must be assessed and managed appropriately by the corporation's officers, directors, employees and representatives in accordance with the internal governance systems in place.
10 Whether it is the corporation that is acting, or the persons who act through or on behalf of it, may have important implications for the nature of its responsibilities, other than legal responsibilities, which a corporation is capable of having.

consequences follow. In this sense, corporate laws that define fiduciary obligations, as well as the laws that are enacted to prevent harm to others, are risks external to the corporation, and its officers and directors.

When a corporation acts and causes harm, the law will treat the corporation and the harm caused as either the legal liability of the corporation itself, that of its officers and directors or both.[11] The law does not factor the behaviour of shareholders into the ascription of liabilities, unless those shareholders are discharging other roles on behalf of the corporation. Indeed, the law explicitly limits the liabilities of shareholders to the monetary value of their shares. The concept of limiting the legal liability of shareholders within Anglo-US jurisprudence has fascinating implications when we turn to identify its potential application in the context of the state-owned company, where the roles of shareholder and director are less well distinguished. Consider, for example, the fact that states have direct international law obligations to protect human rights. In the event that a state-owned entity breaches human rights obligations in its own state or abroad, the state may have difficulty in arguing that obligations are limited only to the monetary value of its shares.

An interesting and important question also arises about corporate responsibility in jurisdictions where corporate behaviour may cause harm but either laws do not exist or the rule of law is not otherwise applied to standardise behaviour that would avoid those harms or provide for liabilities in the event that harms materialise. This is by no means an exotic question, as energy resources are frequently produced and energy infrastructure projects implemented in emerging markets where corporate harms are not adequately covered by the law.

In answer to this question, civil society activists assert that corporations have obligations beyond the law and ethical responsibilities to avoid harms that can arise from corporate activity in these circumstances. Corporate officers and directors, by contrast, have a tendency to analyse this question in terms of risk rather than responsibility, as legal responsibility is, by definition, absent in these circumstances.[12] The external risk to the corporation caused by the actions of third parties which have in turn been harmed by the corporation will likely take precedence over the question of whether the corporation also has an ethical responsibility to avoid such harms.[13] Such external risks cannot be taken without the directors of a corporation taking some responsibility for resulting harms. At the very least, corporate directors will owe

---

11 The evolving jurisprudence on corporate crime (eg, corporate manslaughter) is an excellent example of an area where the corporation, as well as its directors and officers, may be held legally and criminally liable for actions.

12 Note the independent and confidential follow-up analysis to the World Resources Institute study of 190 projects undertaken by international oil companies (pdf.wri.org/development_without_conflict_fpic.pdf). This study found that the time and associated delay costs for the start-up of new projects had "nearly doubled [during] the past decade". The follow-up study of a subset of those projects found that "non-technical risks accounted for nearly half of all risk factors with stakeholder related risks constituting the largest single category". See Report of the Special Representative of the Secretary General of the United Nations on the issue of human rights and transnational corporations and other business enterprises, John Ruggie, A/HRC/14/27 (April 9 2010), Paragraphs 70-1.

13 The question of whether the corporation has an ethical responsibility may be less relevant as long as the corporation's officers and directors undertake ethical responsibilities to act on behalf of the corporation. They do so by making ethical judgements that are in addition to, or perhaps override, their fiduciary obligations.

fiduciary responsibilities to their shareholders to ensure that they are properly appraised of, and assured against, the risks created by the corporation's actions.[14]

4. Joint ventures: risks and responsibilities

While it is possible to organise an energy infrastructure business in any of a wide variety of business forms, energy projects are often implemented through incorporated entities. The main business rationales behind this structuring preference are the desire to distribute risk, mitigate tax obligations, contain legal liabilities and efficiently mobilise financing. The complex legal, tax and financing considerations of critical importance to the successful development and implementation of projects are expertly addressed elsewhere in this book.[15]

Energy infrastructure projects are often strategic in nature, require large-scale investment and the alignment of wide-ranging interests, and confront a daunting array of external and internal risks. For these reasons, investors often seek out joint venture participants bringing complementary advantages to the development of a project or its implementation and subsequent operation. A participant's knowledge of, and connection to, particular governments and markets, sector experience and expertise or technical and financial capability may enable risks and costs to be reduced and opportunities captured. However, the joint venture relationships themselves may also present potential risks and constraints that must be fully understood and assessed. Internal and external risks may stem from the corporate identity of a shareholding partner to the joint venture or the complexity or attributes of their organisational structure, ownership (public, state or privately held), national identity, business culture or scope of operation. The remainder of this chapter examines the internal risks that these various attributes may present to joint ventures implementing energy infrastructure projects.

4.1 Corporate identity

An international energy company, including its many affiliates and subsidiaries, is an artificial legal person. In addition to having the capacity to enjoy rights and undertake legal obligations, corporations, like people, define their identity by what they do and how they go about doing it. Fundamentally, corporate behaviour is measured by performance over time and this defines the corporation's values and shapes its identity. A corporation's reputation can also be burnished by what it says about itself in the marketplace – that is, the efforts made to project and shape its image through brand marketing and assertions about performance quality and standards of behaviour. Size, market capitalisation and jurisdiction of ownership and

---

14   Given the enhanced role that large multinational corporations can play in providing public goods, there is also a solid argument to be made that officers and directors may also owe a duty to prevent harm to people, communities and the environment even where there is no legally binding requirement to do so. Indeed, certain jurisdictions are moving in this direction. Consider, for example, the directors' duties under Section 172 of the UK Companies Act 2006.
15   The analysis in this chapter focuses on the corporate form and, in particular, the incorporated joint venture. However, it is important to note that joint ventures may be, and frequently are, pursued on a contractual unincorporated basis – particularly at the project development phase (ie, before construction and operation).

control enable corporations to assume a status and power that far transcend those of the individual founders and other people that manage, direct and otherwise act for them.

Large international corporations, and the projects that they sponsor, can become significant national, international and even geopolitical actors. For example, at the time that the Baku-Tbilisi-Ceyhan oil pipeline (with a capacity of 1 million barrels per day) was constructed, it was considered to be the world's largest non-military infrastructure project, crossing three countries and responsible for delivering 25% of the world's demand growth in oil between 2007 and 2009. The failure of a large multinational corporation can also have significant societal implications for shareholders and pensions,[16] jobs[17] and the macro economy[18] that may prompt governments to consider certain companies to be too big to fail. Governments can, and frequently do, show preferences for domestic corporations that serve as national champions viewed as fulfilling important roles such as the security of domestic energy supplies,[19] distribution channels to enable goods and services to be supplied to populations and the operation of critically important pieces of strategic infrastructure.[20] Governments may also seek to exert control over particular corporations through their listing on a domestic exchange, transparency and compliance requirements imposed and the national identity of the directors and officers. All of these factors present both comparative advantages and constraints on behaviour for companies to consider when evaluating potential joint venture participants.[21]

Shareholder participants to a corporate joint venture should be aware of the identity and values of their co-venturers. For example, a range of external risks may be associated with the perception of a joint venture participant's corporate identity

---

16   The shareholders of Enron lost $74 billion in the four years immediately prior to the company's bankruptcy (with $40 million to $45 billion attributed to fraud). More than 20,000 of Enron's former employees in May 2004 won a suit of $85 million for compensation of $2 billion that was lost from their pensions. From the settlement, the employees each received about $3,100. See Axtman, Kris, "How Enron awards do, or don't, trickle down", *The Christian Science Monitor* (June 20 2005) (www.webcitation.org/5tZ5aLijg); "Enron's Plan Would Repay A Fraction of Dollars Owed", *The New York Times* (www.webcitation.org/5tZ5cPyj6); Vogel, Carol, "Enron's Art to Be Auctioned Off", *The New York Times* (April 16 2003) (www.webcitation.org/5tZ5cZwye).
17   General Motors lost 27,000 jobs in 2009 as the employee base declined from 91,000 at the end of 2008 to 64,000 when it emerged from bankruptcy (business.timesonline.co.uk/tol/business/industry_sectors/engineering/article6682876.ece).
18   The US Federal Treasury took a stake in AIG in excess of 90%, costing taxpayers $182 billion in order to prevent a financial collapse in which more than $500 billion in credit derivatives had been issued to the market (en.wikipedia.org/wiki/AIG).
19   In March 2005 the Chinese National Offshore Oil Corporation tried to acquire Unocal with a bid that valued Unocal at between $16 billion and $18 billion. Following a vote in the US House of Representatives, the bid was referred to the US president on the grounds that its implications for national security needed to be reviewed. The Chinese National Offshore Oil Corporation withdrew its bid and, shortly thereafter, Unocal merged with Chevron (www.china.org.cn/english/2005/Aug/137165.htm).
20   In 2006, Dubai Ports World agreed to sell the subsidiaries of P&O responsible for operating major port facilities in the United States to AIG for an undisclosed sum of money when the acquisition was referred to the US Treasury Department and Homeland Security over concerns about the alleged negative impacts on port security that would stem from their ownership by a UAE company (.9.%5e%20http//portal.pohub.com/pls/pogprtl/docs/PAGE/POGROUP_PAGE_GROUP/PO%20GROUP%20NEWS%).
21   In the energy sector, additional factors that will shape a corporation's identity in the marketplace include its reserve base, access to critical infrastructure, technical know-how and management systems, global and regional experience, experience and performance record as an operator of complex projects, financial strength, connectivity with governments and strategic alliances (including framework agreements for the procurement of goods and services arrangements).

by external stakeholders with advantages in one context and serious disadvantages in another. Moreover, the perception of certain forms of external risk may vary from one participant to the next, with important consequences for the joint venture. For example, one participant may perceive the risk that arises from societal[22] impacts or the application of extraterritorial compliance requirements in a way that differs from another participant. If these differences are not appropriately accommodated within a governance framework, they may impede the success of the joint venture. These risks require identification and careful management at a very early stage in the formation of the joint venture; they should also be prescribed in the joint venture's shareholders' agreement.

4.2   **International oil companies, organisational complexity and joint ventures**
International oil companies[23] can be enormously complex organisational structures that present large-magnitude internal risks to themselves, as well as to their joint venture participants. They are typically structured into one or more holding companies that own a vast array of subsidiary and affiliated companies incorporated in a wide variety of jurisdictions that, in turn, hold the underlying asset base. Integrated international oil companies operate disparate businesses that range from upstream exploration and production to the development and operation of midstream transportation infrastructure to the downstream management and operation of refining and processing facilities and retail sales outlets.[24] These diverse activities often require distinct business models with discrete strategies, plans and performance milestones.[25]

The many subsidiaries and affiliate companies are often collected under a holding structure and organised into separate business segments administering multiple business units. In addition, it is common for international oil companies to have a leadership presence in each country that may or may not correspond to the leadership of the business unit or the officers and directors that populate the boards of the subsidiaries. The consequence of these overlapping authorities and accountabilities is that the external face of the corporation may not always correspond to actual decision-making power. This sometimes leaves the ostensible senior managers to hold 'steering wheels' that are not attached to an authentic corporate decision-driving capability.[26]

---

22   See the "Guiding Principles on Business and Human Rights: Implementing the United Nations' Protect, Respect and Remedy Framework", presented to the United Nations Human Rights Council on March 21 2011 (A/HRC/17/31).
23   Note that references to 'international oil companies' and 'international oil and gas companies' are used interchangeably throughout the chapter.
24   For example, many large publicly listed international oil companies are listed on multiple exchanges, comprise large numbers of employees (often exceeding 100,000), are present in dozens of countries, have tens of thousands of service stations, operate dozens of refineries and produce oil and gas in many countries. The individual subsidiary and affiliated corporations that are required to hold the many assets and operating capabilities may number in the hundreds, if not thousands, with each subject to external compliance requirements.
25   There is a vast difference in strategic perspective when comparing the oil trader, in a market where commodity prices change in real time, to an upstream exploration and production company developing new upstream reserves where the full cycle from discovery to delivery of first oil or gas may take 10 or more years.

The governance structures utilised to manage internal risks for large international oil companies can be more effective when they are centralised and simplified, and behaviour throughout the organisation is standardised. However, the efficiencies gained in the context of standardised behaviour and compliance may come at a cost. Standardisation can constrain flexibility and creative business initiatives that enable larger profits and more effective operations within a more decentralised structure, where authority is devolved to the local business units. This type of flexibility is also important when operating within disparate business environments and cultural contexts where a business unit may need to accommodate these factors and potentially stretch the governance boundaries set by the head office.

Whether an organisation is centralised or decentralised, its corporate leadership needs to provide a clear and coherently balanced set of values. Indeed, clear values function as a type of glue binding multiple and varied groups into a single coherent structure. Business values enable management to execute decisions that appropriately balance project needs (eg, to deliver on schedule, on budget and on specification) against safety requirements (ie, the need to bring workers home safe and sound). It may be that large international oil companies reach a tipping point in scale where there is a need to focus increasingly on standardised internal systems to manage risk and behaviour, so that operations remain consistent, safe and compliant across the organisation.[27]

An international oil company will typically own its share of an incorporated joint venture through a special purpose subsidiary. The international oil company will appoint employees to each of the boards of the special purpose subsidiary owning its interest in the joint venture, and to the joint venture's own board, and these employees may not be the same individuals. Moreover, the members of the two boards will likely have responsibilities that transcend the business unit in which the joint venture is being developed, especially where the infrastructure in question is cross-border and involves multiple business units and country representations.[28] The joint venture's management team will likely include secondees of the international oil company who may need to work with secondees from the other joint venture shareholder companies, depending on whether the international oil company is the operator of the joint venture.[29]

The internal governance issues that arise within the context of a joint venture organisation can be extraordinarily complex. For example, the individual joint venture participants cannot assume that their own representative(s) or the joint venture representatives of other international oil company participants adequately

---

26 For example, a large integrated oil company may have diverse business activities taking place in a single country or region, including refining and marketing, upstream exploration and alternative energy. Each of these businesses will have a reporting line within the broader organisation and that reporting line may or may not pass through the hands of the company's in-country president.
27 In order to function efficiently across multiple jurisdictions under a common identity, a balance must also be struck between central management and control, and devolved authority.
28 Indeed, as energy infrastructure projects frequently involve exploration and production activities, an international oil company's marketing segment may also overlap with the upstream segment, thus potentially confusing decision making up the chain of the organisation.
29 An 'operator' is a person or company that is either a proprietor, lessee or contractor operating or managing an upstream concession or energy infrastructure. See www.oilgasglossary.com/operator.html

communicate up and down their own management hierarchy. Without uniform communication and buy-in to the proposed actions of the joint venture within the internal structures of each co-venturer, the ability to mandate and subsequently assure the joint venture's actions is greatly diminished. As a consequence, joint venture participants are advised to anticipate and understand this complexity and open multiple channels of communication within their international oil company participant, as decision making can go very wrong.

Forging decisions and providing adequate assurance on risk within a large international oil company may be as challenging as negotiating a commercial arrangement with an external counterparty. The myriad subdivisions and functions often produce conflicting lines of authority, which can complicate internal processes and obscure accountability enormously. Organisational complexity – when combined with confusing statements on business values or ambivalent guidance from leadership – can send mixed signals about the relative demands of cost and responsibilities such as safety concerns. The most tragic and spectacular recent examples of accidents that appear to have resulted from internal breakdowns in organisational behaviour may be the BP 2005 Texas City Refinery disaster and the subsequent BP Deep Water Horizon oil spill.

In the wake of both incidents, the respective investigation commissions appear to have drawn common conclusions on root cause; these point towards, among other things, BP's organisational structure, management systems, values and behaviour.[30] However, BP has also been widely hailed for the exceptional leadership that it demonstrated in leading the development, construction and operation of some of the most complex energy infrastructure projects in recent decades.[31] The disparity in performance may suggest undue emphasis placed upon excellence in individual higher-profile project performance rather than more mundane attention to consistency in delivery across the global organisation. Although the Deep Water Horizon spill involved offshore energy infrastructure implemented by an unincorporated joint venture that included Anadarko and Mitsui, the implications for joint ventures involved in cross-border energy infrastructure are significant. Joint venture participants would be well advised to develop a clear understanding of the internal risks presented by organisational complexity and the potentially wide-ranging external impacts that result when these internal risks are not adequately processed, understood and acted upon by the organisation.

---

30   See the Report of the BP US Refineries Independent Safety Review Panel (BP Texas City Refinery) (January 2007). See also the National Commission on the BP Deep Water Horizon Oil Spill and Offshore Drilling, Deep Water Horizon Study Group Center for Catastrophic Risk Management University of California Berkley (July 8 2010). See also Chief Counsel's Report, National Commission on the BP Deepwater Horizon Oil Spill and Offshore Drilling, Chapter 5, "Overarching Failures of Management" (February 2011).

31   For example, the Baku-Tbilisi-Ceyhan and South Caucasus oil and gas export pipelines from the South Caspian and the Tangguh liquefied natural gas terminal in Indonesia were exceptionally complex and well-executed projects. At the time of financial close, the Baku-Tbilisi-Ceyhan pipeline was hailed as the most complex project financing to have hit the market for a decade. All three projects engaged extensively with local and international civil society organisations and were held by many as models for corporate responsibility.

4.3 State oil and gas companies, ownership and control

Many state-owned or national oil companies were formed as a result of perceived abuses by international oil companies in the Middle East and Central and South America. After nationalising the industry, a collection of oil producing countries formed the OPEC cartel.[32] Other national oil companies were formed within the state-controlled economies of the Former Soviet Union (subsequently the Russian Federation) and China in order to control natural resource extraction, production, refining and marketing at home and abroad (eg, Gasprom (Russia) and CNPC (China) respectively). Still other national oil companies were formed to exert greater national control over the development of newly discovered resources (eg, Statoil in the Norwegian sector of the North Sea).[33]

The 16 largest oil and gas companies in the world are all state owned; they control the overwhelming majority of global oil and gas reserves.[34] As the global demand for oil and gas increases, state-owned and international oil companies are seeking reserves within increasingly complex environments both politically (eg, Iraq) and technologically (eg, within the Arctic, ultra-deep water Gulf of Mexico and, onshore, within shale formations). Clearly, in order for publicly traded and privately held international oil companies to remain competitive, alliances will need to be forged and maintained with state-owned oil and gas companies. A natural fit emerges for the moment between technological capability and global supply, distribution and marketing capabilities of global international oil companies on the one hand, and access to the enormous reserves and financial strength of state-owned oil and gas companies on the other. As a consequence, joint ventures between state-owned oil and gas companies and international oil companies are proliferating globally.

In the case of complex cross-border export infrastructure projects, state-owned oil and gas companies often lack the necessary expertise in project management and project financing. In addition, where the state-owned oil and gas company is owned by the energy producing country, large private sector joint venture partners can help the producing country to diversify and absorb political risks. For example, a neighbouring country may be more reluctant to interrupt the flow of energy resource where large international oil company interests, backed by powerful governments, are also involved. Examples of these alliances, often organised as joint ventures, are developing not only within the jurisdictions in which the state-owned oil and gas companies hold their national reserves, but also in third-party jurisdictions presenting considerations of wide-ranging comparative advantage and internal risk. Understanding the different business characteristics and other rationales behind state-owned oil and gas companies, and the resulting behaviours, is critical to

---

32 The five founding member countries (the Islamic Republic of Iran, Iraq, Kuwait, Saudi Arabia and Venezuela) formed OPEC in September 1960. These countries were later joined by Qatar (1961), Indonesia (1962), Socialist People's Libyan Arab Jamahiriya (1962), the United Arab Emirates (1967), Algeria (1969), Nigeria (1971), Ecuador (1973), Gabon (1975) and Angola (2007).

33 While national oil companies constitute a diverse group, very few are efficient, possess technical and managerial competence and demonstrate transparent systems of corporate governance (see further the discussion below). Notable exceptions to this include the partially privatised Statoil (Norway) and Petrobras (Brazil).

identifying potential emerging risks and mitigating measures that may be required to enable a successful and sustainable joint venture relationship.

To understand the nature of some of the internal risks that may arise within a joint venture relationship between a wholly state-owned oil company and an international oil company, several core differences need to be identified. The respective roles played by management, directors and shareholders in many state-owned oil companies are less clear-cut than in the typical international oil company. In practice, these three roles are discharged by the government in a wholly state-owned oil and gas company.[35] This has fundamental implications for the very purpose of the state-owned oil and gas company, which often differs significantly from that of an international oil company. This is especially true where the oil company is a publicly traded entity – recalling that the overriding objective is to deliver shareholder value. By contrast, the objective of the state-owned company may be to hold and responsibly manage the state's interest in its national resources. The state's interest may not coincide with the objective of maximising the monetary value that the national resource secures in the market, or in responsibly managing reservoir quality. Rather, the state's interest may be to ensure security of energy supplies; use its petro-dollars to subsidise other unrelated sectors of the state's national economy;[36] or use its national resource to secure strategic objectives within international markets.[37] In addition, state-owned oil and gas companies may own and/or seek to control existing and future energy export infrastructure in order to control the monetisation of the state's resource or regional supplies. Often, state-owned oil and gas companies are deployed simply to be involved with foreign companies developing energy export infrastructure across their territory, with the objective of intelligence gathering or knowledge and skills transfer.

The legal, regulatory and fiscal regime in which state-owned oil and gas companies operate may create vastly different individual incentives for behaviour. For example, the manager of a state-owned oil and gas company may be subject to administrative and even criminal sanctions associated with procurement of goods, works and services that exceed budgeted amounts. In these circumstances, where the downside for a cost overrun in the case of an international oil company is simply higher costs, the official of a state-owned oil and gas company may refrain from decisions for fear of administrative or even criminal sanctions due to misallocation of state funds. With this in mind, an international oil company partner need not

---

34 See complete table itemising the top 50 oil and gas companies in the world ranked according to their reserve base (www.petrostrategies.org/Links/Worlds_Largest_Oil_and_Gas_Companies_Sites.htm) (list updated in May 2010).

35 The strategic oversight played by the director in an international oil company is often fulfilled by a government ministry (eg, the Ministry of Oil and Gas) and the management roles are discharged by technocrats appointed by the ministry, with the shareholding function ostensibly played by the government on behalf of the population.

36 For example, Venezuela's economy is 95% dependent on oil export earnings and much of these earnings have been used to shore up funding in unrelated sectors. See *CIA World Factbook* (August 19 2010, retrieved September 3 2010).

37 On February 4 2009 Ukrainian online newspaper *Ukrayinska Pravda* quoted Russian Prime Minister Vladimir Putin as stating that "Russia enjoys vast energy and mineral resources which serve as a base to develop its economy; as an instrument to implement domestic and foreign policy. The role of the country on international energy markets determines, in many ways, its geopolitical influence".

wonder why decisions are delayed as responsibility is appealed up the chain of organisational command (often to ministerial level or above), resulting in far greater damage to joint venture projects. Local law and custom may also force behaviour for both the international and the state-owned company that can create obstacles for the international company before its home government, or for the joint venture itself when seeking international institutional financing.[38]

4.4 Western international oil companies and compliance obligations

Increasingly, a public listing comes with considerable transparency and other reporting and compliance requirements that demand standards of behaviour to which privately held companies or state-owned entities are not subject. The myriad of regulatory and reporting requirements that apply to US and Western European public markets presents formidable challenges to businesses. These requirements include the US Sarbanes-Oxley Act,[39] the US Dodd-Frank Wall Street Reform and Consumer Protection Act,[40] the US Foreign Corrupt Practices Act[41] and the UK Bribery Act,[42] all of which have extra-territorial application. Indeed, the 14th Annual Global CEO Survey (2011) conducted by PricewaterhouseCoopers references overregulation as the number three global risk to business growth, just behind recession/economy and deficit on its list of top nine risks.[43] In addition to legal compliance, the transparency requirements to which Western international oil companies are subject enable accountability to civil society across a broad spectrum of financial and non-financial risk factors.

Compliance and transparency requirements to which Western publicly traded and privately held international oil companies are subject need to be intimately understood by non-listed international oil company and state-owned partners. These requirements will be imposed on the joint venture entity as a whole. In the case of legal compliance, the consequences of failure can be severe and may carry both fines and penalties, including potential criminal sanctions for directors and officers of the non-complying publicly listed joint venture shareholder company. Transparency requirements have ensured that Western international oil companies (especially those that are publicly traded) focus greater attention on the non-financial impacts on people, communities and the environment in which they operate.[44] This can have important governance implications for the behaviour of the joint venture that are best comprehensively understood and addressed in the shareholding agreement at the outset.

---

38   Accessing appropriate information about state-owned partners, their financing and the basis for decision making can be problematic when trying to comply faithfully with extra-territorial application of regulatory and compliance obligations to which listed international oil companies may be subject.
39   PubL 107-204, 116 Stat 745.
40   PubL 111-203, HR 4173.
41   15 USC §§ 78dd-1 *et seq.*
42   UK Bribery Act 2010 c23
43   See www.pwc.com/ceosurvey.
44   "National oil companies (NOCs) concentrated in emerging markets, including China's Sinopec, Saudi Arabia's Saudi Aramco and Kazakhstan's KazMunaiGaz, were among the worst performers when it came to reporting on their anti-corruption programmes, according to research by Berlin-based Transparency International and New York-based Revenue Watch, which rated 44 companies accounting for 60 per cent of global oil and gas production." (www.ft.com/cms/s/0/7f6fb784-435f-11e0-8f0d-00144feabdc0,_i_email=y.html)

4.5 Cultural and behavioural identity factors

The fiduciary duty is a fundamental principle broadly informing both US and UK models of corporate governance. For example, a corporation's board of directors will owe a fiduciary obligation to maximise shareholder value. This is a principle ostensibly shared by other members of the Organisation for Economic Cooperation and Development (OECD). However, the *de facto* ownership structures and historical and cultural characteristics that underlie corporate governance practices in many OECD, as well as non-OECD, countries reveal goals that diverge from the objective of maximising shareholder value and this may result in behaviours that appear opaque to joint venture participants.

The corporate governance objectives of companies formed, owned or controlled in many common law and civil law jurisdictions, as well as OECD and non-OECD economies, are in fact driven by different motives. These may be, for instance, designed to serve:

- growth and market share (eg, the *keiretsu* in Japan);
- the state (eg, in France where a substantial percentage of top industrial, commercial and service companies are either state owned, family controlled or management controlled);
- family interests (eg, Kazakhstan);
- the military (eg, Pakistan); or
- the interests of an oligopoly that may be intertwined with state, security and industrial interests (eg, Russia).

These objectives may be re-enforced by cross-shareholding structures, combined with acute, strategic interventions by the government either directly or through government-controlled vehicles or agencies, or through industrial or family interests.

When examining corporate governance practices in countries that adhere to principles of state capitalism, the drivers may differ further and resulting behaviours may appear more opaque still. As emerging markets expand and continue to absorb more and more commodities – especially oil and gas – the dominance of state capitalism in energy markets, from China on the demand side to Saudi Arabia on the supply side, is an obvious and overpowering factor. As mentioned above, the 16 largest oil and gas companies in the world are state owned; they control the overwhelming majority of these reserves, as well as large foreign currency reserves. As the macroeconomic topography continues to shift, with significant imbalances still holding forth the prospect for future shocks, the overriding strategic interests of states will continue to assert themselves and may result in new forms of protectionism and market manipulation that will complicate how business may be conducted effectively.[45]

---

[45] "Despite massive state interventions in economies around the world, many corporate leaders and investors act as though globalization remains the dominant paradigm. That is a mistake." See "State Capitalism and the Crisis", *Ian Bremmer McKinsey Quarterly* (July 2009). Also, the head of the International Monetary Fund, Dominique Strauss Khan, warned on February 5 2011 that "the economic rebound across the world is built on unstable foundations". (www.telegraph.co.uk/finance/globalbusiness/8296987/IMF-raises-spectre-of-civil-wars-as-global-inequalities-worsen.html)

International oil companies that may be driven by a desire to do business in a manner that strikes an appropriate balance between maximising their shareholders' value and minimising their potentially detrimental impact upon people, communities and the environment will continue to confront powerful forces, including the state, military, domestic and industrial interests, enmeshed in opaque governance mechanics that are almost entirely inaccessible. International companies that choose to enter into joint venture relationships with state-owned companies are advised to inform themselves of these dynamics and take them properly into account when agreeing to the principles that will govern a joint venture relationship – whether those principles are set forth in a joint operating agreement or shareholders' agreement.

5. Emerging trends

A number of contrary forces[46] appear to be converging that place increasing pressure on corporations and energy infrastructure joint ventures to manage their internal risks and responsibilities. Differences in the purpose and philosophy underlying emerging principles of corporate governance will form a quiet backdrop against which many of these disparate forces are eventually rationalised. Will corporate governance principles converge into a global consensus? Will the forces of state capitalism continue to enable an opaque set of standards to be applied to state-owned corporations, while distinct and increasingly onerous standards are applied to Western private and publicly traded companies? Will the state-owned oil and gas companies investing beyond their borders benefit from an absence of post-colonial baggage that international oil companies continue to labour under, especially in North Africa and the Middle East?[47] What will the consequences be for both state-owned and international companies that have invested with autocratic regimes in the Middle East and North Africa in the aftermath of the unfolding 'Arab Awakening'? How these questions are answered and whether these disparate forces are eventually harmonised remain unknown.

In the context of corporate responsibility, a discussion has been taking place between civil society and business over the past decade that has focused primarily upon encouraging responsible conduct among large multinational companies – typically Western in origin and publicly traded. This discussion has arisen against the background of a common set of assumptions, including those relating to the continued free flow of goods, capital and knowledge within an increasingly positive business environment where open markets and the rule of law prevail.

A closer examination of the essential forces behind this discussion reveals important tensions – with civil society activists less concerned with investment climate and competitiveness, and business more sensitive to the risks and costs associated with enhanced regulation. From the perspective of business, there has been an expectation that the right balance between competitiveness and regulatory policy can be maintained so that corporations can operate in a manner consistent

---

46   For example, macroeconomic stability, security of supply, energy poverty, climate change and other detrimental impacts on people, communities and the environment.
47   Point raised in discussion with author by Susan Maples, post-doctoral research fellow at Columbia Law School (March 18 2011).

with the requirements of corporate responsibility without incurring significantly higher costs. However, this expectation may be less important to civil society activists – especially where the impacts of business on people, communities and the environment are at stake. In the wake of the recent financial crisis, the luxury of glossing the cost of compliance and responsible corporate activity with cheap debt has disappeared and the core components of this debate are emerging, perhaps enabling a more realistic assessment.[48]

In the energy sector, the increasing concentration of hydrocarbon production in the former Soviet Union, the Middle East and North Africa, the growing power and global reach of state-owned oil and gas companies from both producing and consuming markets, and enhanced competition for resources will continue to strengthen the hand of geopolitics in energy markets.[49] At the same time, the growing demand for commodities will fuel the economic growth and industrialisation of emerging markets and drive inflationary pressures on more mature OECD economies.

In order for international oil companies to compete effectively and participate in the growth of these markets, alliances will need to be forged between state-owned and international companies. The perception of financial and non-financial risk among Western governments and civil society organisations will continue to force international companies to comply with enhanced regulations and higher standards of behaviour.[50] However, the recent drive — post-Enron and post-2008 financial crisis - for enhanced compliance and regulation is found primarily in the US, UK and Western European contexts. The consequence is that state-owned companies – and to a certain extent private equity funds, hedge fund investors and privately held companies – are subject neither to the degree of financial and other public reporting obligations[51] nor to the forms of pressure to which international companies (and

---

48 At a certain point, the so-called 'virtuous' circle that has enabled the corporate responsibility discussion to thrive may prove chimerical. Other attempts to harmonise ethics and self-interest (ie, enlightened self-interest) demonstrate that the corporate responsibility thesis is neither new nor unique to the business context, nor ultimately achievable. In an unethical world, there would be no profit in ethical conduct as there would be no incentive to pay for it. In an ethical world, people would pay for ethical conduct, but the market would also reward a wide range of other conducts that might be amoral or immoral. In either case, there is no necessary relationship between ethics and markets and, therefore, no reason to believe in the sustainability of corporate responsibility's virtuous circle. While corporations will be held to account for conduct that causes harm, the efforts of civil society activists will likely transform this responsibility into a legal, rather than a self-declared ethical, obligation or corporate responsibility. Point raised in discussion in January 2006 with John Gardner.
49 See *Forbes*, www.forbes.com/lists/2010/18/global-2000-10_The-Global-2000_Rank.html. See also www.petrostrategies.org/Links/Worlds_Largest_Oil_and_Gas_Companies_Sites.htm). Note that while Exxon-Mobil is the world's largest publicly traded company measured by sales, profits and market capitalisation, it is only the 17th largest company measured by reserves of barrels of oil equivalent.
50 Higher standards of corporate behaviour are required in order to minimise the dramatic impact that this explosion in population, corporate activity and economic growth is having on the global macro economy, as well as on people, communities and the environment. There is also a danger that uneven and aggressive extraterritorial enforcement actions may be perceived to betray strategic and political motivations, prompting emerging powers to launch countervailing enforcement measures in the coming decade. Accordingly, an international effort, perhaps led by the United Nations, might be preferable.
51 Consider, for example, some of the many compliance requirements imposed on international oil companies by the US government: the 2002 Sarbanes-Oxley Act, the 1977 Foreign Corrupt Practices Act or the 2009 Energy Transparency Through Security Act. However, state-owned companies should be more directly subject to legal obligations in their activities (eg,, human rights obligations are directly binding upon states and hence state-owned entities).

especially publicly traded international oil companies) are subject by regulators and by civil society activists.[52] Interesting questions arise as to whether governments in Asia and China, in particular, will choose to become more aggressive in their own policing, perhaps on an extra-territorial basis, of corporate behaviour over which they may choose to assert their jurisdiction. Additional questions emerge in the context of whether international oil companies will continue willingly to assume ethical responsibilities where doing so may place them at a strategic disadvantage in attempting to partner with state-owned companies, or whether they might prefer the greater certainty associated with more straightforward regulation.[53]

Perhaps concerns about climate change and other potentially harmful business impacts may promote sufficient consensus to precipitate the introduction of forms of regulation that will minimise these impacts, ensure security of supplies and further contribute to a level playing field between state-owned and international companies. As regulatory obligations become more onerous, the remaining gap between legal requirements and ethical responsibilities may well diminish as the external risks to business increase.[54] Accordingly, a shift in focus from a more academic question concerning the ethical responsibilities of corporations towards an assessment of risks that they present to people, communities and the environment, and the risks that are in turn posed to corporations when ignored or inadequately managed, may ultimately forge a more pragmatic merger of interests between civil society and business.[55]

The implications of these trends for international oil companies seeking to partner in the development of energy infrastructure are significant and constitute a space to be carefully monitored. From an internal governance standpoint, state-owned and privately held international oil companies that have little to no exposure to US markets, but a strong commercial need to access technology, management systems and distribution markets, may attempt to structure around these requirements. This would present significant compliance challenges and risks at best, and potentially loss of competitive advantage as state-owned companies grow in wealth and sophistication.

6. Conclusion

Internal risks can often cause a joint venture to fail. Of those joint ventures that

---

52 The recent flotation of Glencore should have fascinating transparency, reporting and governance implications for a leading company operating in the opaque world of global commodities trading.
53 While free market principles can and should prevail in a globally competitive economy, it is naïve to assume that implicit and explicit biases are not acted upon by governments to support corporations that coincidentally fulfil strategic objectives. The many state and private interests of Chinese origin that are paying above market value to secure vast commodity supplies globally are less concerned about returning shareholder value than achieving Chinese state strategic objectives.
54 In this regard, see the recent report by John Ruggie, special representative of the UN secretary general, on the issue of human rights, transnational corporations and other business enterprises, entitled "Guiding Principles for the Implementation of the United Nations 'Protect, Respect and Remedy Framework'", posted for public comment on January 31 2011. While the Ruggie framework does not represent a 'hard' legal requirement, it does form 'soft law' to which companies will increasingly be held to account.
55 When corporations ignore or inadequately manage the external risks created by their behaviour, the corporation has failed to address internal risks. The emphasis away from responsibility and towards risk creates a much more pragmatic approach to addressing potentially harmful corporate behaviour and is the preferred approach counselled by Ruggie.

manage to deliver their business objective in spite of internal risks, many are dysfunctional and fail to deliver on schedule as well as on budget. Proper management of internal risks is a core issue of corporate governance. A wide range of contrary and conflicting global forces referenced in this chapter pose compliance, regulatory and corporate responsibility demands on joint ventures implementing energy infrastructure projects. Requirements for higher standards in corporate behaviour appear more rigorous in US, UK and Western European contexts. This stands in marked contrast to applicable requirements from markets that have fastest growth in hydrocarbon production, as well as the largest reserves. The ability of privately held and publicly traded international oil companies to compete effectively, and/or join forces with state-owned companies to deliver secure supplies to developed markets, remains largely a question of policy. In addition to maintaining competitiveness, global policy makers will need to strike an effective balance among competing factors. These include market, environmental and strategic factors on the one hand, and the requirements of responsible corporate behaviour on the other. The sustainability of this balance will depend, in part, on the ability of policy makers to forge a universally applicable normative framework that can accommodate the pragmatic demands of joint venture arrangements.

Project finance and risk mitigation

William E Browning
Infrastructure Development Partnership LLP
Alexandre Chavarot
William J Clinton Foundation

Transboundary energy infrastructure investments are exposed to a number of political, commercial and technical risks, as discussed in the previous chapters of this book. Finance certainly is an additional risk, in that the proposed investment may not attract sufficient capital to meet expected project costs. It can, however, also be successfully utilised as a risk mitigation tool, both by ensuring that risks are contractually allocated to those entities best able to mitigate and absorb them, and by involving international financial institutions whose umbrella effect often helps to ensure that the terms of the relevant host government agreements and intergovernmental agreements are respected.

This chapter provides an introduction to project finance, discusses the risk mitigation benefits of project finance and presents an overview of the different participants in a project financing, as well as the main sources of project debt. The chapter also highlights key criteria for a successful project financing, with an emphasis on managing the project financing process. It finishes with a discussion with representatives of the Baku-Tbilisi-Ceyhan, Nord Stream and Blue Stream pipeline projects on key finance considerations.

1. What is project finance?

'Project finance' refers to a financing scheme in which a specific finance plan is put in place to fund a single investment or project, with the future cash flows generated by the project asset(s) being the main, if not the only, source of funds available to service the project debt. Project finance is also called non or limited-recourse finance, to reflect the fact that lenders have no or limited recourse to project sponsors once a series of project completion tests are met.

Below is a simplified structural diagram of project financing.

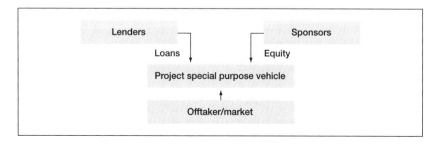

The project sponsors or equity investors invest their *pro-rata* share of equity in the project company. The company raises debt from a variety of sources, with lenders having no or limited recourse to the balance sheets of the project sponsors.

Project finance, therefore, differs from corporate finance in that the debt is raised specifically by (in most cases) and for the project company, as opposed to a scheme where a project investor funds its share of project costs from its corporate funds.

The nature of sponsor support is an essential component of negotiations between equity investors and project lenders. Forms of sponsor support typically include completion and/or performance guarantees, debt service undertakings to cover specific technical or non-commercial risks, as well as ongoing warranties that suppliers, contractors and/or sponsors provide to lenders during part or all the operational phase of the project.

2. Why use project finance?

Project finance is used for two main reasons:
- to access capital, both for the weaker project sponsors which may not be able to fund their share of project costs on their balance sheet and for the stronger ones which may want to make sure that their creditworthiness is not affected by the sheer size of the project; and
- to ensure that project risks are properly mitigated and allocated.

2.1 Access to capital

Accessing capital through project finance is particularly relevant for transboundary energy infrastructure projects, which tend to be very capital intensive. Few oil and gas companies or utilities can fund multi-billion dollar investments on their balance sheets. Some companies have a simple rule of thumb that says that they will seek to fund their share of a project on a limited-recourse basis in case it amounts to more than 5% of the value of their balance sheet. This figure is easily reached for most transboundary oil and gas pipelines or power grid interconnections.

In the case of multi-sponsor projects, project finance is also a way to allow the smaller sponsors and/or the least creditworthy ones to mobilise the funds necessary to pay for their share of project costs without explicitly relying on the support of their more creditworthy partners. A company with a $100 million net worth and no credit rating would find it very difficult to pay for, say, a $150 million share of a large oil and gas infrastructure project unless a significant portion of it is funded by debt raised at the project level.

In this case, the larger and more creditworthy sponsors may also prefer to undertake a project financing than to have to carry their weaker partners (eg, by funding their partners' share of project costs in exchange for a disproportionate share of future project revenues). Similarly, companies with a minority investment in a large project may prefer to raise debt at the project level in order to avoid increasing debt on their balance sheets.

As a result, project finance tends to be the technique of choice for large energy infrastructure projects being developed by groups of multiple sponsors with varying creditworthiness and/or financial strengths. This is particularly the case for projects in emerging markets where the risk mitigation aspects of project finance add another level of benefits.

2.2 Risk mitigation and allocation

Project finance can potentially remove risks from a proposed investment from the point of view of the project equity investors or sponsors through the combination of three elements:

- an independent review of contractual arrangements and market studies by lenders and their advisers;
- the benefits of a detailed risk allocation embedded in the project commercial contracts; and
- the deterrent effect that some lenders provide.

We review each of these factors below.

(a) Lender contractual and market review

This is largely driven by the fact that lenders have little or no other methods of recouping their loans than through project cash flows. Lenders will seek to understand and quantify most project risks: they typically ask an independent engineering firm to review and opine on the construction and maintenance contracts; they are likely to insist on contingent funding plans in case the risk of cost overrun is material; they will similarly review the permits and seek independent legal opinion(s) to confirm that the project is in compliance with domestic and international laws; and they will validate market studies and confirm that prospective revenues are more than sufficient to meet debt service obligations. Last but not least, they will commission a stringent environmental and social impact assessment to ensure that the project breaches no international environmental performance standard.

Such detailed lender due diligence should ensure that most commercial, technical and legal flaws of an energy infrastructure project, to the extent not already addressed by the project sponsors through various support mechanisms, are identified and adequately remedied. The process is akin to having a second opinion of a project's feasibility and expected economic benefits, which can often result in an improved contractual structure and risk allocation.

(b) Risk allocation

Risk allocation is at the core of project finance. Each of the project stakeholders is expected to bear the risks that it is best able to manage, with no or limited recourse to others in case some of the risks materialise. Risks are often divided between the pre-completion and operating periods, on the one hand, and between commercial and political risks, on the other.

Pre-completion commercial and technical risks are largely borne by contractors, with a key element being the existence of an engineering, procurement and construction contract. Such a contract means that there is one point of responsibility for all construction work among the various suppliers and contractors. The engineering, procurement and construction company is committed to delivering a project meeting pre-agreed performance tests on schedule and, in case of a lump sum turnkey contract, under an agreed budget. The contractor is 'on the hook' for any delays or performance shortfall, through the existence of an engineering, procurement and construction wrap (in which the contractor provides a 'wrap around' guarantee covering the performance of all project subcontractors) and/or liquidated damages. Lenders will typically be the indirect beneficiaries of such liquidated damages, although they will rarely rely solely on them as a risk mitigation tool (primarily because the size of liquidated damages in most instances will be lower than the amount of loans). In addition, the ability to shift risk significantly to the engineering, procurement and construction contractor may vary dramatically with market conditions.

An alternative to the existence of an engineering, procurement and construction contract often is for the sponsors to manage project construction and to provide completion guarantees to the project lenders. In this situation, the sponsors are typically responsible for interface risks between the various project contractors and suppliers. They provide lenders with the option of paying back their loans in case the project is not completed by a negotiated backstop completion date. The point of these various pre-completion risk allocation options is that there is clarity on who bears such risks – which in and by itself is a strong risk mitigation factor. A project funded through a corporate financing may be equally successful and its construction could indeed be managed in the same way as with a project financing; however, the sponsor(s) will not necessarily face the same scrutiny as the one imposed by project lenders.

The same logic applies to the risks of the operating period. Under a corporate loan, lenders will not necessarily perform a detailed due diligence of the performances of all the cash-generating assets of a company; however, with a project financing they have no choice but to anticipate all likely developments and to ensure that their loans will be repaid in most reasonable downside cases since they are looking solely (or almost solely) to the project cash flows. This can lead to the debt package being resized, through partial and/or accelerated prepayments or cash sweep mechanisms, in case a project fails to meet certain completion tests and/or encounters certain material events.

Political risks are an important aspect of risk allocation for transboundary energy infrastructure projects. As explained elsewhere in this book, host government and intergovernmental agreements are very effective tools to mitigate such risks. These agreements create a legal framework that binds together the various countries in which the project is located. The involvement of development finance institutions, through a project finance scheme, often adds significant weight to this framework.

The table on the right summarises the typical risk allocation for a transboundary energy infrastructure project.

(c)   *Development finance institutions*
Development finance institutions can have a deterrent effect on host governments. These institutions include multilateral organisations such as the World Bank and its

	Pre-completion risks	Operating risks
Commercial risks (technology risk, cost overrun, delay etc)	• borne by project sponsors and engineering, procurement and construction contractor(s) • lenders largely protected through completion guarantees and/or lump sum turnkey contract with pre-committed cost overrun facility	• shared between project sponsors and lenders
Political risks (expropriation, restrictions on exports/imports, currency transfer and convertibility, breach of contract, political violence etc)	Typically allocated throughout project life between: • project sponsors (exposure often limited to their equity, with political risk carve-outs of completion guarantees) • development finance institutions (through a combination of direct loans and export credit guarantees or political risk insurance made available to commercial lenders) • host governments through their obligations under the host government and intergovernmental agreements	

sister institution the International Finance Corporation, regional organisations such as the European Bank for Reconstruction and Development, the European Investment Bank and the Asian Development Bank, as well as bilateral organisations such as export credit agencies and development agencies. They each maintain political relationships with host governments, either directly or through their country's embassies. Some of them require host governments to acknowledge their involvement in a project financing, often giving them additional legal protection should a host government not fulfil its obligations under the terms of a host government or intergovernmental agreement. Importantly, they monitor project implementation and often pre-empt issues from becoming serious impediments to the successful development and operation of a project. Development finance institutions can exercise political leverage by threatening to reduce their involvement in a country – which can be very effective in emerging markets with limited funding alternatives and reputational sensitivity.

These institutions can often fund corporations of a host country or government institutions, but are unlikely to lend directly to a large international company on balance sheet. Most of them will, however, directly finance infrastructure projects in which international companies are involved to the extent that such projects meet their development criteria. These financings will typically be done on a project finance basis.

3. **Project participants' roles and sources of project finance debt**

The participants in a project financing can be divided between equity investors, lenders, host governments, contractors and offtakers or buyers of the product(s) sold by the project. The table below summarises their main roles.

Equity investors
• Identify commercial opportunity.
• Define project scope.
• Commission and pay for feasibility studies.
• Commission environmental and social impact assessment.
• Develop indicative finance plan.
• Develop and implement contracting strategy.
• Select lenders and negotiate debt package.
• Pay for lenders fees, consultants and legal expenses.
• Invest equity.
• Monitor construction activities.
• Often operate project.
• (Hopefully) generate a return on investments.
Lenders
• Review and test economic analysis.
• Confirm project debt capacity.
• Review and validate feasibility studies.
• Review and validate environmental and social impact assessment.
• Negotiate term sheet and loan agreements.
• Seek credit committee approvals of agreed finance documentation.
• Ensure conditions precedent to drawdown are met.
• Disburse loans.
• Monitor project construction and operations.
• Receive interest payments.
• Obtain loan repayments.
Host governments
• Approve investment proposal made by sponsors.
• May provide financial and/or fiscal benefits.
• Could be an equity investor.
• Agree environmental and social requirements.
• May provide direct or indirect guarantees to sponsors and/or lenders against specific risks.

Contractors
- Agree contracting strategy proposed by sponsors.
- Negotiate construction contracts.
- Construct project facilities.
- Typically pay liquidated damages in case of cost overrun or delay.
- Receive payment during construction period.
- May be asked to make a small equity investment in project.

Offtakers or buyers
- Buy output of project.
- Could help mitigate market risk, assuming they are
- investment grade entities.
- Could also be minority investors in project.

Consultants
- Review technical, construction, process, legal, environmental and other pre-construction, construction and operating aspects of a project on behalf of lenders.
- Provide ultimate third-party advice for lenders.
- Can provide enhanced risk mitigation by challenging project assumptions, planning, process and implementation.

Project finance loans are provided by a combination of commercial banks, development finance institutions including export credit agencies, and multilateral and bilateral development banks, as well as by capital markets investors in specific circumstances.

However, capital markets are seldom tapped to fund greenfield projects. This is mainly because bond investors lack the flexibility that is often required by project sponsors to amend debt covenants during the construction phase of a project. There is also typically a 'negative carry' cost of raising the full amount of a bond at financial close versus a regular loan drawdown reflecting construction cost schedules. Lastly, few greenfield energy infrastructure projects would meet the initial investment grade rating that is required by the largest pools of bond investors, thus limiting the potential size of such bond financings and significantly increasing their costs.

The table on the next page summarises the key characteristics of the primary sources of project finance debt: commercial banks, export credit agencies and multilateral development banks.

4. **Key criteria for success**

As alluded to above, a pure project financing is not an appropriate vehicle or financing methodology either for every project or for every sponsor or set of sponsors. It is a costly (usually 2% to 4% of project capital costs, corresponding to arranging, advisory and legal fees, but excluding interest) and time-consuming undertaking. A decision to take the project finance route should be done cautiously and only after

	Commercial banks	**Export credit agencies**	**Multilateral agencies**
Purpose	• Provide loans to sound investment projects and/or companies. • Maximise remuneration through upfront fees and loans margins. • Minimise risk exposure. • Develop commercial relationships with sponsors.	• Promote exports from their country. • Minimise risk exposure.	• Finance the development of countries and/or specific economic sectors. • Promote sound economic principles. • Help mobilise commercial bank funding. • Promote best environmental practices.
Products	• Corporate or project senior loans.	• Insurance or guarantees provided to commercial lenders (European export credit agencies). • Direct loans (other export credit agencies).	• Corporate or project loans. • Corporate or project equity investments. • Guarantees.
Constraints	• Maturity. • Size of unsecured loans.	• Cover typically limited to 85% of export contract plus local costs. • Sourcing constraints. • Average life of loans.	• Direct loans typically limited to $100 million to $150 million per institution per project.

careful analysis of an array of considerations. Among the primary goals is effectively to transfer risk from the sponsors to the lenders, but achieving this goal comes with associated complexity and thus can affect overall the cost and schedule.

While in terms of timing and cost project finance does not compare favourably to corporate finance, there may be other valid reasons generally set out above that make project financing the preferred approach. Below are some indicative durations of the project financing process for recent transboundary energy infrastructure projects:

Project	Capital cost	Debt raised	Time from initial lender approach to first drawdown
Baku-Tbilisi-Ceyhan	$4.05 billion	$2.6 billion	31 months
Nord Stream (phase 1)	€5.5 billion	€3.9 billion	17 months
Blue Stream (initial financing of offshore section)	$3 billion	$2 billion	26 months

In order to support project financing successfully, a project should meet a few key criteria and/or display a number of the following characteristics:
- require significant capital investment and/or be capable of attracting substantial leverage – at a high level any project financing follows roughly the same formula in terms of requiring advisers, independent consultants (significantly, paid for by the sponsors), site visits, ancillary commercial analysis (eg, potentially upstream reserve reports) and the like; therefore, a potential project must have a critical mass to support these kinds of transaction cost;
- have multiple sponsors – project finance is often a way to reconcile the different constraints (eg, capital requirements, creditworthiness), risk appetite and requirements of a number of sponsors; it is rarely put in place by a single sponsor as such an arrangement would do little to shift risks to lenders;
- have a common approach to the financing – the sponsor group should develop a common set of financial objectives, including risk allocation, and agree on a target finance plan. Without this a diverse sponsor group can find itself disadvantaged in dealing with a perhaps larger, equally diverse lender group that can make varying demands across the sponsor group, increasing the complexity of the financing and driving up short-term and long-term costs;
- include a political/regulatory risk element if development finance institutions are to be involved. As described above, the development finance institutions bring with them a halo effect of political risk mitigation, but commensurate with that positive is the potentially onerous level of scrutiny and due

diligence which exceeds that of the commercial banks (although the difference between the two is closing as non-governmental organisations and similar activist organisations increase pressure on commercial banks); therefore, utilising development finance institutions in the debt mix may make sense only where there is a need to mitigate political risks and/or development finance institutions provide more advantageous finance terms; and

- have insufficient equity – a common issue with a multi-sponsor group is a mix of credit and attendant balance sheet issues. In many instances it may well be that some of the sponsors cannot fund the project with equity alone and the proposed project may be the sole or primary asset of that sponsor.

5. Project financing process

The financing process can be divided into a financial structuring phase and a finance implementation phase, as shown in the diagram below.

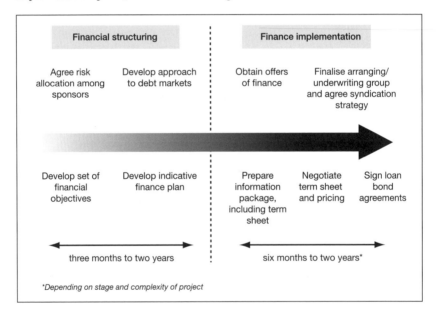

Financial structuring activities are inward looking: the focus is on the sponsors developing a set of common financial objectives and a target risk allocation, as well as determining an indicative finance plan. These activities are often facilitated by a financial adviser acting as a broker between the different investors.

The finance implementation phase involves detailed discussions with lenders, from the selection of arranging banks to the negotiation of debt term sheets and pricing. During this phase lenders also provide feedback on the various draft project commercial agreements.

Each phase can take up to two years, depending on the complexity of the project and the degree of advancement of its commercial aspects.

6. **Managing the process**

The project financing process can be quite daunting and takes on a life of its own for large projects. From a project perspective, the effort should be viewed and approached as a primary and fundamental aspect of the overall successful development of the project. It should be staffed accordingly. Yet it is often treated as an after-thought or ancillary activity. Within the project development, the primary internal areas that will need to be coordinated in order to work in parallel are technical aspects/construction, commercial elements, finance and legal considerations.

Externally, stakeholders requiring hands-on management include lenders, consultants, contractors, affected/host governments and other stakeholders, including non-governmental organisations.

The diagram below gives some indication of the complexity in managing external stakeholders.

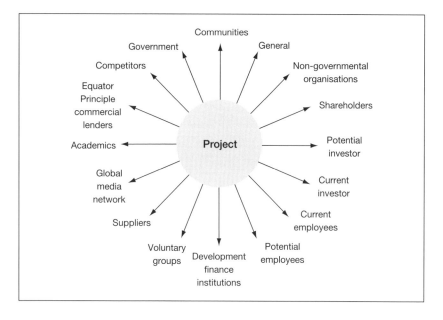

Orchestrating a coordinated effort among all the various internal and external stakeholders is a key activity, and one that should be recognised at the outset and given consistent and proper attention. There is no doubt that an over-zealous and unconstrained set of lenders can adversely affect the core project. Conversely, if the technical team does not embrace general lender requirements and involvement, then the lending side (and thus, ultimately, the source of money to move the project forward) could slow to an unacceptable level. Both are most certainly unintended results, but they nevertheless represent risks that can be avoided and/or managed.

Internally, the financing process and the project's commercial and technical development must work hand in hand. A project's technical team will have a set of cost and schedule goals from which it will be loath to deviate. Therefore, the project's

Project finance and risk mitigation

management must be able to articulate a digestible internal message of cooperation with lenders and their representatives. Similarly, the project finance team and the technical team must work together to frame technical sideboards and a common negotiating position for use during the finance negotiation. This common view becomes particularly critical during site visit/monitoring negotiations. Site visits by the lenders' independent consultants are a (recurring) primary tool in the lenders' compliance arsenal and must be acknowledged as such. However, compliance site visits are disruptive, take personnel away from critical tasks and can be expensive (remember that the sponsor pays for everything) and so need to be agreed to occur on a schedule that is appropriate and relevant. To achieve that goal requires a high degree of internal coordination. A modicum of planning and coordination will pay dividends as the project construction finishes and the project gears up for financial completion when the lenders' recourse will be primarily, if not solely, to the project and not the sponsors.

The roles of the financial, commercial and legal teams must also be carefully coordinated. Commonly, the commercial envelope for a project will have been agreed by the time lenders get involved; however, in the effort to secure the project further, lenders will often layer on additional burdens that could adversely affect the agreed commercial envelope. The same sensitivity can be said for lender-mandated legal enhancements. Therefore, the project's finance, commercial and legal teams must work in lock-step to avoid a significant deterioration in project commerciality and achievability. If suggested lender requirements approach unrealistic levels, it is imperative that the project teams – finance, commercial, technical and legal – be able to put forward a reasoned, countervailing view to support the project. Only with such a coordinated approach will the most effective finance package and contractual structure, both for the lenders and for the project and its sponsors, emerge. The

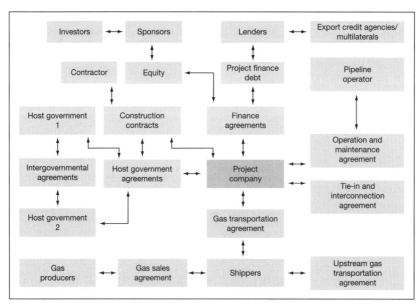

diagram on the previous page summarises the multiple parties and types of agreement that need to be developed by the project commercial, finance and legal teams, and approved by the project lenders (in this instance, a gas pipeline).

Given the limited world capacity for significant project financings, the inter-related nature of banking today and the ability of information to proliferate via the Internet, it is difficult to switch horses during the negotiation of most project financings, particularly the largest ones. There simply are not enough experienced banks engaged in the sector, especially following the financial crisis of 2007-2008 during which a number of commercial banks exited the sector altogether. And bad news travels fast. Therefore, there comes a point in any finance negotiation on large financings when negotiating leverage shifts and the project becomes much more subject to the demands of a potential lender. It may be that, as discussed above, some or all of the sponsors are unable to equity finance or the project does not want to reveal to the market that some banks were unwilling to lend to the project, or any combination of the above. In any event, there is a tipping point when, for a variety of reasons, the requirement to keep the banks in the deal obscures the ability of the project perhaps to negotiate as favourably as it might otherwise wish. Much like the internal tensions outlined above, this risk is mitigated to a large degree by having a well-informed, aligned, multi-disciplinary negotiating team.

Taking a longer look at practical management of relevant external stakeholders (lenders, their consultants, contractors, affected/host governments and non-governmental organisations), we see much the same picture. Each constituent group requires a devoted manager inside the project. Some are quite obvious: the finance team will liaise with and manage the lenders; the technical team will handle the contractors and, to a certain degree, most of the consultants. A trickier area is the overall issue of non-governmental organisations, wherein a stakeholder identification and follow-on management plan is mandatory. The influence of non-governmental organisations is commonplace and well known today, but the lender group will nevertheless continue to look to the project for a successful, timely approach to identifying, analysing and responding thoughtfully to these organisations' concerns. From the project's perspective, this should be a dedicated and consistent messaging and management effort, not a piecemeal, reactive part-time activity. Lastly, assigned, consistent liaison with host government counterparties is critical to success. No matter how robust the underlying investment regime or agreement may appear at present, every infrastructure project exists at the behest of the state. Accordingly, a well-managed project will be one that has developed close ties to the relevant host government stakeholders.

Educating the entire project team as to the nuances of the project financing process from lender identification, to financing negotiation, to the drawdown process and, ultimately, to financial completion and loan maintenance is key to smoothing out the upheavals that will naturally occur in any project. It appears to be common sense, but in reality in any large-scale project people have multiple activities, many of which are primarily geared towards the successful start-up of the project – the challenge, thus, is to work as an integrated team.

7. Case studies

It is instructive to examine the implementation of actual project financings across a range of questions. To that end, a series of questions was put to:
- John Wingate, finance manager of the Baku-Tbilisi-Ceyhan oil pipeline;
- Paul Corcoran, finance director of Nord Stream; and
- Ruslan Vazetdinov from Russian Project Finance Bank, who was financial adviser to OAO Gazprom on the Blue Stream financing.

By way of background, below is a very brief description of each of the referenced projects:
- Baku-Tbilisi-Ceyhan – this is a 1,768-kilometre oil pipeline running through Azerbaijan, Georgia and Turkey, completed in 2006. The pipeline starts at the Sangachal terminal near Baku, Azerbaijan and transports oil from the Azeri, Chirag and Gunashli offshore oilfields in the Caspian; the terminus is at a newly built marine terminal in Ceyhan, Turkey. The pipeline company is owned by international oil companies, including the State Oil Company of the Azerbaijan Republic. Construction costs were approximately $3.7 billion, of which $2.6 billion in debt was provided by a combination of development finance institutions, export credit agencies and shareholder loans. Financial close took place in early 2004.
- Nord Stream – this is a 1,224-kilometre gas pipeline linking Russia and the European Union via the Baltic Sea. It will consist of two parallel lines that will transport 55 billion cubic metres of gas per year when completed in 2011 and 2012. The pipeline starts in Vyborg, Russia and terminates at Griefswald, Germany. The pipeline company is owned by Gazprom, BASF SE/Wintershall Holding GmbH, E.ON Ruhrgas AG, NV Nederlandse Gasunie and GDF SUEZ SA. Construction costs for the pipeline are expected to be €7.4 billion; in the Phase I financing €3.1 billion in debt was secured from export credit agencies and a further €800 million from commercial banks. Phase 1 financial close took place in May 2010.
- Blue Stream – this is a 760-kilometre gas pipeline project (370 kilometres onshore and 390 kilometres offshore) from Russia to Turkey under the Black Sea, with a capacity of 16 billion cubic metres of gas per year and a maximum depth of 2,150 metres. The pipeline originates at Izobilnoye, Russia and terminates at Samsun, Turkey. The onshore pipeline is owned by Gazprom and the offshore pipeline company is owned equally by Gazprom and ENI SpA. Construction costs for the offshore portion were approximately $3 billion, of which $2 billion in debt was provided by a combination of development finance institutions, export credit agencies and private banks; the onshore portion raised debt of $573 million. Financial close for the initial financing took place in 2000, with a refinancing in 2005.

The interview questions were as follows:
- Why was project finance used to fund the project?
- What were the finance objectives of the project sponsors?

- What pitfalls did you try to avoid when you started the financing process? How successful were you in doing so?
- What were the key project financing challenges that you faced?
- Did the geographic location of the project (multi-jurisdiction, offshore, onshore) present any specific obstacles/benefits?
- What are the key lessons learned from the financing of this project?
- What would you implement differently?
- What additional challenges/opportunities would the project sponsors of this project need to consider if they were undertaking the same project today?
- Was your project affected by the financial crisis?

Below are their answers to each question.

Why was project finance used to fund the project?

Wingate (Baku-Tbilisi-Ceyhan): With a very diverse sponsor group, the financing was based on attracting the most debt, especially given that we were working with a significant, high-profile project in a jurisdiction in which banks had not previously lent on any serious basis. We were also focused on mitigating perceived political risk.

Corcoran (Nord Stream): I believe the shareholders' objectives were to minimise their equity investment and to improve their equity return.

Vazetdinov (Blue Stream): Blue Stream is a hybrid financing, with elements of a project finance structure, such as a project company with ring-fenced assets and tariff payments secured by dedicated project revenues, and aspects of a corporate financing such as full sponsor guarantees throughout the life of the loans and non project-related gas export contracts used as security. Gazprom credit (as well as Russian sovereign) was not acceptable to international lenders in post-1998 crisis time; therefore, a joint venture structure with support of a strong international partner (ENI) was the only possible solution.

What were the finance objectives of the project sponsors?

Wingate: Much as in the question above as a basis, but also our sponsor group was particularly focused on avoiding any inter-sponsor credit support issues; hence we wound up with a very complex borrowing structure. In addition, insulating the upstream (which consisted of sponsors' affiliates) from the pipeline project financing was a goal.

Corcoran: The objectives were to mobilise long-term finance, to utilise blended shareholder credit rating in order to have a stable long-term transportation tariff, and to achieve a solid equity return.

Vazetdinov: The availability of external financing was a precondition for the project to fly since Gazprom simply did not have its own funds to finance the project in an extremely low energy price context. Obviously, the sponsors were keen to maximise the amount of debt and to minimise finance costs to benefit from the financial leverage. Commercial sources were not available for Russian borrowers, so one of the objectives was to swap export credit agencies supported financing for the offshore section of the pipeline for untied funds provided by ENI to finance onshore section, which was the sole responsibility of Gazprom.

What pitfalls did you try to avoid when you started the financing process? How successful were you in doing so?
Wingate: We were concerned that the deal, which was integrated across three countries and was the product of several years of hard-fought, coordinated negotiation, did not deteriorate with lender influence. In addition, we wished to maintain the envisioned borrowing structure as it achieved many of the sponsors' goals, albeit in a perhaps complicated way. I think we were very successful and the sponsors were happy with the result. We were concerned with the size and manageability of the lender group – meetings in full flower might reach 70 or 80 in number. We were partially successful in dealing with lender-nominated spokespersons, but at some point everyone must be at the table.
Corcoran: We wanted to avoid too much interference from lenders into project contracts – that is, avoid contracts that conflict with financing objectives and would be renegotiated by lenders. We tried to anticipate objections from credit committees to details of the deal, so as to avoid reasons for rejection at credit committee level. We were very successful.
Vazetdinov: Since the underlying offtake contract with Botas placed firm time obligations on the supplier, there was a risk that we might have to make unnecessary concessions to the lenders given the time pressure. But we managed to ensure that the financing did not affect the commercial arrangements, including the existing gas export contracts, and that the financing was concluded in time and under acceptable terms.

What were the key project financing challenges that you faced?
Wingate: I think our three primary challenges were the novelty at that time of the countries in which we were working, the credit mix that spanned the entire spectrum, and the fact that (believe it or not) we were in a low oil price environment, which made the project look like a high (perhaps unacceptably high) priced alternative! Times change – that in itself is an important lesson.
Corcoran: We were going to market in the middle of the 2009 financial crisis, in a context where reduced capital allocation and shrinking country limits (for Russia), as well as low liquidity, made financing difficult. It required a high sweet-to-sour ratio between the export credit agencies-covered and uncovered debt components, and a clean deal (documentation, risk allocation, etc) for success.
Vazetdinov: The key challenge was to raise significant debt in the international market at competitive prices in an unfavourable post-crisis market. Also, we managed to put together a proper project financing structure that allowed the refinancing in 2005 of the export credit agencies' facilities by unsecured bonds covered partly by SACE (the Italian export credit agency) and partly by Gazprom, but without non-project-related gas export contracts.

Did the geographic location of the project (multi-jurisdiction, offshore, onshore) present any specific obstacles/benefits?
Wingate: Yes, as mentioned above, because we were in a new area it was a challenge to get the lenders to look realistically at the risks. We had a few government upsets

in the interim – the Rose Revolution two months before financial close, to name one – which, we thought, made the project look riskier than it is. Time has proved us right. Perception can be a powerful obstacle. While not at all geographic, the fact that the project is a long, large-diameter oil pipeline posed its own set of risks or perceived risks that had to be managed. We had a significant stakeholder management team to look into and respond to questions from non-governmental organisations.

Corcoran: Not especially; however, the multi-country permit process, as well as the international waters jurisdiction, needed some work with the banks' legal advisers.

Vazetdinov: The project was supported by an intergovernmental agreement between Russia and Turkey that established a preferred stable tax regime. Actually, the multi-source export credit agencies financing package covered both the offshore section (joint responsibility with ENI) and the Russian onshore section (sole responsibility of Gazprom). As a result, the requirement for Gazprom's own funding was minimised.

What are the key lessons learned from the financing of this project?

Wingate: To me, the main ones are: have a robust project, plan for contingencies and be prepared to explain your financing objectives clearly and convincingly. While some things are out of your control, others can be managed, explained and successfully negotiated through preparation, and a dedicated, integrated team/sponsor group is key.

Corcoran: The key lesson is that if the components of the deal are right, a deal can be done no matter how bad the market is – which in itself is a very encouraging message.

Vazetdinov: The financing was structured twice. In August 1998 – two weeks before Russia declared default – we signed a detailed term sheet with the arrangers that envisaged Gazprom's sole responsibility and a big commercial bank tranche. After that, we spent about two years agreeing the joint venture terms with ENI and structuring the financial package without commercial lenders. The main lesson was that the right form of credit enhancement should always allow access to debt markets, even in hard times.

What would you implement differently?

Wingate: It might not have been possible given what I have outlined above, but reducing the size of both the sponsor group and the lender group would have streamlined our efforts. But as I say, that is probably just wishful hindsight – in reality, everyone on both sides of our deal was there for a very good reason. No group was expendable.

Corcoran: We could have tried to simplify the insurance arrangements, which led to unnecessary complications. Otherwise, I don't think we would have implemented this financing very differently in hindsight.

Vazetdinov: I am not sure we would have done anything differently, given the constraints we faced at the time and the terms we achieved, and the ability we had to refinance should conditions change significantly.

What additional challenges/opportunities would the project sponsors of this project need to consider if they were undertaking the same project today?
Wingate: Azerbaijan and Georgia, and to a lesser degree Turkey since many financing have occurred there, are different places now than in the late 1990s when we started seriously investigating financing opportunities. Oil is headed back up, the Baku-Tbilisi-Ceyhan project has demonstrated the doability of a large-scale transnational oil pipeline financing – so I would say a financing could be done faster and with fewer lenders. The countervailing view, I suppose, is that oil pipelines continue to attract a heavy dose of attention from non-governmental organisations.
Corcoran: We might consider the opportunity to involve the European Investment Bank in the Nord Stream finance plan, given that bank's renewed appetite for energy infrastructure assets.
Vazetdinov: Things would probably be much easier today as Gazprom has been recognised as a creditworthy standalone entity, as shown in the successful Nord Stream financing. The new accounting regulation in the European Union does not support the tailored transportation tariff profile that was originally used in Blue Stream. Now the tariff should be flat and stable to avoid trapped cash.

Was your project affected by the recent financial crisis?
Wingate: No, not really. We reached financial close in February 2004 and financial completion mid-2007. So basically we were unscathed. However, as with everyone, the crisis has limited the scope of any possible refinancing opportunities.
Corcoran: Very much so. In February 2009 a market sounding of 30 banks revealed only one bank able to commit. The deal was improved (better sweet-to-sour ratio) and deal terms ended being much more favourable to lenders than the market standard a year previously. The export credit agencies recognised their deal-maker status and were much tougher in negotiations than was usual previously.
Vazetdinov: As I understand you mean the 2007-2009 crisis. No, the project was not affected, given that its revenues and debt service are relatively stable and fixed.

While the finance managers and financial advisers each come at the questions a little differently and are obviously informed by the complexities and characteristics of their own projects, the themes remain the same and are informed by the basic structure and drivers of any project financing as outlined in the initial part of the chapter, whether a small or large project, cross-border or in a single jurisdiction. In addition, especially true for exceptionally large projects, they touch consistently on managing the process, both internally by aligning sponsors goals and with respect to coming up with effective ways to deal with some very large, disparate lender groups.

8. Conclusion

While project financing may not suit every circumstance, it has proved an effective tool in situations where a project/sponsor(s) wished to access capital, whether for reasons that arise out of an already weaker credit picture of some or all sponsors, or a desire to ensure that a project does not negatively affect existing credit of one or more sponsors. Given a clear view of the intended result, project finance's other attribute is as a mechanism for effective risk allocation and mitigation.

Project financing has many components that are working in parallel, sometimes seemingly at odds with the underlying technical project. Therefore, sponsors must give due attention to harnessing and managing the various activities: a successful project financing must be proactively managed by a dedicated team. Project financings (and therefore the associated individual lenders) result in long-term relationships, and project personnel and the lender group should strive to work in a partnership, based on a full understanding of each other's vital role in the success of the overall project. Setting and maintaining the correct tone and thus productive relationship is a critical feature of the management of the process from the lender negotiation period through to loan repayment.

If a project meets the criteria and a sponsor group is inclined and properly managed, a project financing can be the key to a successful financing result and overall project implementation.

Technical and operational risk

Deborah L Grubbe
Operations and Safety Solutions LLC

This chapter examines the nature of technical and operational risks by considering the economic value that may be placed on these risks in a due diligence process prior to a merger and acquisition (M&A). Ironically, these risks are referred to as non-financial risks. They appear here in three broad categories: process (technical), plant (facility) and people (workforce, leadership and culture). Sample questions provided for each category should help M&A teams assess pragmatically this important dimension of risk.

In most business enterprises, the nature of the discussion relating to non-financial risk has changed over the past 30 years. With the advent of instant, global communications and markets, the obligation of business leaders to learn from the errors of others has grown exponentially. Market capitalisation can be quickly and severely affected by operational issues, as in BP's precipitous drop in value over the Deepwater Horizon incident in May/June 2010. Whole industries can also become affected – for example, the decline of the nuclear industry in the United States after the events at Three Mile Island (1979) and Chernobyl (1986). Companies can be wiped out with one serious operational error. Union Carbide, once one of the top five US chemical firms, no longer exists due to an extremely serious incident in Bhopal, India, in December 1984, in which thousands were killed and many more injured.

This chapter approaches non-financial risk by considering what a prime M&A team would look for when buying or selling a world-class asset set. Unfortunately, today's economic conditions mean that only a small percentage of the global asset base is first rate, which makes the whole M&A process much more challenging for the purchaser. Such purchaser is rarely willing to overpay for an asset – even though, in this author's opinion, it usually does.

Before getting into the details, let us take a moment to investigate a broader standard of measure – that is, a concept of what a world-class asset consists of. Simply put, a 'world-class asset' can be codified through its vision for itself, its physical plant, its market penetration, and its internal and external work processes, including, but not limited to, business, marketing, research and development, manufacturing, engineering, financial, supply chain, information technology, and legal and sustainable outreach (environment, health and safety included). First-rate standards usually show up in the form of benchmarked data, which helps steer the M&A team toward a more appropriate valuation. In some circumstances, the M&A team may not know what it does not know, which can easily result in a less than optimal assessment and perhaps – even worse – a broken deal. In addition, it is quite possible that the seller's valuation may even be missing some of this data. A discussion of whether to include non-financial risks and how to judge initial negotiating positions is beyond the scope of this chapter.

Typical industry benchmarks focus on narrow areas such as individual policies, maintenance, staffing, effective use of capital, safety and environmental performance, reliability, flexibility and asset life. In a successful M&A process, it is important to look beyond these benchmarks and to delve deeper into the details. Failure to do so could result in being burdened with an asset that is both ill suited to purpose and a value detractor.

This chapter explores the technical, facilities, workforce and cultural aspects of non-financial risk assessment.

1. Valuing the technical asset base

Standard M&A processes are already well developed to evaluate a firm's patent estate and a research and development/new product pipeline. These areas will not be addressed in detail here; however, it is critical in any M&A activity to give intellectual property due consideration. The balance of this chapter addresses the technical organisation, or soft assets, and the technical aspects of the manufacturing equipment, or hard asset base.

Contrary to popular opinion, an engineering or scientific brain is not a commodity; it is another form of intellectual property, and one that can be fairly mobile. In fact, it can leave and go to work for a competitor relatively easily. Experience in this area indicates that one gets what one pays for. If you need top technical talent to stay productive throughout your M&A process and if this same talent is critical to the future success of the new firm, it is crucial to plan for that in advance. Some firms have decided that their value play is not in the technology area; that is certainly to be expected in some markets. However, possible questions to ask during M&A activities include the following:

- What are the key technologies in the firm being acquired?
- How commonplace are these technologies when compared with competition and market?
- How competent are the technical personnel? Do they hold any notable certifications, licences and technical recognitions? How many are there?
- How much training have the individuals received? How much was through formal education? How much on the job?
- Does the firm have a process of formal technical progression? How well has it been executed? What benefits have been realised?
- What kinds of technical expertise are being purchased? To what degree?
- How many technical members of staff are there? What is the level of embedded contractors versus full-time employees?
- How much is spent on purchasing technical services externally? How does this spending compare to internal technical costs?
- Are there specialised technical resources leveraged across the various business and manufacturing units or does each unit have its own? Or is it a blend of

the two models?
- Is the intellectual property managed and developed by these resources?
- Is this intellectual property codified into engineering, design and operating standards? Are such standards available to every employee?
- Is the codified intellectual property up to date and on a regular and timely review/refresh cycle?
- Will these standards and other documents be included in the acquisition?
- What percentage of the overall corporate intellectual property is captured explicitly? Does intellectual property include application or does it cover only scientific knowledge?
- How tightly is the intellectual property managed and protected? Is anything proprietary?
- Is the technology in use to manufacture products inherently safe? If not, what are the processes in place to assure that the right levels of risk are being considered at the appropriate management levels, up to and including the board of directors?
- How easy is it for the technical people to raise their concerns directly with the senior executives?
- What are the respective safety and environmental performance records? How are the technical people supporting the environment, health and safety organisation and the operating organisation?
- How active is the technical organisation outside of the firm? How do they interact with customers? With contractors in the capital project development teams?

2. **Valuing the facility**

In addition to technical resources, the facility itself, or hard asset, must be properly evaluated. Sometimes, along with the financial and litigation records, this is all a potential buyer may see. The buyer may not be given additional opportunities to tour the plant or to interview employees until after the deal is consummated. How then does one gain the insights needed for an appropriate valuation?

If the buyer is granted one facility tour, some tricks of the trade can be useful in getting a sense of how disciplined the organisation is. The purchaser will desire a high level of discipline if the firm under consideration either handles hazardous chemicals or performs technically demanding work. The quickest way to assess the operating discipline of any organisation is to evaluate how neat its facilities are, and to sense the kinds of chatter and demeanour of the people on its manufacturing floor. Simply put, if the workplace is clean and tidy, and the employees are focused and working as a team, there is a higher likelihood that the operational discipline is high and that the factory is managed well. A workplace can be neat even if the asset is old.

Additionally, the M&A team may choose to hire an operations professional to assist it in making this evaluation. In addition to housekeeping and morale, what would this expert look for? Possible questions to ask in the various areas to investigate appear below.

General:
- What is the location and age of this facility? Is the ground owned or leased?
- Where is the nearest large metropolitan area? What are the road and transport system conditions?
- Is there a characterisation of the ground underneath the facility on record?
- At what depth is the nearest drinking water aquifer?
- Are there any regulatory issues or outstanding requirements in government records?

Environment, health and safety:
- If hazardous materials are involved in the products' manufacture, how are these being handled and stored?
- Has there ever been a loss of containment? If so, how often?
- What has been the local authorities' reaction to such incidents?
- What is the general community tolerance of the facility and its management? How does the facility management formally or informally interact with the community?
- Are all safety incidents reported/logged? How are they investigated?
- Are the key learnings from these incidents shared among the workforce? How and by whom?
- Are all environmental incidents reported? What is the performance level of this facility for liquid waste, solid waste and gaseous emissions? (One should verify the environmental permit compliance, if possible.)
- Does the facility have the ability to track excursions outside the normally safe operating envelopes?
- How many fires have broken out in the facility? Why are they occurring?
- What is the level of theft from this facility?
- How are the raw materials stored? What kind of access is provided around the hazardous materials and finished product?
- How often does the facility practise emergency response to a disaster?
- Who on site keeps the required process safety information and drawings up to date? Can a copy of the information be reviewed?
- Are all drawings up to date and have been found generally to represent the process equipment, piping, instrumentation and electrical equipment found in the facility?

Production and quality:
- What is the source of raw materials for this site? How rugged is the supply chain? Are there alternative sources of supply? Have there ever been any supply interruptions that lasted more than two days?
- How rugged is the manufacturing process? How drastically does raw material variability affect finished product quality?
- Is there an automated process control system? Who is authorised to make changes to the safety control systems?
- Is there an active quality programme? Is the site certified under the

International Organisation for Standardisation? Which certifications does the site hold?

Maintenance:
- How is maintenance performed on this facility? Who employs the maintenance workforce and what level of training has that team had over the past five years?
- Is the maintenance history of every piece of major equipment involved in the purchase documented?
- What is the maintenance philosophy? Do the maintenance workforce and supervisors know the difference between preventive and predictive maintenance?
- Is there a computerised maintenance/work order system that is integrated with other site systems? What percentage of the maintenance work is planned?
- Is there an active management of change process?

Co-occupancy and power:
- If the purchase involves only one unit on a multi-unit site, who will provide the utilities? How will demand changes be addressed and what is the cost/price philosophy?
- If services are shared, how will capital improvements need to be addressed?
- Will the purchased unit be a supplier for a manufacturing unit under other ownership?
- Does the site have its own power generation or does it run on purchased power? What is the level of momentary outages per year that occurred each year over the past five years? Is there an uninterruptible power supply on site? How often has it been used? How often is it tested? (One should ask to see the records.)
- Can the site be isolated from the grid?

3. **Valuing the workforce, leadership and culture**

In most if not all businesses, human capital is a key contributor to profitability. How that human capital works and what kinds of requirement need to be placed on it vary from industry to industry and from business to business. For example, some manufacturing needs to be done by people with specific knowledge, while other manufacturing can be wholly contracted out. Both could yield similar results, yet hold different levels of risk. The same dynamic applies to maintenance and engineering. However, experience shows that one must be mindful when addressing areas that require specialised technical knowledge.

Since the Columbia Shuttle Accident Investigation Board issued its report in 2003, a lot has been said in the media about safety culture. Safety culture is a microcosm of the larger organisation's culture and, as such, it plays into the M&A discussion. When bringing together two distinct cultures, one must pay attention to the key differences between them. It is important always to remember the often stated truism that culture eats strategy for breakfast. If the finances look fantastic, the owners are willing to sell and the cultures are vastly different, then prudence suggests a very deliberate and detailed approach to address the cultural differences directly.

The M&A team must also take into account the desires and needs of the new owners, as their desired corporate culture and level of profitability can affect staffing decisions. The amount of change and the time required to make a successful transition from the established culture of an existing asset to the newer environment should not be underestimated. Below are some questions to consider in an evaluation of the workforce, leadership and culture of the project to be purchased.

Questions relating to the workforce:
- From where does the majority of the personnel at this location come? Are they locals, travellers, nomads? What is the average length of service?
- Where does the supervision come from? Does it come up from the ranks? Is it long-tenured or younger?
- What are the hiring requirements for the hourly employees? Are these requirements rigorously followed? Is a test required? Is literacy verified? If so, to what level?
- Are many people on the site related to each other?
- How have more senior managers obtained their positions? Are they from the area?
- How mobile is this workforce?
- Is the economy of the local area strong or distressed?
- Is this manufacturing site one of many in the area? Is it the largest? Is it high profile?
- What is the workforce turnover?
- How long do people normally stay in position?

Questions relating to leadership and culture:
- Does the firm being purchased have stated values? Are these values known to everyone? Are these values lived out in the daily workplace? Is it palpable and consistent?
- Are the leaders sending a consistent tone from the top?
- Do the leaders periodically measure their culture? In what way?
- Is there a general level of awareness of, or satisfaction with, the culture and what constitutes a leader in this culture?
- How is the existing culture different from the culture of the acquiring firm? Are there gross dissimilarities?
- What is the general heat map for communications? Up/down? Across? Down/up?
- How does the manufacturing facility management communicate with its employees? How often? What topics are covered?
- How is the manufacturing unit staffed? What is the average age of the workforce? Is this a 24-hour-a-day/seven-day-a-week operation? What is the average level of overtime pay per employee per year?

- What is the proportion of contractors in the daily operation?
- Is the workforce represented by a third party? Is the union governed by an international agreement? What is the bargaining process? How frequently does it take place? Have there ever been any labour actions?
- How often do the union and management meet? What is the tone of these meetings?
- What are the subjects of typical grievances? How quickly are they resolved?
- Is this an organisation where rank has privileges? Are there special parking spaces for the bosses for instance?
- What is the leadership turnover? Why did people leave? Are any patterns noted?

While these questions are not exhaustive, the answers will be helpful to a typical M&A team. The information will complement the team's investigation into the level of risk that it is addressing. The data can support a strong valuation and negotiation.

Assuming that negotiations are successful, this non-financial risk assessment data comes into play again when the new owners take control and begin to develop strategies for the future. The data can serve as a foundation for the planning. With a deeper knowledge of the inherent risks and a broader based understanding of the condition of the wider asset, the transition team can lay better informed plans around the staffing, direction and technical skills required, among other things.

In the end, it pays to begin any M&A process with the ultimate goal in mind. No one likes surprises and many non-financial risks can have a seriously negative effect on valuation. Due diligence across a broad array of operating parameters such as technical personnel, facility, leadership and culture can make a big difference to whether a correct valuation is obtained, and whether the M&A is successful.

Holistic project management and risk

William E Browning
Yashar Latifov
Infrastructure Development Partnership LLP

A great deal has been written on project management philosophy, engineering concepts and execution strategies, and many manuals, procedures and systems have been produced in connection with the project management of energy infrastructure development projects. This chapter neither attempts to replicate this work nor does it try to re-state what are widely recognised as the core concepts of classical or contemporary project management. Instead, the chapter offers the author's perspective on the successful management of risk by organising, planning, managing and running the technical and construction aspects of projects. While the management of different types of risk and uncertainty are comprehensively addressed in other chapters of this book, no understanding of the intricacies of project development is complete without an examination of the basics of project management. This includes exploring the role of project management, first, in informing and understanding the various subject matters addressed elsewhere in this book, and, second, in underpinning a successful overall business strategy.

1. **Risk: the holistic project management perspective**

   Traditional functions of project management such as engineering (concept design, front-end engineering design or detailed design), planning, contracting, procurement, construction, project controls, health and safety, environment and security, and so forth form the bedrock of successful energy infrastructure projects. However, one must not overlook the fact that the projects themselves are critical to the success of the business, in terms of both their long-term net present value and associated reputation factors. In other words, failure is not an option. Energy infrastructure projects, including their development and execution, inevitably form a key part of the business value chain by connecting important energy supply sources to large consumer markets; they can span countries and, at times, continents. Consequently, successfully managing the execution of a large energy infrastructure project is an imperative; processes for such management must rightly become embedded in a company's or developer's corporate culture.

   The challenges that face contemporary projects increasingly involve global competition, shareholder expectations, complex markets, regulatory regimes and practices, and many other factors. Accordingly, when examining the issue of project management in major infrastructure projects, it is critical that business leadership and management also be examined. A conventional definition of 'project management' must be complemented by commensurate business leadership skills and qualities. Only through this combination of a deeply embedded, accepted, understood and rigorously applied project management system – coupled with strong, determined and knowledgeable leadership and governance – will successful technical project delivery be assured. Project leaders, essentially deemed to be (and regarded as) project chief executive officers (CEOs), should look at any project in the context of this sort of comprehensive, interdependent environment and in the context of a hierarchy of risks (and opportunities) in which projects originate, evolve, mature and operate; project leaders should then address project challenges methodically, continually and holistically.

A definition of 'risk' within the complete environment of project management may be set forth as follows: "the level of *exposure* to *uncertainties* that the enterprise must understand and effectively *manage* as it executes its strategies to achieve its business objectives and *create/protect value*".[1]

Key issues or factors will influence project performance, including the classical balance of time versus cost versus quality. Accordingly, all relevant uncertainties and the extent of their exposure must be properly understood and effectively managed. This is the role of the successful 21st century project CEO and his or her project

Graphic 1[2]

Project risks	Industry/ competitive risks	Institutional/ regulatory risks	Country risks	World price risks	World system risks
Construction	Industry evolution	Regulatory stability or intervention	Political stability	Commodity prices	Political conditions/ war
Operations	Demand/ growth rates	Contract enforcement	Financial, economic stability/ inflation	Exchange rates	Trade regimes
Partner/ally	Supply conditions	Legal stability	Expropriation/ taxation	Interest rates	Global demand
Contract negotiation	Costs		Repatriation policies	Risk premium	
Project management	Distribution		Terrorism/ civil unrest		
	Prices				
	Infrastructure				

---

1 Deloach, James, *Enterprise-Wide Risk Management* (2000).
2 Professor Lessard, Don, Massachusetts Institute of Technology (2004).

management system. Once created, the value should then be protected, as well as enhanced on an ongoing basis.

Graphic 1 above illustrates the broad range of risks that any project CEO must bear in mind when applying the company's customised project management system to the proposed development at hand. Any project management system of any long-term and intrinsic value will take into account the sets of relevant risks that stem from traditional project risks – that is, the risks beyond those that were once thought to be the sole responsibility of a project CEO. A project management system and a delivery team must examine risks and mitigants in a holistic manner. For better or worse, today's project CEO has a very broad remit to identify, understand and mitigate relevant risks, and his or her project management system and personnel must stand up to that challenge.

Graphic 2[3]

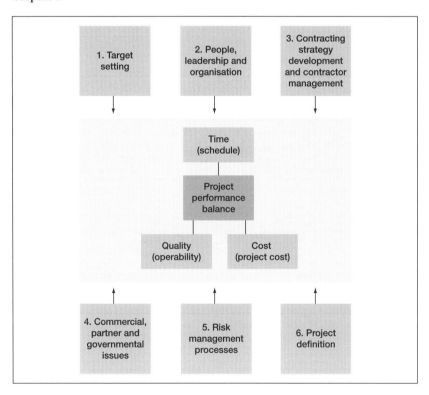

2. Organisationally managing risk

Graphic 2 above gives an idea of the influences on the time-cost-quality equation. This chapter does not look especially deeply into the drivers, influences and

---

3 Professor McRae, Gregory, *Dealing with Uncertainties in Project Management*, Massachusetts Institute of Technology (2004). This graphic has been modified from the original.

relationships that inform a project and affect its deliverability. However, Graphic 2 illustrates that regulatory, industry, corporate, commercial, personnel, as well as traditional risks and activities such as accurate project definition, schedule and target setting must not only be identified but actively managed. In a manner of speaking, Graphic 2 shows, from a tasks and activities standpoint, how and by whom the risks from Graphic 1 can be dealt with – that is, it essentially describes the great breadth of management activities. To a degree the graphic also illustrates the notion cited above that a project management system should be deeply ingrained in an organisation, so that all risks and the commensurate activities intended to manage those risks are commonly accepted and agreed. In this way, the company as a whole should pull together toward an agreed goal in support of the project CEO and the delivery team.

As indicated above, it is not the purpose of the chapter to analyse each risk and the factors and/or activities pertinent to it. The risks and uncertainties are numerous and specific to each infrastructure project, its geographical location and the particular factors driving the project. It is important, however, to analyse in more detail the structure or, to be more precise, the architecture of a typical project management system.

3. **Structure and goals of a project management system**

Typical project management systems are called capital value processes, stage gate or spiral processes, or define/appraise/select processes. These typical systems feature many common elements and are designed to get to the same place, largely through the same sequential process. The overarching goal of a project management system, and perhaps its single most important feature, is to enable and facilitate, in a systemic and systematic manner, the identification and management of the components of a problem that will contribute most to uncertainties in outcomes. An appropriate architecture for a project management system will assure a company that it is doing the right project and doing the project right. While this sounds like a platitude, it is ultimately and fundamentally correct. It is also extremely tricky to get right.

In simple terms, the essence of a project management system can perhaps be summarised by two words: holistic planning. A process based on the continuous and methodical examination and challenge of situations, and then the determination of appropriate responses will address planning risks and uncertainties at different levels, from the straightforward to the chaotic, including the following aspects:

- Coordination of all tasks and functions, such as planned tasks, task management and, most importantly, interface management between tasks and functions;
- Variations, such as uncertainties in task durations and in external requirements and expectations, and the consequential requirements to allow for schedule and budget buffers;
- Known unknowns, such as which risks can be managed by planned iterations among tasks and coordination among functions. This leads to determining the requirements for larger contingency planning;

- Unknown unknowns, which lead to unplanned iterations and, as a consequence, to the requirement for dynamic planning; and
- Chaos planning,[4] where plans evolve daily as the result of, among other things, typical emerging market conditions, external interface management changes, project dynamics in the construction process and a construction site that is not static (eg, a pipeline construction site can span hundreds or thousands of kilometres and thus have continuously changing circumstances). This requires accommodation of change in the project goals and plans. Accordingly, options must be kept open.

Each major infrastructure project, in its own time and place, represents an evolving process. These projects are not static subjects; they have their own unique, specific features and drivers, risks and uncertainties, and threats and opportunities. The challenge (or at least one of them) is to ensure that a project – and a project management system – has a robust design that can adapt to the near constant changes endemic to any large project. Although we have discussed at some length the soft risks that lie outside traditional project risks, the fundamental design of a project management system must ensure that the project's robustness will be neither overlooked nor underestimated.

It is, therefore, vital to examine and to understand the design of, and the thinking behind, an appropriate system. There are many possible process designs to choose from. The considerations expressed in the previous paragraph apply not only to a major project management process in its entirety, but essentially to every stage (or phase) of the project management process. Each stage works by building on the previous stage, so it is imperative that the analysis be built on a firm foundation. Such staged/phased value and risk analyses are employed by more or less all major industrial corporations.

Graphic 3 overleaf illustrates in a skeletal fashion the basic architecture of the staged (or phased) process used in various forms by many companies.

In the context of a staged or phased project management approach as shown it is essential to understand, consider and continually manage how risks and uncertainties affect project performance, and to appreciate and understand the value created at every stage of the project development, design and execution.

In general, certain key considerations/opportunities arise during the sequential process. These include the following:
- Synchronising project design and business case evolution;
- Maintaining optionality well into the execution phase;
- Managing external stakeholder expectations;
- Managing supplier/service provider performance risk; and
- Selecting the optimal time to freeze the variables that most affect the ultimate value of the overall project.

---

4   Professor Lessard, Don, Massachusetts Institute of Technology (2004).

Graphic 3

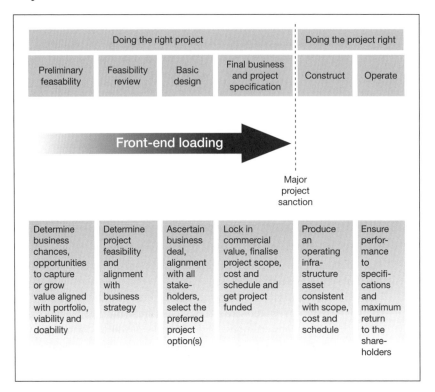

4. **Period of maximum influence**

It is important (and somewhat axiomatic) to draw the reader's attention to the reality that while the project management system is designed to identify and manage risks across the natural sequential development of a project, the project CEO's influence and that of his or her team decrease as the project is defined and expenditures are incurred. Graphic 4 below illustrates this point.

The primary message of the above representation is that the delivery team's influence is materially more significant at the early stages of the project management process, while the expenditures are low. However, the team's influence reduces equally significantly with time, as the definition and expenditure levels increase. It is crucial, therefore, to ensure that the initial phases (ie, prior to the sanction of the construction phase or final investment decision) have been robustly undertaken and challenged. Bluntly put: the only way at a later stage to increase influence is to make unplanned, sometimes catastrophic expenditures. This is also a red flag for determining that the project management system has not been successful.

While it is a commonly acknowledged rule, the influence-expenditure relationship in the context of this book's thorough discussions of risk remains an interesting observation: the more uncertainties there are, the bigger the influence is and, therefore, the bigger the value a delivery team can bring to the project and the company's

Graphic 4

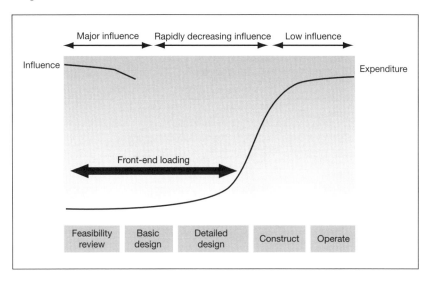

business. In other words, it makes sense to use a project management system to look at project risks and uncertainties not just as hazards to be avoided, mitigated or compensated, but also as properly managed opportunities to enhance project value.

5. **Key questions in the project management process**

    In terms of practical analysis, certain key questions require robust answers during the project management process. The answers to these questions will form the basis of a healthy iterative process that will guide change when the project team has the greatest influence (ie, at the outset and through to the point when the decision to proceed with the project has been taken). Any number of factors can be looked at, but the following illustrate the basic broad range of challenges that the experienced project manager should continually review:

    - What types of uncertainties are relevant at each stage of the project management process and what risks and/or opportunities do those uncertainties create for the project's success or failure?
    - What types of (planned and unplanned) iteration will there be in the process of project development and project design?
    - How is each stage's content, both technical and non-technical, integrated into the complete system?
    - How and when are the project stages tested, and what are each test's pass criteria?
    - At what points in the project management process can progress, and the ultimate likelihood of success, be assessed?

    These questions should be addressed in a iterative planning review within or across stages or phases, as illustrated in Graphic 5 below.

Holistic project management and risk

Graphic 5[5]

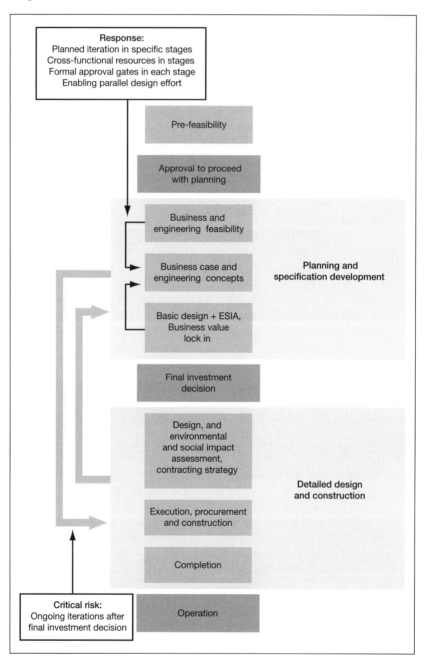

---

5 Professor Eppinger, Steven, *Risk-Based Product Development Planning*, Massachusetts Institute of Technology (2003). This graphic has been modified from the original.

6. **Successful project management system: an iterative process**

    The project management process must be controlled both within each phase and across phases; it must also be effected by cross-functional holistic reviews – that is, reviews that cover all types of risk and uncertainty relevant for each specific phase, as discussed above and extensively throughout this book.

    End-phase reviews using some of the questions set out in section 5 above should be structured so as to avoid backtracking. Once a subsequent phase is sanctioned, backtracking becomes the single most critical cost risk element for the project. In the engineering context, for example, critical problems (those risks most costly to mitigate) occur in each instance of re-iteration after engineering specifications have been approved and major commitments made for detailed design and construction – that is, final investment decision (these process risks are represented by red loops in Graphic 5). As indicated above, throwing money at a situation that was effectively missed in the stage gate project management system process basically represents a failure of that process.

    The more appropriate process will therefore look as represented by green loops in Graphic 5. The risks are identified, prioritised and assigned in each specific phase. The necessary iteration cycles must be held within each stage to address the assigned risks, and phases must be structured in a way that very much discourages backtracking to the maximum extent possible. Scheduled reviews held at the completion of each phase and before sanctioning a subsequent phase is a key control component of a successful process.

7. **Conclusion**

    An efficient and successful project risk management process is a sequence of comprehensive risk identification, analysis, response planning, monitoring and control, which is then repeated continuously as the overall business, together with relevant project development and execution, evolves and pushes to delivery. Every project management system has a leader, but the process must be ingrained in, and championed by, the entire organisation. As aptly put by Dwight Eisenhower, "in preparation for battle, I always found that plans are useless, but planning is indispensable".

    Acknowledgements: Massachusetts Institute of Technology – BP Projects Academy, 2004.

A strategic approach to environmental and social impact assessments

David Blatchford
Martin Lednor
Infrastructure Development Partnership LLP

1. Introduction

Environmental and social impact assessment (ESIA)[1] reports are traditionally the main vehicle for predicting the environmental and social impact of development projects, mitigating their effects and managing their consequences. However, recent observations of a number of large energy infrastructure projects, all requiring project finance, suggest that the ability of ESIAs to fulfil these needs effectively is questionable, and that their cost has not been commensurate with value. ESIAs have become expensive procedural hurdles with the objective of regulatory compliance, rather than practical tools to identify and manage risk.

Stakeholders today are far better informed and connected. The combination of modern communications and an increasingly active network of non-governmental organisations has significantly increased the reputation risk for projects, as well as the expectations that their sponsors will not only behave appropriately, but also have clear strategies and systems in place to be able to demonstrate good performance. With the increased external focus on projects, it is critical not to outsource risk to construction contractors or neglect the importance of temporary or associated facilities as both aspects have the potential to increase the non-technical risk of projects.

These observations lead us to the view that since their conception, environmental and social assessment processes in general and ESIAs in particular have not kept pace with the now accepted need to manage environmental and social performance throughout a project's life cycle. Moreover, we contend that many ESIAs are excessively long documents that lack focus and relevance, as well as methodological rigour.

This situation presents a range of potential risks for project sponsors and results in inferior environmental and social outcomes. Such risks are manifested in a variety of forms, all of which ultimately have the potential to require redesign or retrofitting, cost overruns and schedule delays – not to mention project sponsor/construction contractor tensions, arbitration and dispute resolution.

In order to address this situation, we advocate a considered approach to

---

[1] 'ESIA' is the generic term used here to refer to environmental impact assessment, environment social and health impact assessment, environmental impact statement and environment effect statement.

environmental and social risk assessment, impact prediction and mitigation management based on a more strategic, timely and holistic view of environmental and social impact assessment. With this approach, ESIAs occupy an *intermediate* position in a continuum that commences with project conceptualisation and proceeds through project design, construction, commissioning, operations and decommissioning. This results in a process that is more closely aligned with a project's life cycle – a situation that is entirely appropriate for issues that are now (or should be) central to project definition and execution, as opposed to peripheral window-dressing to achieve regulatory compliance.

This chapter describes how a strategic approach to environmental and social assessment can result in better environmental and social outcomes, reduce non-technical risk and, in so doing, deliver better value to all stakeholders. We do this by first outlining the concept of the approach (section 2), followed by a description of a modified approach to preparing ESIAs (section 3). We then explain how this change of emphasis can lead logically and seamlessly to the next step in the project's life cycle – a stage that involves the most risks and greatest potential for impact (ie, the construction phase) (section 4). We conclude by demonstrating the considerable potential for achieving better environmental and social outcomes at reduced cost and within a shorter timeframe.

2. Strategic approach

A strategic approach implies timely recognition, integration and management of a range of environmental and social issues that collectively represent non-technical risks to projects. Traditionally, many of these risks emerge only during the ESIA phase of a project, and often with costly scheduling consequences (eg, land access and compensation, route and site selection of plant and infrastructure, waste management).

Earlier consideration of such non-technical risks is likely to be a natural outcome of a strategic approach, but this is not to say that all activities simply need to be dealt with earlier in a project schedule. In reality, a conventional project's life cycle minimises upfront expenditure until the project is fully sanctioned.

In addition, elements of project design are often not at a suitable level of detail at early project stages to allow specific impact analysis and management. So the key is to be able, on the one hand, to identify the environmental and social issues and risks associated with projects, and on the other, – generally through experience – to determine:

- the information required to characterise the issues to an appropriate level at each phase of the project's life cycle. The scope of this phase typically starts out broad and gradually narrows. The level of detail required tends to be superficial at the outset and gets more detailed as the project evolves, with each issue requiring individual consideration. The objective is to collect systematically and build on the right information through each phase of the project and in the process screen out extraneous information. Equally, it is imperative that potential fatal flaws are identified early and, where contentious issues are identified, the minimum data sets required to decide

whether the project should go ahead;
- the time, resources and methods required to acquire and analyse the information (desktop research, and census and surveys – recognising seasonality or cyclical issues, statistically representative data sets, etc), and the bearing that this has on internal project milestones (eg, financial investment decision, ESIA preparation and regulatory approvals, project financing, preparation of construction contracting bid documents);
- the opportunity for a phased, incremental approach to data collection, with effort scaling up as various design elements and site options become clearer, and different levels of internal project approvals are obtained. Here, the overriding objectives are building data sets commensurate with the specific questions that need answering (which pre-supposes that the right questions are being asked), while anticipating future requirements and avoiding data replication and redundancy. In this sense, more is not always better; and
- project interdependencies. These can be important at multiple levels and in varying situations – for example, where:
  - multiple regulatory ESIA approvals are required, as is the case with transnational projects such as pipelines;
  - land access agreements influence construction start-up dates;
  - ESIA commitments influence contracting bid documents; and
  - environmental and social baseline data collection surveys influence impact prediction verification, mitigation monitoring and contractor performance assessment, among other things.

The other key benefit of a strategic approach is the potential to improve considerably the effectiveness of the ESIA step in the overall process.

ESIAs are written primarily to serve regulatory needs; their scope, timing and execution often present significant shortcomings, particularly for complex, large-scale projects. Rather than being focused on impact prediction and providing information on mechanisms to prevent or minimise impact (with impact and mitigation being clearly linked), there is a danger that they become excessively long documents that are difficult to digest, are not well focused and require considerable rework to be useful in subsequent phases of the project. This is particularly true for the construction phase, as described in section 4 below.

This observation can partly be explained by the tendency to scope ESIAs to meet regulatory needs, while failing to recognise the considerable benefits of addressing the post-approval aspects of the environmental and social assessment process. A strategic approach prescribes that ESIAs are but one step in a series. The ESIA process should be viewed by sponsors as an iterative and living risk management tool, as opposed to a tick in the box, whether for regulatory approval or financing.

3. Redefining ESIA

3.1 Design basics

The basis of a good ESIA is a number of discrete foundation stones and interdependent components. When these issues and topics are well defined early in the ESIA process, unnecessary expenditure of time, resources and budget is avoided.

Whereas the foundation stones are static entities, there are also a number of dynamic components to ESIAs that must be recognised and managed. After all, one of the original purposes of ESIAs was to allow stakeholders such as communities and governments to make informed decisions on proposed developments; this can only be achieved through consultation, review, reformulation, monitoring and evaluation – all dynamic, time-dependent processes.

These (dynamic) components, referred to here as 'interdependent components' to reflect their interrelated characteristics, evolve during the impact assessment process and are typically integrated with other project disciplines. The interdependent components, in combination with the foundation stones, result in an ESIA that is well planned and positioned within a strategic environmental and social framework, and therefore a useful tool serving multiple uses over and above meeting regulatory requirements (see Figure 1 below).

Both the foundation stones and the interdependent components need to be set within an integrated schedule to ensure that the aims and objectives of the ESIA, and more importantly the mechanisms to achieve them, are integrated with the overall project schedule.

Figure 1 depicts the complexities associated with the evolution of the project through various phases. In particular, it highlights the special importance of defining

Figure 1: The strategic ESIA concept

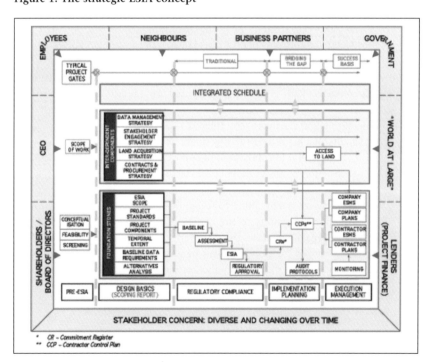

the 'design basics'. The project is framed within eight categories of stakeholder who collectively have different views on a project, all of which have the potential to change over time. One indication of a successful project would be when it is accepted by the majority of all categories of stakeholder.

Large-scale infrastructure projects invariably attract a lot of scrutiny from stakeholders. The ability to manage non-technical risks to an acceptable level hinges on identifying the foundation stones, and particularly the interdependent components, early in the project's life cycle and with the clear end game in mind. This allows non-technical risks to be minimised; a position on fundamentals associated with the overall impact assessment and management process can then be articulated to stakeholders.

3.2 Foundation stones

Six foundation stones have been identified:
- ESIA scope – the scope of the ESIA and particularly the disciplines and subjects to be covered to ensure that project risk is managed;
- Project standards – the standards to be adopted for the disciplines defined in the ESIA scope;
- Project components – the primary components, as well as associated facilities that collectively define the spatial extent of the project;
- Temporal extent – the phases of the project;
- Baseline data – data requirements needed to answer specific questions. This avoids the trap of collecting everything that can be collected just because it exists; and
- Alternatives analysis – a process of rationalisation and optimisation, particularly for linear projects where options exist for route alignments and facility locations.

The following discussion addresses each of these in turn.

(a) **Scope of ESIAs**

Over the past 40 odd years, the scope of impact assessments has grown from environmental impact assessments (in its most literal sense) to the position today where there is no accepted nomenclature and accepted scope of the assessment. The International Finance Corporation's Performance Standards provide a benchmark, but many companies believe these to be excessive requirements important only for projects seeking to attract financing. Consequently, impact assessments are variously described as ESIAs, environmental impact assessments (EIAs), environment social and health impact assessments (ESHIAs), environmental impact statements and environment effect statements.

Many sponsors struggle to articulate clearly what they mean by 'social' (the 's' in ESIA). Why is health mentioned in an environment social and health impact assessment, but safety ignored? Do these assessments equally cover the workforce, as well as the local population? Land acquisition has the significant potential to affect livelihoods and is often one of the biggest challenges for a project, however, it is lost in the 's' of the impact assessment nomenclature and treated separately under the resettlement action plan (RAP) process. Confusingly, even the RAP process can apply when people are not resettled.[2] Whatever the nomenclature used, the ESIA scope must be clearly defined at the outset and provide a clear plan for ESIA data gathering and baseline assessment.

(b)   Project standards

One of the complexities of large-scale infrastructure projects is the plethora of potentially applicable standards to consider. This is especially true if social standards are included.

Standards applicable to large-scale projects are generally dispersed among a range of technical, legal and policy documents. Usually, these form a pyramid with the intergovernmental agreements (for transboundary projects) at the apex:

- intergovernmental agreements;
- host government agreements/production sharing agreements;
- specific national regulatory requirements, including laws that enable international conventions ratified by the host country;
- specific requirements (eg, references to the European Union or the US Environmental Protection Agency);
- corporate requirements;
- financing requirements;[3] and
- best practice/guideline.

The requirements of each need to be identified and a decision made as to what constitutes "applicable project standards". This is often an onerous exercise as like-for-like comparisons are often difficult between different sets of requirements. Emissions and discharge numerical standards are easier to line up and directly compare. Execution requirements, or the way the project will be managed to minimise impact, are far harder to clarify – hence the concept of 'contractor control plans' (CCPs), discussed in section 4 below. Ultimately, however, standards need to be collated and a decision made as to the most stringent standard that the project is either willing or required to apply. Such standards then need to be communicated to stakeholders.

Different company approaches have been observed: those driven only by regulatory requirements to a prescribed format (often the bare minimum, on the assumption that 'minimum' equates to 'least cost'), or those by the desire to try and minimise impact identified in the impact assessment process (and therefore minimise non-technical risk). Increasingly however, corporate requirements are driving project sponsors to compliment national regulatory requirements with corporate requirements.

---

2   There are eight International Finance Corporation Performance Standards on social and environmental sustainability. The resettlement action plan process defined in Performance Standard 5 applies to households that are economically displaced, as well as requiring physical relocation.

3   Even if a project is not seeking external financing, the Performance Standards are increasingly becoming the *de facto* international benchmark for environmental and social performance; many oil and gas companies are aligning their own corporate standards to the Performance Standards.

(c)   **Project components**
Traditionally, impact assessments have concentrated on the key project components – that is, what is the basic project that is being constructed and operated? This is generally described in the project description section of the ESIA.

Many projects evaluate and predict impact arising from the construction or operation of the main or core project components. Impact from temporary facilities such as accommodation camps, quarries and borrow pits, pipe lay-down sites and temporary or permanent use of third-party waste facilities are often not incorporated into the assessment, even though they can have more impact than the main project components themselves. Such facilities are often left to the contractor to manage, frequently on the false assumption of reducing costs or managing risk. This is a common tactic of companies having contracting philosophies based on the premise of making the contractor responsible for as many issues as possible: "Leave it to the contractor – it can deal with it!" is a typical refrain. Consequently, the contractor becomes responsible for issues that have the potential to cause both the greatest impact and, hence, the most damage to the reputation of the sponsor.

Another outcome is that the contractor creates a legacy for the sponsor to deal with well after the contractor has demobilised, or after the point where the cost to the contractor to fix the problem exceeds completion payments. This favours the 'walk-away option' and invokes the legacy.

The International Finance Corporation Performance Standards[4] recognise the need to consider 'associated facilities' and provide some guidance as to what constitutes such facilities. Associated facilities add a dimension to the definition of the project that is often both underestimated and difficult to address. Fundamentally, projects must recognise the implications of shared or dependent third-party facilities early in the project definition stage and classify all operational sites – both from an impact assessment perspective and an execution (operational control) perspective. In reality this is not an easy task; criteria need to be developed to characterise sites with respect to the need for impact assessment or contractual stipulations prescribing management requirements over existing operations. This step not only defines the spatial extent of the ESIA *per se*, it also has an important bearing on the extent of the applicability of the management plans that flow from the ESIA. This can raise a range of complex issues such as site stewardship, occupational health and safety, incident recording and reporting, third-party liability, security and insurance.

(d)   **Temporal extent**
This is not new to projects where the recognition of various project phases is well understood. Generally, construction activities for large-scale infrastructure projects occur over a relatively short duration and create more significant impact than that associated with the operational life of the project. ESIAs need to consider clearly both

---

4   Performance Standard 1 entitled "Social and Environmental Assessment and Management Systems" addresses the requirement to consider risks and impacts in the context of the project's area of influence, recognising that significant impacts may arise from non-core associated facilities.

construction and operational impacts. The intermediate step of commissioning can create its own impact that needs attention. Finally, due to the uncertain nature of these activities at the time of project development, decommissioning or abandonment is generally addressed as a series of commitments to national regulations/best practice guidelines.

(e) *Project baseline*

Given that one of the main objectives of ESIAs is to assess impact, it follows that the quality of the information used to make these assessments (the project baseline) will have a fundamental bearing on accuracy of these assessments. Herein lies one of the major flaws of many ESIAs: the description of the project baseline is a voluminous assemblage of easily acquired material (eg, cut-and-paste material from earlier ESIAs) or misdirected survey work that is often of little or marginal relevance to the prediction or management of impact.

This situation arises when little thought goes into determining what data are relevant to the proposed activities, and when the inextricable link between baseline condition (including inherent natural variabilities) and impact assessment is not established (see text box below). The result can be redundant data, incurred at an unnecessary cost and over an extended period of time, often without any significant understanding of project impact or the potential for its mitigation. It is also important to scrutinise data derived from secondary sources, as often data are lifted from sources with no consideration of quality or completeness.

The key to developing the project baseline foundation stone lies in preparing a carefully planned and clearly articulated project baseline study design. This can either be a stand-alone document or part of a broader scoping report (see section 3.3). It is essential that it serves as a basis for the characterisation of potential impact and its management, and not merely provide a context for the project (see text box). Furthermore, the project baseline should provide a repository of information capable of being used to answer multifaceted questions. Most notably it should allow:

- prediction – the impact of activity A is likely to be B; and
- confirmation – the impact of X was Y, as outlined in the text box below.

In simple terms, the project baseline study design should:
- identify key issues;
- delimit the likely extent of the study area;
- develop specific questions that need to be answered with respect to each issue;
- determine the information required to answer each question; and
- outline the method(s) to be used to obtain this information, and the timeframe.

Baseline description – a pre-condition to impact assessment

In the context of ESIA, the term 'impact' is taken to mean a change in a variable or component of interest caused by a perturbation or intervention related explicitly to some aspect of the proposed development. In a strict technical sense, impacts must be calculated as differences – that is, prior to the initiation of a development activity, potential impact is calculated as the difference between the forecast condition of a variable of interest with development, and the forecast condition without. During and after development the actual impact is ideally calculated as the difference between the measured condition of the variable with development (including mitigation), and the expected condition without. This encompasses the important concept of the likely future condition of a receptor, as well as the existing condition.

Impact assessment is essentially synonymous with impact prediction, and specifically prediction of changes from a baseline condition, as demonstrated by the results of post disturbance monitoring. The key here is not so much making predictions *per se*, but making accurate predictions, recognising that this is likely to be easier to accomplish with physical systems (eg, transfer and fate models) than biological, ecological or social ones. And a pre-condition to making accurate predictions is a well-defined and understood baseline.

Making accurate predictions also implies that they can be tested during the post-disturbance monitoring phase and this aspect should influence baseline data collection, as well as the form of the prediction. In other words, it is the process of refining a generalised question into a form that can be falsified which is important.*

*Refer to Duinker, PN (1989), "Ecological Effects Monitoring in Environmental Impact Assessment: What Can it Accomplish?", *Environmental Management* Vol 13, No 6, pp 797-805.

(f) *Alternatives analysis*

Linear projects, such as pipelines, can offer much greater opportunity for considering project alternatives than single location projects. Constraints analysis is a specific form of alternatives analysis that is particularly important for linear projects. For example, pipeline route and facility site selection is the process of selecting an optimised route. The term 'optimised' relates to identifying particular issues (often termed 'constraints') in the area of interest. As these are increasingly understood, a specific alignment (route) or location can be selected depending on the nature of the constraint. Linear developments often use the 'corridor narrowing' concept where baseline information is collected in increasingly narrow corridors as the final alignment becomes better defined.

The use of constraints mapping, from an initial desktop exercise to detailed field data gathering, is a continuum and should be aided by the development of a project geographical information system (GIS) initiated in the pre-ESIA phase of the project (see section 3.3).

3.3   **Interdependent components**

Five interdependent components have been identified:
- a comprehensive stakeholder engagement strategy;
- a land acquisition strategy;

- a contracting and procurement strategy to ensure that the terms and conditions within tender documentation inform prospective construction contractors of project requirements;
- a clear data management strategy based on the development of a GIS; and
- an integrated schedule of the ESIA work within the overall project schedule.

These components, which all complement the foundation stones, are further discussed below.

(a) ***Stakeholder engagement strategy***

Figure 1 is framed by the range of stakeholders who may have an interest in the project; although often overlapping, interests will vary from group to group and may change over the life of the project.

A good stakeholder engagement strategy is a key tool facilitating free, prior and informed consultation aimed at building and maintaining positive stakeholder relationships. The stakeholder engagement strategy should broadly identify:
- the overall objectives and corporate/project goals;
- all stakeholders who may have an interest in the project – individuals, organisations and communities;
- the areas on which the engagement activity will be focused;
- the nature of the information to be disclosed, and the process and formats suitable for each stakeholder group;
- the communication tools to be employed;
- the schedule;
- the resources required and responsibilities;
- the information feedback mechanisms; and
- the monitoring and reporting requirements.

True engagement involves two-way communication and information sharing, consultation and participation – not just telling stakeholders about project plans and activities. Engagement is also an ongoing process throughout the project's life cycle and is partially directed by specific milestones such as the disclosure of the ESIA.

For more on this subject, please refer to the chapter of this book devoted to the subject of investor risks.

(b) ***Land acquisition strategy***

Acquiring rights to land can be a complex process, especially in regimes where the regulatory process requires strengthening. Many large-scale infrastructure projects fall into the category of 'in the state interest' and, therefore, through the government, have the ability to invoke laws on expropriation (or involuntary resettlement).

The purpose of RAPs, addressing impact on landowners and users, is to minimise project risk by ensuring that a robust and transparent process is followed. The need to understand the regulatory process and gaps (or even conflicts) with other requirements such as performance standards is crucial and requires a significant amount of upfront work. The extractive industry has traditionally found that land acquisition is a problematic issue that often becomes critical, with potentially significant reputational issues. Many of the non-technical risks (and consequential delays) for large-scale projects, in particular linear projects such as pipelines, arise from land acquisition issues, particularly when there has been a lack of clear strategy and understanding of the issue early in the project's life cycle.

For more on this subject, please refer to the chapter of this book devoted to land issues.

(c) *Contracting and procurement strategy*

The contractual terms and conditions for construction contractors are critical to a project's success. The issue is the timing of the implementation of the procurement and contracting strategy with respect to the development of the ESIA. There is often a need to provide contractual requirements regarding environmental and social management before the ESIA is completed. In an ideal world, the ESIA will be completed before the construction contractor procurement and contracting phase begins. In reality, this is not feasible and so the two processes are dovetailed to ensure that sufficient detail is provided in tender documentation to guide the contractor as to a project's environmental and social requirements. The best strategy is to lay the ground rules in the tender documentation and refine them as impact assessment activities evolve.

Where responsibility for particular issues sits is another important decision that needs to be reflected in bid documentation for construction contractors. Major engineering contractors are not necessarily specialists in areas such as waste management, biorestoration, community investment or cultural heritage. Sponsors should evaluate carefully whether it is better to contract directly with specialist companies rather than via a major contractor for specific services. In turn, this is linked to the overall contracting mechanism with, on the one hand, lump-sum contracts and, on the other, reimbursable options; both have implications with respect to achieving environmental and social objectives.

(d) *GIS/data management*

One of the key tools in any large-scale infrastructure project, particularly a linear one, is the development of a GIS, a spatial data management system to support many different end users and act as a management tool throughout the project. By storing, interrogating and presenting a large and varied data set, the management system can significantly assist in a range of project activities from planning and design to impact assessment activities and documentation, and construction and operations management.

Where mapping is either inadequate or non-existent, the use of high-quality satellite imagery is a pre-requisite to form the GIS base mapping. Acquiring imagery can appear costly despite dramatic improvements in the resolution of imagery and significant price decreases over the past 10 years); but having imagery upon which constraints and land parcels can be mapped, and pipelines routes developed is a major benefit to a project resulting in hidden savings. The GIS should also capture constraints associated with geotechnical conditions and geohazards and in doing so, serve as the project's single repository of environmental and social data,

Additionally, a GIS can be linked to other specialised data management packages that will manage the processes and datasets required for mitigating and managing the social and environmental issues associated with the project, as well as fulfilling a reporting function.

(e) *Integrated schedule*

Integrating environmental and social management activities with other project aspects is fundamental to a successful project. However, integrating elements that have not traditionally been seen as mainstream is a challenge and results in the following scheduling pressures:

- Developing engineering definition as an iterative process, incorporating mitigations for environmental and social issues into project design where feasible. In addition, there needs to be sufficient development of the project design to allow impact evaluation;
- Ensuring that the sponsor-agreed project standards and commitments match the contracting and procurement documentation given to contractors, and that contractors are aware of the requirements before their bids are fully assessed (this is covered further in section 4 below);
- Integrating the land access schedule, the acquisition of rights and production of RAPs;
- Integrating a coherent stakeholder engagement strategy, with key messages evolving over time in response to project definition and status of the ESIA studies and stakeholder interest; and
- Developing a project GIS and, where appropriate, acquiring satellite imagery.

3.4 Scoping report

'Scoping' can be defined as "the process of determining the content and extent of the matters which should be covered in the environmental information to be submitted to a competent authority for projects which are subject to EIA".[5]

A well-reasoned scoping report that, at its core, draws together the foundation stones and interdependent components described above can save significant effort and resources by articulating those issues pertinent to the project and therefore minimising obsolete data gathering and assessment.

Scoping involves broadly considering issues and alternatives, and deciding what is likely to be significant in the context of the particular project and its proposed location(s) (covering both measurable and intangible, perception-based issues) and determining priorities. As prioritisation takes place at a relatively early stage in the understanding of issues, it has to be based on the experience of individuals well versed in the relevant industry sector.

---

5   European Commission, *Guidance on Environmental Impact Assessment: Scoping*, Office for Official Publications of the European Communities, Luxembourg (2001).

3.5 ESIA scope of work

The scoping report typically forms the key document of the ESIA scope of work – that is, the package used by consultants to prepare their proposals to undertake the work.

Scopes of work do not always accurately portray the specific project situation and assessment requirements identified in section 2. Often, they are adapted from scopes of work prepared for other projects. They can also be unnecessarily complex. Just as there is opportunity for ESIAs to be more focused, relevant and effective, a clear and concise scope of work along with a scoping report that captures the elements described above will help consultants produce better bids.

In order to ensure an adequate assessment, genuine project experience is required by the ESIA consultants and project sponsors in full-cycle environmental and social scoping and execution. Too often cut-and-paste ESIA scopes are used in tender documentation without regard for the particular project in question. This situation is often exacerbated by a tendency for ESIA consultancies to produce 'recipe' products, as well as the absence of any incentive to think progressively and identify clients' needs. To be fair, the ESIA tendering process generally favours both lowest cost options determined after a pre-qualification process and consultancies with the longest pedigrees. These can preclude significant original thought and lead to the continuation of poor quality ESIAs with little regard to purpose.

4. From ESIA to pre-construction

Contractor performance is critical in managing potentially higher environmental and social impact and unforeseen events arising during the construction phase.

As described in section 1 above, ESIAs have traditionally been technical assessments designed to predict and inform regulators and the public of the potential environmental and social effects of both the development proposals and management options. The level of effort, however, directed towards presenting and packaging the results of the assessments in a form conducive to effective implementation in the field by contractors is not commensurate with risk. As a result, contractors may well commit to a project's environmental and social standards and policies – in some cases hundreds of company specifications (generated over years to ensure that all potentialities are covered) – or undertake to meet commitments detailed in ESIAs. However, many will lack a detailed appreciation of what is required.

This presents additional risks to the project sponsors as they need to be able to manage contractor activities, while demonstrating to regulators, partners, the corporate entity and potential lenders that mechanisms are in place to ensure compliance with commitments made, including effective monitoring and report of outcomes.

The more strategic approach to environmental and social planning and assessment advocated in this chapter recognises the critical role played by contractors when it comes to managing risks and mitigating impact. This needs to be reflected in the scoping of, and approach to, ESIAs and in the way that the results of ESIAs are communicated to ensure that the ESIAs are recognised as a critical step in the overall process of understanding risks. Ultimately, the effective performance of the construction contractors will have a large bearing on the success or otherwise of the project. Whatever the result, it will generally affect the sponsor more than the contractor.

4.1 Construction contractor bid packages

The performance expectations of project sponsors must be clearly articulated and incorporated into bid packages. ESIAs rarely meet this need, either due to the sequencing issues identified above (section 3.3(e)) or the focus being limited to obtaining regulatory approval.

In the absence of a completed ESIA, the approach applied has been to include as many standard environmental and social measures in the bid package as possible; the ESIA is sometimes added as a late inclusion to the bid documentation. This has resulted in poorly understood requirements and, consequentially, poor performance and extended contractual disputes due (in part) to the requirements not being more clearly spelt out.

The bid package is therefore a critical document in the strategic ESIA process as it provides the contractual leverage to ensure that commitments made in the ESIA are both understood and implemented by the construction contractors. The package should include a high-level description of expectations (to explain the requirements for both a robust management system and management plans), followed by increasingly detailed requirements – for example, based on specific data gathered from site specific surveys or through seasonal requirements.

4.2 Commitment registers

One of the initial products of an ESIA should be a commitment register. This is relatively well understood by a number of environmental consultancies; however, it is not always applied particularly well. Commitments must be unambiguous, non-repetitive and written with a practical application in mind. This is not an easy task for consultancies without a first-hand and practical understanding of the way projects are implemented on the ground.

The registers themselves should be collated into a database to allow sorting and inclusion into CCPs (see below) and the identification of applicable commitments to specific project phases, issues and geographical areas. It is an advantage if commitments can be sorted by stakeholder interest.

The strategic ESIA approach focuses on the practical applications of the impact assessment and the conversion of a traditional academic document into something with a clear end game: to provide the teeth to manage the effective implementation of the commitments. Many regulatory agencies are primarily interested in the identification of impact and mitigation measures; however, our contention is that the enlightened sponsor will continue past the regulatory requirement and use the ESIA as an effective environmental and social management tool through the development of a commitment register and, subsequently, a set of CCPs.

4.3 CCPs

The concept of CCPs was first developed and implemented for the Baku-Tbilisi-

Ceyhan pipeline in 2003, whereby the minimum performance requirements to be met by the contractor (often described in the form of mitigation measures) were detailed in a series of thematic plans. Contractors were then required to develop contractor implementation plans and procedures that contained procedures and method statements specifying how they would implement the measures included in the contractor control plans.

The main benefits of contractor control plans are that they:
- clearly lay out the contractual requirements by discipline;
- use a common format;
- translate the many commitments made in ESIA documentation (and other source documentation) into specific actions;
- provide a means to monitor contractor performance by providing an audit protocol;
- avoid approval of contractor plans by a lender group (in a financed project);
- assign responsibilities to the contractor (differentiated from the sponsor); and
- allow transfer of a template between projects, with a combination of common and specific sections.

The concept of CCPs, and the way it can be derived from the ESIA, is possibly the biggest single outcome of the more strategic approach to ESIAs – the vision from the outset of the ESIA to develop the CCPs to ensure that the contractors are provided with every opportunity to understand the issues, the legal basis for the issue and what they are expected to do. Such a common transferable template is a bonus for sponsors as it can become the blueprint for managing a particular issue. This approach also allows for sections to be customised – for example, to meet corporate or local regulatory requirements.

4.4 Sponsor plans

In parallel to developing CCPs, sponsors are required to develop their own company plans spelling out how they will manage a particular issue.

The key here is the early development of the overall strategy with respect to contracting and the approach to be taken for ESIA issues, recognising the advantages to the sponsor for some activities to fall directly under their responsibility rather than being pushed onto the contractor.

5. Conclusion

A strategic approach to environmental and social risk assessment, impact prediction and mitigation management is required for large-scale infrastructure projects in order better to manage non-technical risk, to produce better environmental and social outcomes, and to deliver better value to all stakeholders – all at a lower overall cost to project sponsors. The relative benefits of this approach over existing practice are illustrated in Figure 2.

These benefits can be achieved by taking a more holistic view of environmental and social impact assessment, better integrating the assessment process into the conventional project's life cycle sequence, reinvigorating the ESIA component of the

Figure 2: A cost-benefit comparison

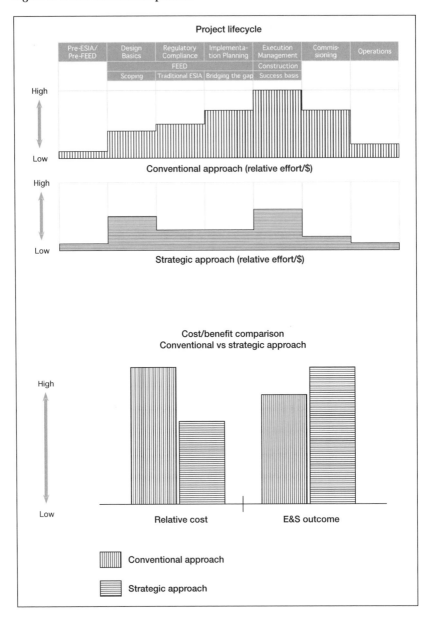

assessment process to make it more focused and relevant to contemporary needs, and better managing the potentially high-risk pre-construction phase by adopting the concept of CCPs.

A strategic approach requires earlier and more deliberate planning than current approaches to impact assessment. It also requires a different approach to scoping and preparing ESIA reports, which have tended to become staid and cumbersome, as well as recognising the bearing that contracting strategies and the form of contractor bid packages can have on non-technical risk, and environmental and social performance.

The concept of foundation stones and interdependent components acknowledges the need to establish the ESIA fundamentals early and manage an evolving process without losing sight of the end game. This allows a project to manage risk to an acceptable level and from the point of view of multiple stakeholders.

The strategic approach requires adjustments by all parties. Here are four key suggestions:

- Sponsors should spend more time and effort on specifying the need to ensure that the design basics are clearly defined and relevant to the project. Such specifications should be part of the contractual scopes of work developed for environmental and social consultancies;
- Sponsors should view the ESIA as a step in an overall process to manage non-technical risks, rather than a regulatory requirement;
- Consultancies should place greater emphasis on the rigour behind the assessment process and regard the end point as the development of plans to manage the identified risks, rather than an assessment process only serving a regulatory requirement; and
- Regulators themselves should likewise review their scoping processes and guidelines, and consider the benefits of a strategic approach to managing project risks.

The likely result will be reduced cost, enhanced environmental and social performance, greater stakeholder acceptance and reduced project risk.

Security risk management

Antony FS Ling
LPD Strategic Risk Ltd

1. **Introduction**

   Brian Wellcome's job was just to get the stuff out of the ground. Yes, he was in charge, but someone else would look after the politics: the locals, guerrillas – whoever. Some security guy had said something to him once, but he knew they always exaggerated. Anyway it was now all too late – just guns, screaming and death. He knew he was going to die. But Brian did not die – he was far too valuable as a hostage.

   In our globalised world, Wellcome's desperate situation is just one of the increasingly common security problems that companies deal with. While this case may be extreme, impacting as it does on duty of care, liability, reputation and cost, it is just one of the wide range of interrelated security, social and political risks confronted by companies today.

   Since 9/11 companies have begun to recognise security as an integral part of business,[1] particularly since the search for commodities is taking them into increasingly volatile environments. However, what businesses have yet to grasp is that by giving priority to tackling the causes of security failure, they can minimise the consequences and cost of a security incident. Security is more than just concentrating on protection. Companies should understand how socio-political and environmental concerns become security problems. Implementing this holistic approach is critical to mitigating the variety of asymmetric risks encountered by energy infrastructure projects, particularly transborder projects, in the developing world.

2. **Scope**

   This chapter examines the wide range of tools, processes and techniques available to manage the causes of security failure that are likely to be encountered in international and transboundary energy infrastructure projects. While the focus is on the host government, civil society and community dysfunction or hostility; reference is also made to how environmental and health issues can become politicised. The direct or indirect threats (or risks)[2] and possible impacts are discussed with the mitigating actions required to counter them holistically; the objective of

---

1   Eddy, Kester, "Security is linked to all aspects of business", *Financial Times* (July 12 2010).
2   This chapter makes no distinction between 'threat', which is sometimes used to describe unmitigated risk, and 'risk'. 'Risk' is generally used in preference.

this process is to help a project work in peaceful and profitable surroundings.

All the chapters in this book discuss risks that, one way or another, could cause security problems if left unmitigated.

Some key business security risks, such as fraud, internal theft, information security, information technology, product contamination and counterfeiting, are not covered. While important, these risks are outside the scope of this chapter, which concentrates on what can broadly be called socio-political security issues. Physical and electronic protection, while a key final defence, is not discussed here either.

3. **Holistic security – managing the causes of security breakdown**

A popular definition of 'security' is safety against attack. 'Attack' does not exclusively mean violence. Criminal, political or economic attack – often supported by propaganda – is the preferred form of hostile action and violence which in this context, becomes a continuance of the politics by other means[3] (eg, a breakdown in the wider national political context). The project developer is often no more than a high-profile target through which an adversary can affect the host government. Three examples of this are the attacks on the Bougainville copper mines in pursuit of independence from Papua New Guinea, the wider threat and attacks against international extraction projects by Al Qaida and attacks against the oil industry by guerrillas, particularly in Colombia.

The security impact on a transboundary energy infrastructure project or operation can be caused by problems in just one country. For instance, in 2005 the risk affecting the transportation of oil and gas by pipeline or tanker was mitigated by managing security problems in Iraq rather than in Turkey, Jordan or the Gulf.

'Holistic security' is best defined as the recognition, identification and mitigation of the causes of security breakdown. If a company can minimise the causes of security failure, it can minimise the chance of it occurring.

Possible security threats range from petty theft to serious political attack. Petty theft is often the result of a community's alienation from a project.[4] In such cases, traditional leaders feel no compulsion to exercise discipline and often even encourage that petty theft. This was the case in the diamond mines of southern Namibia when, leading up to independence in 1991, theft was seen as duty in the struggle for freedom of the South West Africa People's Organisation. Examples of operating among serious political violence, at the other end of the spectrum, are oil projects in Colombia or Nigeria and gold extraction in the Democratic Republic of the Congo.

Host governments, often with limited capacity, have a duty to protect those investing in their country. But companies involved in projects also need to manage the security risks that they encounter. Companies fail, totally or in part, in that task for three broad reasons:

- They might fail to recognise and then mitigate security risks in their project's operational region;[5]

---

3   von Clausewitz, Carl, *On War*.
4   But it can also be extreme poverty, where everything has a value.
5   For instance, Total in Myanmar in 1996.

- They might alienate local communities or international and national civil society;[6] and
- They might not recognise the security limitations of host governments and their security agencies. These include organisational and training weakness and host governments' inability to adhere to international human rights standards.

Ironically, the companies that have experienced problems have subsequently provided the benchmark for the management of these security risks.

Problems often arise either because a company fails to take qualified advice early enough in the process, or as the result of cultural or political communications breakdown – such as, in particular, failure to recognise that a seemingly minor issue or risk can quickly become critical if not managed properly.

Successful avoidance or mitigation of these problems starts with proper identification of the risks.

4. Risk assessment processes

4.1 Identifying risks

One of the initial tasks in the pre-entry process for any energy infrastructure project is the identification of potential security risk. There would seem little value in starting a project if the company does not know whether it can keep its people or investment safe.

The region or countries in which the company hopes to construct or invest may be unstable or lawless. The potential cost of security or the reputation risk, such as in Myanmar, should be carefully analysed. Looking ahead over the life of the operation, there may be a dormant or potential crisis that could put the whole project at risk. Possible high-risk security issues will need monitoring.

Project developers should assess any growing regional tensions, particularly where the project itself could increase political impact. A good example of foreseen security risk was with regard to the construction of the main export pipelines carrying Caspian oil (Baku-Tbilisi-Ceyhan) and gas (South Caucasus) through Azerbaijan, Georgia and Turkey to the Mediterranean; there was little doubt that the pipelines would stoke tensions between Russia and US-backed Georgia. Since the project recognised early those risks, the project could put in place actions to monitor emerging pressure. An example of failure to recognise a problem was in 1988 when operating and investing companies did not foresee the full extent, or cost, of the potential security problems along the Cano Limon pipeline running inside Colombia but close to and parallel with the Venezuelan border. Among other security issues, the project companies should have recognised at the outset that routing a major export pipeline close to an international border carries the risk that attackers can mount quick attacks from the neighbouring country without fear of hot pursuit.

---

6   For instance, Monterrico Metals in Peru in 2005.

Analysing the threats long term is particularly critical pre-entry: risks should be identified as early as possible and logged on any project viability and risk matrix.

4.2 Security risk assessments

Once the decision has been made to build, or invest in, a major international energy project, a series of detailed security risk assessments need to be carried out. Since risks change, these should be revised frequently. Project security and political advisers should identify and assess security risks under five general headings: regional, country, reputational, impact and project risk assessment.

(a) *Regional security risk assessments*

These should consider the long-term threats to the project in the light of regional nervousness, particularly if existing stresses could be exacerbated by the project, such as in the case of the Baku-Tbilisi-Ceyhan and South Caucasus pipelines mentioned above. Regional risk assessments should look at tensions between countries and at any emerging long-term geostrategic threats. They should also consider how these might affect project security – for instance, any possible negative effect on a project resulting from the recurring tensions between Colombia and Venezuela, or the impact of any worsening situation in the Democratic Republic of the Congo or Angola.[7] A risk for the Baku-Tbilisi-Ceyhan pipeline was its close proximity to the war in Nagorno Karabakh; the risk itself could not be mitigated, but the impact on the project could be minimised by routing the pipeline out of hostile artillery range. An example from the developed world would be illegal immigration into the southern United States, where tensions could affect a project.[8] Project developers should assess whether the relevant governments or the international community could easily resolve regional issues, or whether the situation will inevitably worsen, escalate into violence which will affect the project.

(b) *Country security risk assessments*

These should look at national cohesion and examine tensions between local and state government, as well as the potential for pressure to turn violent. Law and order issues such as corruption and staff safety should be assessed.

An important aspect of a country risk assessment is the extent of the rule of law. Functioning rule of law is critical to national security, the security of the project and future operations; arbitrary justice is not conducive to long-term stability. This threat is linked to reputational risks discussed in subsection c below. Pakistan, where tribal tensions combine with politicised Islam to destabilise parts of the country, or Indonesia facing a growing push for Papuan independence, are both examples of country risk. An example from the developed world was the Provisional Irish Republican Army's threat to UK strategic east coast oil facilities.

---

7 For instance, refugees squatting around the Soyo liquefied natural gas facility.
8 Particularly immigrants in search of work.

(c) *Reputational risk assessments*
These examine the possibility of a project's unwitting link with either corruption, or the media associating the operating company with alleged human rights abuse carried out by the host nation's security forces while deployed in defence of a high-profile project. This is exemplified by BP's reputational problems in Colombia in the mid-1990s. Association of a company with human rights abuse could make it legally liable under, for instance, the US Alien Torts Claims Act. Any possible impact of untrained host government security forces on a project should be assessed, especially how such impact might affect a project's local community. Examples of this include police overreacting when confronted by a hostile crowd at the construction camp for a liquefied natural gas plant in Papua New Guinea in 2005, and the arbitrary arrest and alleged torture of a community leader in Turkey opposed to the Baku-Tbilisi-Ceyhan pipeline also in 2005. That same year Burmese villagers sued Unocal for complicity (with the Burmese army) in forced labour, rape, torture and murder. The case was settled out of court.

Reputational risk assessment should consider whether people have the chance of a fair hearing in court. In the absence of the rule of law, corporate ethics become critical in mitigating risk to the reputation of a company.

(d) *Impact risk assessments*
These examine the influence of the project on the country – particularly local communities. For instance, the provision of project security by private, paramilitary or state entities should be assessed for any possible negative impact on communities. This should also be part of the social component in the project's environmental, health and social impact assessment. Experienced security advisers should be involved in writing the environmental, health and social impact assessment, especially in countries such as the Democratic Republic of the Congo, Colombia and Nigeria where the security threat to large projects and their neighbouring communities correlate.

(e) *Project security risk assessments*
These should consider impact on construction and related activities including transportation and construction logistics and should be based on the prevailing security environment assessed in the regional and country assessments. Project security risk assessments, taken together with the other project assessments such as geological or commercial, could influence pipeline route selection, as was the case in the Baku-Tbilisi-Ceyhan pipeline. The locations of border crossings need special consideration in this regard. In certain security environments a facility might need to be sited so it would not be overlooked by high ground that could be occupied by people hostile to the project. A site-specific security risk assessment for every location should consider risks from, or to, the local community from the project, as well as any ethnic tensions, including migration or immigration, that could affect the project's security.

Companies should follow the general principles set out below when assessing their projects' security risks:

- Be open. Project developers should publish risk assessments on the project website. Only information likely to put people, including communities or the project itself, at risk if published should be excluded. For instance, responsible non-governmental organisations approached the developers of the Baku-Tbilisi-Ceyhan pipeline to publish their risk assessments; this was done after agreeing that anything that could increase the risk to people or the project, if made public, should be excluded.
- Ensure that construction team leaders and other influential project engineers understand and accept the security risks and how the safety of their teams could be affected. 'Buy-in' is vital when security considerations are likely to influence the position of facilities or the route of a pipeline. Project developers should also ensure that security risk is coordinated with other project functions such as environmental, health, political and social.
- Reflect the high security risk in the overall project risk assessment, if relevant.
- Bring potential show stoppers to the immediate attention of the project leader.
- Provide early warning of worsening risk as a project progresses – for instance, of a new security threat to project employees or to safe transportation of equipment to site.
- Involve engineering, procurement and construction contractors in the security risk process as soon as they are contracted.
- Update assessments as required and redistribute them as consultative papers for comment to all interested parties. Feedback should be revised and incorporated.
- Ensure that there is no conflicting analysis in any of the five security risk assessments or with other non-technical risk assessments such as social assessments. Assessments are interdependent and can be published together as one document.
- Do not be mesmerised by, or over-assess, risk (even though sound risk assessment is critical to the success of a project). Be prepared for the unknown.

4.3 Information providers

All risk assessments should be the result of wide consultation, including with foreign embassies, other international companies, host country security agencies, parent country foreign ministries[9] and commercial information providers,[10] and responsible non-governmental organisations.[11]

---

9 For instance, the British Foreign and Commonwealth Office or the US State Department.
10 For instance, LPD Strategic Risk Ltd, Control Risks Group and Kroll.
11 These include Human Rights Watch, Amnesty International, WWF, Oxfam, Greenpeace and those in the field of regional conflict analysis, such as International Alert. An irresponsible non-governmental organisation might be:
 • a single issue organisation set up exclusively to oppose the project;
 • an organisation operating as a front;
 • an organisation hostile to the project *per se*; or
 • an organisation that attacks the project as a publicity vehicle for some unrelated political objective.

The involvement of project staff, particularly locals, in the security risk assessment process can be constructive. It also gives them the necessary ownership.

Those compiling the assessments should visit the ground to talk to communities and to reconnoitre potential locations to confirm their security suitability. An example to the contrary was a last minute, and reluctant, visit to a potential gas facility site in Indonesia, which revealed that the ground on which the project was to be built was in fact the village football field, on which the person who sold it to the project had no rights. Unfortunately, no background checks had been run on the man selling the site. Thus, it is key to ensure that due diligence is carried out in support of all contracts. Failure to do so can have legal, as well as financial and reputational, consequences.

4.4 Due diligence

'Due diligence' is a term used for a number of concepts involving an investigation of either a business or a person prior to signing a contract. Hiring people without knowing their past history is an unacceptable risk. Failure to check the background of security employees or even government security agents employed by the project may make the operating company legally liable if these individuals commit a crime or attract hostile attention.

Many governments penetrate international companies working in their country, usually by recruiting an employee. This may be inevitable, and companies should assume that this is the case and operate accordingly. Rather more problematic is the recruitment of an intelligence source by a third country; this is most likely to happen in the former Soviet Union and can be detected by sound vetting of staff, including consultation with trusted host and home government agencies. It is also important to carry out background checks on major service providers such as the project's engineering, procurement and construction contractors. Subtle considerations, such as a director being deemed unacceptable by the host government, need to be known by the operating company before the issue becomes public and politicised.[12]

4.5 Measurement and presentation of risks

Mathematical or statistical measurements of risk, also known as quantification, do not generally lend themselves to political, security or health risks. There are too many variables and unknowns. Who can put a percentage value on someone's motivation, intention, methods or capability to carry out an action, directly or indirectly, against a project?[13] Qualitative risk assessments are the usual method of choice.

All relevant available information is collected and analysed in light of the answers to the following questions
- Has the source of the information proved reliable in the past?
- Is there collateral information?
- Does it sound credible?

---

[12] Examples might be an Armenian director of an international company operating in Azerbaijan or vice versa.
[13] For related risk theory, see Bernstein, Peter L, *Against the Gods – the remarkable story of risk*.

Security risk management

Once the information has been analysed, it can be graded from 'very low' to 'very high' for both impact and probability.[14]

The diagram below shows how impact and probability are compared to calculate risk.

		Probability			
		Very low	Low	Medium	High
Impact	Very high	Medium	High	Very High	Very High
	High	Low	Medium	High	Very High
	Medium	Very Low	Low	Medium	High
	Low	Very Low	Very Low	Low	Medium

A risk table can be developed using the following headings with a typical example from an unnamed developing country.

Example of country risk assessment

Risk description	Comment	Impact on project programme	Probability	Manageability or influence by company	Risk rating
Security deterioration related to elections	Parliamentary elections scheduled for 2008. Presidential elections possible for 2009. Electoral registration process completed to reported satisfaction of all parties involved. No major irregularities reported. Low risk of security incidents involving former combatants	Medium	Low	Low	Low

---

14   There are many presentational systems; some are limited to just 'high', 'medium' and 'low'. It is not the presentation but the intellectual rigour of the analysis that is important.

264

As in the fifth column of this example, project companies should consider whether they can control or manage a major incident or whether such incident would be outside their sphere of influence. For instance, if civil war breaks out project companies cannot usually expect to influence the war's course or outcome, but they can manage the impact on themselves by having in place, for example, a functioning and rehearsed evacuation plan.

Assessments need to be reviewed frequently since risks alter when the political, security or local community situation changes. Security risk can also reduce as a project develops because the company has carried out various mitigating actions to lower the assessed risk.

In summary, risk assessments enable a company to forecast a project's potential security problems. Once those are identified then plans should be made to prevent them from happening by mitigate their causes.[15]

5. Mitigating risks

As a rule, plans will be needed to mitigate any identified risk graded 'medium' or above. These plans should include details on managing the causes, as well as the consequences, should prevention fail.

In the developing world, and for transboundary projects, the cause of these risks[16] is likely to belong in any of three overlapping areas:

- the social area, which includes the local community, and national and international civil society (including non-governmental organisations and the media);
- the political area, which covers a wide range of issues that are either regional, national or host government-related; and
- the project area, which includes employees, contractors and workers.

5.1 Social risks

A security incident often happens because the company has failed to implement or even make plans to manage social risks. Problems arise when companies fail to recognise that the cooperation of their local community is key to their security. An example of this is not setting up functioning consultation, disclosure or grievance management processes. Principles such as free, prior and informed consent should be standard. Communities are important stakeholders in any project. Sins of commission can be even more damaging. Examples include taking too much water from local rivers, or project logistic convoys producing continuous and pervasive clouds of dust as they drive through villages.

Projects should work with cultural experts such as anthropologists, social impact specialists, non-governmental organisations and, particularly, local community

---

15 This process is sometimes called intelligence-led security. However, the word 'intelligence', when applied to business, has been used by non-governmental organisations and the press to attack companies for engaging in espionage (eg, BP in Colombia in 1996).
16 Some key business security risks such as fraud, internal theft, information security, information technology, product contamination and counterfeiting are excluded from this chapter (see Section 2 above).

leaders and local people in order to understand community issues, and work to minimise impacts and maximise community benefits. For example:
- local peoples invariably want jobs with the company and contractors. High expectations should be managed by never offering anything that will not definitely be forthcoming;
- it may be possible to help the community, particularly through training and initial funding, to set up a commercial organisation to support the project with transportation, provision of security guards and primary health care staff. Provision of healthcare for project workers and the community increases productivity by reducing sickness, absenteeism and employee turn over;
- project drivers should be encouraged to give lifts to locals. A mining company in Yemen, Thani Dubai, has a policy to pick up people on the road because "it's what we do because they're our neighbours and friends". The company operates peacefully in what is seen by many as a hostile environment. Nothing is more likely to alienate local people than failure to extend what in most cultures is the common decency of a lift; it is such issues that fester and grow in peoples' minds. Prior planning can negate any liability risk from carrying locals in company vehicles; and
- large and expensive infrastructure projects, such as building hospitals, should be avoided; inevitably maintaining medical staff, equipment and medicines will be an endless cost to the company despite any promises to finance from host or home governments, or non-governmental organisations. Blame for failure will inevitably be with the company.

The local community's security is of equal importance to the security of those working on the project. At worst, the project could become unsustainable if it is perceived, directly or indirectly, that community security has been compromised by the action or inaction of the company. The security environment of a project often worsens because the presence of foreign workers is exploited by anti-government groups' misinformation machinery supported by intimidation and subversion.[17] A delicate situation can be exacerbated by the host nation's security organisation overacting – typically, to a demonstration. Detailed plans to deal with this risk should be made under the second of the overlapping areas: 'political'.

5.2 Political risks

The host government's security forces may be of doubtful effectiveness and may even be operating without legal authority. Any overreaction by troops or police can become a major challenge for a company.

Over the past 15 years a great deal has been learnt about working with less mature military organisations. This is partly because courts are increasingly often hearing allegations that a company has breached either international[18] or host nation human

---

17   Foreign contracted semi-skilled workers will inevitable cause ready propaganda by chasing local women.
18   Such as the US Alien Torts Claims Act.

rights law through its association or complicity with abusive host government armed forces.[19] Apart from the legal issue, the resulting bad publicity can damage both reputation and brand, and – in some cases – even the company's share price.

Companies in the extraction industry have come together with governments and non-governmental organisations to establish the Voluntary Principles on Security and Human Rights.[20] These address the concerns of companies, governments and human rights groups by setting standards. The voluntary principles have several notable and distinguishing characteristics, particularly, that they are operational rather than aspirational in nature.

The voluntary principles comprise three sections:

- Risk assessment – this section recognises that risk assessments should also consider human rights-related factors;
- Interactions between company and private security – this section asks companies to enforce the principles and guidelines by including them in contractual provisions with private security providers; and
- Interactions between company and public security – this section addresses one of the areas most frequently scrutinised by the public and civil society, namely, the security cooperation between public forces and companies.

It is the relations between company and public security that are the most problematic of the provisos of the voluntary principles. A practical issue is that when a company relies on local troops for protection, close cooperation becomes critical. Inevitably, host government troops in most developing countries are badly equipped, undertrained, immobile, unsupported medically and underfed. The company does not want them scrounging or stealing food off the locals; it is a relatively easy and without political risk to feed them and support them medically. But if the company is going to help to improve the troops' military capability, then there are risks of association with any possible acts of abuse. The voluntary principles do not rule out aid to host government troops, so long as such aid is ethical and in line with international human rights standards. As a guide to giving assistance to host government security forces, the principles consider that:

- no lethal aid should be provided to host governments;[21]
- any assistance should be fully documented, transparent and duly witnessed;
- wherever possible any support provided should be with equipment and not money;
- the reason for providing assistance should be made public;
- wherever possible, the use of any assistance should be monitored and (publicly) documented, and any evidence of compliance, or non-compliance, should be retained;

---

19   For more detail see Dimitroff, Tom, *Human Rights Norms & Potential Foreign Direct Liability Risks to Companies*.
20   On December 20 2000 then US Secretary of State Madeleine Albright and UK Foreign Secretary Robin Cook launched the Voluntary Principles on Security and Human Rights together with non-governmental organisations and major companies in the extractive industry.
21   Some equipment is dual use – for instance, radio sets can be used for directing fire onto a target. Generally, dual-use gifting should be avoided.

- security equipment should be procured in line with company standard procedures; and
- the equipment provided should contribute (directly or indirectly) to the project's security.

Companies naturally want their armed state protection to live close to their project in order for these armed forces to provide protection. Building barracks to achieve this is acceptable, as long as the guidelines above are followed.

The provision of training by project companies to security forces protecting the project is in line with the requirement of the voluntary principles that "[c]ompanies should use their influence with host governments to ensure that... force should be used only when strictly necessary and to an extent proportional to the threat" and that "the rights of individuals should not be violated". In other words, sponsoring military training, carried out to the highest professional standards, in line with international human rights norms is acceptable. Commercial, not-for-profit or intergovernmental organisations often deliver this training.[22]

It is generally agreed, at least by European companies, that building close ethical ties with host government military and influencing them for the good improves the companies' legal protection and therefore decreases their liability.

Other key voluntary principles requirements include making a thorough in-country assessment of human rights records and the rule of law. Such a review enables a company to identify pitfalls that it may fall into, or bad practice with which it could be associated.

Under the voluntary principles, companies should use their influence with host governments to ensure that:

- individuals credibly implicated in human rights abuses should not provide security services for companies;
- force should be used only when strictly necessary and to an extent that is proportional to the threat; and
- the rights of individuals should not be violated while exercising "the right to freedom of association and peaceful assembly".

While an extractive company does not assume the responsibilities of government, it can promote such principles and protect itself from media and legal attack. But more importantly, a company should ensure that host government forces protecting the project do not mistreat their local communities.[23] Companies such as BP use the voluntary principles practically as rules of engagement to facilitate this.

Host governments have the sole responsibility and mandate for the use of force. However, companies should consider embedding certain aspects of security in the project's prevailing legal regime. This is particularly relevant for transboundary projects.

Each host country and the operating company should consider signing a joint

---

22 For instance, the International Committee of the Red Cross, the Organisation for Security and Cooperation in Europe, Equity International, LPD Strategic Risk Ltd.
23 For further details, see the Royal United Services Institute's Presentation on the Baku-Tbilisi-Ceyhan pipeline by Tony Ling, March 2004.

statement noting their commitment to international human rights norms and their determination to make the project a model in all respects. A joint statement can make the voluntary principles part of the law governing the project.

The benchmark on this subject was the signature in 2003 of a trilateral security protocol between Azerbaijan, Turkey and Georgia to govern the Baku-Tbilisi-Ceyhan and South Caucasus pipelines. The protocol enshrined key principles such as sharing of information, risk assessment and security training cooperation. The parties agreed to comply with international human rights protocols including the voluntary principles. There was an agreement to "pursue any credible allegations of human rights abuse". A Joint Pipeline Security Commission was established and signatories undertook to meet at least annually and to be in regular consultation.

Complimentary to the voluntary principles are the recent UN Guiding Principles for the Implementation of the UN Protect, Respect and Remedy.[24] This business and human rights document elaborates and clarifies how companies, states and other stakeholders can operationalise the UN Protect, Respect and Remedy framework by taking practical steps to prevent any negative impact by businesses on the human rights of individuals. One of the advantages of these guiding principles is that they unite all human rights, including labour, discrimination, legal rights and security, and break down the silos into which these had become isolated. The need to handle overlapping and complimentary human rights disciplines together reflects the requirements, referred to elsewhere in this chapter, for projects in the emerging world to manage the government, socio-political, environmental, security and health environments as a single, interrelated function. These principles also aim to increase a state's legal and non-judicial capability to deal with human rights abuse by a business, be it direct or by association. The state can be the business's host or home country.

A company should explore the possibility of agreeing a combined strategy between involved nations to assist with the training of security agencies that are responsible for the protection of the project.

Independent monitoring of the community policing of a project will reassure the international community (and lenders) that the project is being managed to the highest standards.

The provision of private security is best restricted to unarmed locally employed watch-keepers. Companies should still check that they are not hiring the local criminal gang or anyone with a record of human rights abuse.

Companies should assess the capability of the local prosecuting authority and judiciary to hold accountable those responsible for human rights abuses and violations of international humanitarian law. Lack of acceptable standards in enforcing the rule of law can exacerbate hostility to the host government and, indirectly, the company. Hostility from the international community can affect a global company at its home base.[25]

---

24   Finalised in June 2011.
25   Companies previously working in Myanmar came under sustained political attack, especially in California.

A company's association with state tyranny is the major security risk. But there are several other complex and difficult politically based security risks – for instance, understanding the likely nature of a developing and escalating regional or national threat. Early warning is essential. A company should develop support mechanisms with home governments and embassies. A business should obtain and analyse relevant information to keep ahead of a worsening political environment that might put its workers and the project at risk. A typical example is ensuring that a company has the information and plans in place to cope with any instability, including rioting, during an election.

5.3 Project risks

Good holistic security practice depends to a large extent on employees, contractors and workers owning the social and security environment. Like safety, security should be seen as the responsibility of all.

Every member of the company's work force, including contractors, should understand that they are guests in the country in which they are working, that they are in equal partnership with the local community, and that promoting this concept is an important part of their job.

Managers in the field should understand the business necessity of winning the battle for the hearts and minds of the community and their own role in it. Formal training is necessary if everyone in the project is to embrace this. Non-local staff may have to be encouraged to visit villages and eat local food. Sometimes a cultural leap is required to understand what the project's neighbours need, their aspirations and the importance of sensitivity in the relationship.

Project companies should insist on best environmental practice from all employees. Failures in this regard, probably more than any other, can lead to the breakdown of relations and exploitation by hostile groups.[26]

6. Conclusion

Wellcome's captors released him after nine months of captivity and the payment of a considerable ransom. This did not happen until Wellcome's family, frustrated but ill advised,[27] involved the press. There was a damaging amount of publicity targeted at the company for reportedly not caring about employees; staff morale suffered and the chief executive officer was forced to fly out to the operation and promise expensive, short-term and ultimately ineffective protection that benefited nobody but the contractors providing expatriate guards.

But the final irony in all this was that it had all been so preventable. An independent enquiry ordered by the company's board revealed that, far from being a long-planned terrorist attack, Wellcome's capture had been a result of village

---

26 In 1996 accusations that BP Colombia's environmental failings had resulted in the poisoning of fish and cattle with accompanying heath scares reached the UK national press. The ensuing media attack damaged the company's reputation. More importantly, ELN (*Ejército de Liberación Nacional* or National Liberation Army) guerrillas were able to exploit these reports by organising anti-BP demonstrations and issuing threats against the company.
27 Media involvement in delicate ransom negotiations always muddies the water, complicates communications and ultimately inflates the cost of release.

revenge and guerrilla opportunity. The company had upset its neighbours; the week before Wellcome's kidnapping, a distressed pregnant woman and her husband had come to his fenced rig site asking for help from the clinic, only to be turned away by the guard, who came from a different tribe with a long-running feud with the surrounding people.[28] Wellcome had then refused to meet a deputation from the village. The woman died in childbirth, but the baby son lived.[29] The evening before the attack on the rig the talk in the village was still about this rebuff; the hatred of the company was palpable. There happened to be a stranger passing through this previously peaceful village. He immediately saw an opportunity for political gain and offered revenge that was immediately accepted. It took him no more than a few hours to gather a dozen well-armed comrades from the hills.

The guerrillas were able to capitalise on their popularity and quickly took over the area, the army moved in to counter them and the company never operated there again.

---

28 It is a fallacy that guards should be from tribes not coming from the project neighbourhood. This bad advice is often justified by arguing that locals will conspire together to steal or will turn a blind eye on crime. On the contrary, hiring local guards provides well-needed employment and community involvement, and engages the guards with a sense of project ownership.

29 The son was raised to think only of revenge and hatred of the foreigners who had stolen his land and his mother.

Risk, project-affected communities and their land

Henry Thompson
Oxania Ltd

This chapter examines land in terms of the people who live on it and from it. The users and owners of land are the people mostly affected by the development of primary extractive industries: history shows that, in general, indigenous populations incur significant losses to livelihoods, status and stability as a direct result of major investments in oil and gas. Experience also shows us that indigenous people also constitute a significant risk. The poorer people are, the less educated, resourced and legally empowered they are; the more likely they are to incur loss and create conflict. These patterns of risk can be broadly mapped out. The first part of the chapter addresses the risks and the case for communication and mutual cooperation, and the second briefly examines the issues around land tenure and land rights.

1. Indigenious and local communities

There is an ice factory on the dock at Gloucester, Cape Ann, Massachusetts – a 70-year old black wooden barn with few windows. When a fishing boat comes alongside to be filled with chunks of ice, someone goes to the third floor to haul, in the freezing gloom, blocks of ice the size of upright pianos onto a short but very slick steel hopper, at the far end of which is a rotating wooden drum as big as two oil barrels laid end on end. This massive cylinder of old pine is set with rows of pointed iron spikes as thick as sausages. As the drum spins, driven by a belt and motor deep below, the enormous slabs of ice are crushed and the chunks swirl down a set of steel chutes and out into the boat's hold. In conversation with the genial factory owner, an old employee asks:

"You don't still have that big crusher up on the top floor, do you? The thing with the spikes in it?"

"Yeah, sure we do, that's never going to break."

"I have nightmares about that thing."

"So does my insurance company. But you know what? It's probably the safest piece of machinery we got. Everybody's scared to heck of it."

If risk is as much about perception of it as it is about risk itself, the same can be said of companies and communities. What is seen and understood of them is as important as what they actually are, because in many senses the key element in the relationship between people and companies is trust and belief.

1.1 The context: the gap

Project development gives rise to the following three sets of question, among others:

- In whose interests is the company working?
- Under whose authority and interests does the company have access to the resources?
- How do the government's and companies' interests bring benefit to the affected communities? That is, where do local people come into this?

While the answer to the first question is 'the investors' and that to the second question is 'the government', the answer to the third question is less clear. The stage for the development of resources features three actors – governments, companies and communities. In a developed country, the aim of the community is to maintain and improve local people's livelihoods, the aim of the company is to return a profit to shareholders and the aim of the government is to raise rents and stay in power. The government achieves its goals through taxation, and by representing and protecting the interests of its constituents. In a developing country, the aims of the communities and of the companies are exactly the same as in a developed country. However, while the aims of the government may be the same, the means to achieve those aims are often very limited. The critical differences lie in the quality of governance.

'Governance' covers various meanings and affects a wide set of activities, but its core consists of an interdependent relationship between people and government. This relationship is best understood as a social contract that exists, to a greater or lesser extent, between communities and government. Without working democratic political mechanisms, it is very hard to achieve accountable and effective regulation. To work effectively, this social contract must be based on trust and norms of behaviour that are developed and tested, and re-tested. Where this social contract does not exist because the government is weak or lacks either skills or popular support, the company operates outside the norms of its regulatory background and there is a governance gap. In this environment, different industry sectors react in different ways.

The oil sector is characterised by relatively high speed and high cost investments that are balanced by high margins in the event of success. Licences awarded for onshore exploration typically run to five years, with two-year extensions to follow. Within that time the onus is on the investor to analyse existing data, run some seismic tests and drill one or two wells. This comes at a cost of many millions of dollars and uses teams of specialised contractors who may be on site for only a few months. The work is quite highly skilled, involving a mobile workforce and an international supply chain. If successful, a producing oilfield can be run by a just few skilled technicians.

By contrast, the mining sector works much more slowly and on typically smaller margins. Although exploration drilling programmes can run for well over a decade, the costs of mining exploration are far lower than in the oil sector and, critically, fewer contractors are involved. The work is less pressured and the workforce tends to be more local. If successful, an operating mine usually employs a lot of local people in semi-skilled jobs.

In the oil sector, the emphasis is on making money work, and quickly. In the mining sector, the emphasis is on making people work in a predictable manner – and hoping nobody undercuts the business model. The mining sector, although far poorer, enjoys some significant advantages: the mining developers have a decade to develop a relationship with local people and to train them up to a standard at which they can be active stakeholders in the mine. Mining companies spend a lot of time slowly developing a working relationship with local people that draws people into the companies' decision-making processes on the basis of proximity and mutual benefits. The oil sector has three to four years of part-time activity before it moves into production, which is not enough time to train people to the required standards. While in many countries oil companies have bought their way into trouble with local communities, few have bought their way out of it: lavish social programmes and handouts do not equate to a working social contract.

In developed countries, companies establish the social contract with the communities who live and work around their area of operation within the framework of effective regulation. This is applied through the government and will reflect policies that protect the interests of both communities and companies. In developing countries with little or no effective regulatory capacity, the onus lies squarely on the company to create mechanisms that self-regulate and maintain confidence in the process. This requires a level of effort on par with that provided in developed countries' governance structures. In essence, as the incoming parties, companies have to create a new social contract with the relevant communities and establish new norms, which may be understood as the laws that will frame the evolving social contract.

1.2 The Golden Rule

Following the golden rule – "do unto others as you would have them do unto you" – is straightforward in theory, but in the context of a company making an overture to and with a local community, it requires tremendous dedication to task among a myriad of competing interests and distractions. The local community, when faced with a company, has scant basis for trust. Accordingly, the philosophy around the Golden Rule needs to be shifted past the Shavian acknowledgement that "tastes [aims] may differ", to Popper's "do as they would wish".

Communities tend to resist change where they fail to see what their personal and collective gain is. Similarly to European communities resisting an airport expansion, communities in developing countries often resist a mine or oil exploration as an immediate reaction. European communities generate impact studies, engage in lengthy planning processes and tribunals, and in doing so weigh up the costs and benefits. Communities in developing countries have no recourse to those mechanisms.

Corporate social responsibility is a way of encapsulating the concept that a company's activities can be a 'force for good'. Many companies have carefully worded corporate social responsibility policies that contain references of support to the Universal Declaration of Human Rights or the United Nations Global Compact. This is no simple undertaking. Article 17 of the Universal Declaration of Human Rights states that "everyone has the right to own property alone as well as in association with others" and that "no one shall be arbitrarily deprived of his property", while the United Nations Global Compact requires companies that support it "to uphold freedom of association and the effective recognition of the right to collective bargaining". The implications are clear: the onus is on the company to develop a process to persuade the community to accept a central place in the dialogue and make informed decisions about the proposed developments. This takes considerable time and effort, and communities are frequently poor, uncertain and preoccupied with the business of making a living off their land. As with the opening plays of any relationship, the first contacts are critical in establishing the essential value of trust. However, there is tension between the corporate driver of risk reduction (ie, the cost of getting it wrong) and the stated corporate aims and ideals (ie, the benefits of getting it right). Within the spectrum that lies between least cost/least damage and higher cost/most benefit, companies veer to the former for obvious reasons. The decision itself carries a risk because the costs of getting it wrong, lest there be any doubt, include:

- the depreciation of asset value as the communities that live within the area express and project their dissatisfaction and/or raise rents;
- a breakdown in security and the attendant costs of trying to maintain security in a deteriorating environment;
- class action lawsuits;
- devaluation of the company brand and/or reputation; and
- management time in the company in attempting to mitigate the negative effects of the above.

Controlling that risk simply means that community engagement is not an add-on; it is a core component that needs to be addressed both before and in parallel to any technical considerations. Beside the understanding that positive engagement is both an aim and an ideal, two things stand out:

- the development of the relationship between company and community needs to be planned and built into the project development from the onset; and
- the company needs to accept a significant overall level of effort to manage and react to the relationship as it develops.

In an ideal situation in a developing country where the company proposes to work among unbiased and unaffected communities, the company could begin with a process of engagement and mutual enlightenment. Most projects start slowly and build up. On the way, basic questions need to be asked:

- Is it in the communities' direct interests to accept and support the project?
- If not, how does the company change the project to make it an attractive proposition?
- Can the company accept the risks that may arise if, as in many cases, communities do not want the project to go ahead?

A new context and an unaffected community may represent an ideal scenario to work in since the company can develop a relationship on mutual terms with no prescribed bias. However, it is more likely that the community has prior experience, knows the risks to livelihoods and established norms, and starts from a largely negative stance. Either way, empathy is invaluable. In order for a positive relationship to develop, companies need to understand and appreciate the perceptions of communities, and vice versa. This is hard because it involves the meeting of very different minds, languages and value systems.

1.3   **Case study**

In an area with a 35-year history of oil and gas developments, an oil company is preparing to install an export pipeline that runs from a field development to an existing central processing facility and export hub. The company will spend around $1 billion developing its own central processing facility and the pipeline. The company may subsequently expect to recoup its entire costs in three to four years of operation. The new 135-kilometre pipeline runs broadly east-west through linear sand dunes stretching for tens of kilometres in every direction; the nearest inhabitants are in a Bedouin village over two kilometres from the planned pipeline route in a section that already has a track and pipeline. The route runs over territory that is claimed by a number of powerful tribes. These groups are host to many businesses that compete fiercely for contract work with the oil industry. Following persistent roadblocks and armed intimidation, some existing operators in the immediate area pay a significant number of tribespeople to do nothing. Despite this, roadblocks, kidnaps, sieges, feuds and murders are not uncommon events. (The murder of foreign workers remains rare.)

Experience shows that family-based contracting companies from the local tribes will fight over access to work and on the basis of real or imagined disputes with each other and the company. Behind these incidents appear politicised actors who may direct tribal groups to create situations that may be resolved only through a high level of intervention – and through which new business deals may be established.

The tribe living at the eastern end of the proposed pipeline was asked where its territory ended to the west. This location was marked. At the western end of the proposed pipeline, members of the opposing tribe set their eastern border around 55 kilometres into the first tribe's territory. A previous operator had defined an agreed border about midway between the two points. The tribes were given the following choice:

- work side by side in the overlapping territory;
- agree on a border point midway between their unilaterally designated borders; or
- accept a third party to work in the contested territory.

Although remaining delightfully genial and humorous, the tribespeople roundly refused all three options on the grounds that "nobody knew where the border was" – a proposition somewhat undermined by the naming of individual sand dunes (on tribal lines) by each tribe (and toponymy prescribes primacy of usufruct, if not ownership). What the tribespeople wanted, ultimately, was a grey area in which to compete for both territory, and a small gain in contracting and value at the cost of their tribal neighbours. This is, in its element, a component of the nature of tribal relations and there is a risk inherent in operating through this ontological geography of tribal competition. If that risk is loss of life, that loss carries a cost; compensation for murder in that country is around $35,000.

Between the (international) company and the (national) civil engineering company based far away in the capital that has won the bid to construct the pipeline, the main contract is designed and detailed by engineers and project managers. The contract is detailed on technical matters, but is less prescriptive on how the main contractor shall manage and work with (local) subcontractors. Yet it is widely understood that this is the precise area that contains the greatest risk to the project. The contract was designed with no input from the local groups – that is, the main civil engineering contractor is not contractually obliged to work with local people on the basis of clearly thought-through processes that negotiate each step to the mutual benefit of all parties.

The drivers of the likely stoppages can be mapped out on the next page.

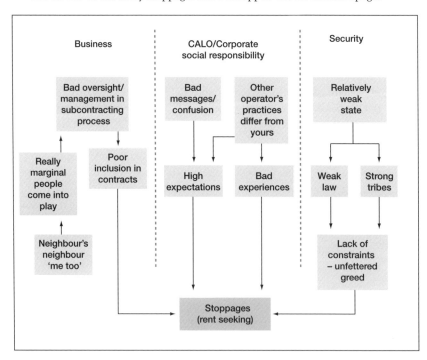

The areas in blue in the chart above are elements that the company can directly address to some extent since they represent the legitimate concerns and interests of people who seek benefits from the work. The areas in brown can be addressed by working at local level with security agents, lawyers and community affairs specialists – that is, those responsible for human security – in developing the confidence and comfort of all concerned.

The communities just outside the field development had felt aggrieved for some time over what they saw as the low level of local hire. To redress this, they had approached the provincial government labour office and purchased a letter of authorisation to supply labour and vehicle hire to the oil company. Despite this piece of paper, the company sent the petitioners away. The communities' reaction was to plan an armed roadblock and stop the oil export trucks. Past experience showed that this would rapidly result in a face-off with the army and the grievances would be relayed to a relatively high authority, with the involvement of senior sheikhs and even the president's office. A payout might or might not be forthcoming. This sort of direct action is both dangerous and costly, and erodes any confidence between the company and the community. When asked if they would bring a civil case against the office that had accepted a bribe to issue a worthless document, the tribespeople said that:

- they would not consider legal action because they themselves had no lawyer; and
- even if there was a lawyer from their tribe and whom they trusted, the legal process was slow and they considered the judges notoriously open to diversion.

When asked whether the situation warranted running the risk of injury or death – two of their tribe had been killed in a similar incident a couple of years previously – they said yes, because in a situation that they create they have both the dignity of self-determination and a rapid result. When asked if the widows and orphans of the two dead men would say the same thing, the tribespeople were affronted.

The company correctly asserted that it acted wholly within the law; it also bemoaned the legal vacuum prevalent in the area of operation. It was then suggested to the company that it simply hire a lawyer to do two things: first, educate the communities about their rights in a weekly series of visits and simple handouts (starting with land law and water law) and, second, represent, at its cost, the community in any case that the community may wish to bring.

The company rapidly refused on the grounds that the community would immediately bring a case against the company. The company was then asked to consider the relative risks and costs arising from legal action, armed action or both, and the connections and trends that may lie between them. The communities' ire was not addressed and remains largely unresolved.

The company has established a network of community affairs liaison officers hired from each of the major tribal communities. These individuals are tasked with acting as the company's spokespersons and gathering information about the communities that may lead to the company designing small-scale development interventions. Typically for oil companies, the budget for community development grants is never more than a fraction of a percent of the company's capital expenditure in that country.

During a strategic planning session with a group of senior managers examining the security issues around the pipeline project, a manager, quite properly (and perhaps bravely) made the following observations: Of a total population of 20,000 people, possibly 10% of males of working age constitute a risk to the operation – that is, 2,000 people whom the company could employ as guards, essentially paying them to be on-side during the construction phase. The cost of paying these men $5.70 a day would equate to under $3.5 million. This in turn equates to under 0.5% of the projected capital expenditure. Surely the simplest thing to do is just to pay these people off?

In response it was pointed out that:
- one should consider the long-term and corrosive rent-seeking behaviour that is reinforced by paying people to do nothing except ostensibly protect the pipeline construction activities from being attacked by themselves; and
- it was somewhat difficult to discriminate between people directly affected by the project and those around them; the 'me too' principle would likely draw in a far greater number of potential protagonists. How would one stop paying them simply because the construction work was finished?

Despite the rent-seeking and history of banditry, what is apparent to anyone in contact with the communities is that people simply want meaningful work.

It is one thing to encourage collaboration between contractors and to develop a dialogue with local communities; it is another to consider the worst case scenarios and act proactively to match the negative drivers with strong influence chains. Much of the armed aggression is driven by competing and malign business interests led not by local people, but by businesspeople in the capital city. The simple identification of the leading protagonists opens great scope for drawing influence into the company's and community's favour but under control of the company. Bridging the gap can be achieved in a number of ways.

1.4   Bridging the gap

Companies are not development organisations. The oil and gas sector is led by highly trained engineers, geologists and financial analysts who collectively hold different skills and perceptions to those found within development organisations, as well as to those found in relatively poor rural communities. It is hard for an oil company to learn to empathise and integrate a learning process in collaboration with local people. As a globalised industry, the oil and gas sector is predominantly English-speaking, while the upstream is based predominantly in non-English speaking countries. Without communication, there is no relationship and without a common language, there is no communication. Thus, the relatively poorly educated and poorly resourced local communities are at a disadvantage in their capacity to communicate with the company. Ostensibly, the onus should be on the incoming entity – the company – to learn the language of the people among whom it operates. In practice this does not happen.

As in any relationship, trust is not created immediately or easily between peoples with very different core values. For example, the relative view of land is very different between communities and companies: Local communities are predominantly agrarian, stable, traditional in outlook and ostensibly conservative. In the most direct and immediate sense, the land from which they derive a living belongs to them and they to it, as farmers and pastoralists; the land is quite literally the foundation of their life and any impact upon it changes and challenges their human security. Primary extractive companies are not so much connected to the land as to the resources that lie beneath it; in many ways the surface is an obstruction since it stands between the company and the resource. The company is at an immediate disadvantage because the community will not easily trust any entity that seeks to change the land on terms that are not those of the community. However, a company can engage in a careful negotiation. A social and environmental impact assessment carried out in 2004 for a major pipeline in the same country as the case study provided above noted that:

> The socio-cultural importance of territory and land ownership is linked to collective identity and personhood in the tribal system. Tribal peoples draw strong links between the concepts of personhood and a sense of belonging, which are anchored for each tribe to a discrete territory or area of land. It is in defending that territory that 'honour' is earned – a concept central to the identity of a tribesman, and that which confers the authenticity of 'descent' and collective heritage. 'Honour' for tribesmen is understood as the essence or quality that distinguishes them from non-tribal, weak people and must be defended…This point is reiterated by another tribal term mu'awwarat, meaning 'things on the inviolability of which honour depends, and hence things in defence of which one's brothers' should offer support against outsiders'. A tribesman, when asked for a list of things that are defined by mu'awwarat will often mention land, first and foremost. The term and the concept of land itself are thus often virtually synonymous with the notion of 'ard' or honour defended.

Just before this report was published in 2007, the general manager of the project was interviewed in the presence of the author of this chapter:

"How are you going to manage the issues of cooperation and compensation with local communities?"

"There is nobody there."

"That's odd, because the pipeline will cross several wadis which are intensively farmed, and passes in proximity to quite a lot of rainfed farming on the plateau."

"Under our development agreement with government, the company makes a single payment to government and they handle that."

"Do you think that any of that funding will reach the farmers out in the fields."

"That is their problem."

"You must know that if your contractor's bulldozers start to cross the land of farmers who have not been compensated, the farmers are well armed and will stop all work - and that's your problem."

"We have the army to protect us."

"Have you looked into the costs of the military security cordon around the existing oilfields in terms of the costs to the operators and to the political stability of the region, and compared those ongoing costs to the potential costs and benefits of paying proper compensation?"

No answer was forthcoming. Four years later, in another interview, the same general manager said that he was spending 60% of his time dealing with local conflict issues.

Within the structure of government licence, private company interest and community ownership and rights, the company is typically placed in the role of wishing to develop an asset within or around a community, the company's asset being seen, by that company, as an entity distinct from the community. If a company can tacitly acknowledge that the government has scant or negative agency within the community, then it should also invert the concept of rights and ownership. The government gives licence; however, the real stakeholders are the community. How would the situation evolve if the community wanted to develop the asset? This question challenges the assumptions behind management and control. The communities are the *de facto* owners of the asset and may lose livelihoods in the event that the asset is developed. However, if the communities themselves wanted to develop the asset, they would likely need a company to provide the financial backing and expertise. Imagine a company operating at the behest of a community and on the community's terms. If this seems unlikely or unfeasible to European norms, consider that in the United States, the landowner holds the mineral rights; accordingly, mines operating on tribal reservations pay all their royalties and production taxes directly to the tribe. This does not resolve all the problems, though, because financial gain does not equate to control over resource management or positive social change.

In the absence of the structures of effective government regulation and socio-political equity, both sides have to forge levels of trust in transparency. For agrarian communities, their livelihoods are perfectly apparent and transparent: an external observer can see land, crops, livestock; the primary assets of a rural community are visible and easily understood – at least by each other. Faced with an external observer, most rural people feel that their lives are stripped bare. By contrast, the workings of an exploration company are diffuse, and the financial flows within it are relatively opaque and the hierarchies largely invisible. Again, the onus is on the company to make its structure and intentions as transparent as possible. For obvious reasons, this can take considerable time and effort. How can people who may be illiterate or have no direct experience of major infrastructural or commercial developments become competent stakeholders?

The architecture that defines the interface between people and company is important. To describe this, one may outline two exploration assets (a gold mine and an oil exploration site) that sit within 200 kilometres of each other.

The goldmine was worked for eight years in the 1970s and 1980s and is undergoing a resurgence of exploration drilling in the hills behind the original site. The drilling programme is expected to run for seven to 10 years. The drilling camp itself is a set of tents and awnings set beside the single track that leads up the valley. The camp is equipped with two television sets (one in Arabic, one in English), both placed outside or in communal spaces. The camp also features a satellite uplink with wireless broadband internet available throughout the camp. The drillers, predominantly plain-spoken South Africans, also have small patches of land outside their tents or below the latrines, where they grew maize, sunflowers and pumpkins. Local people clearly understand the farming element and the drillers' itch to use waste water. Despite the tiny patches of green, there are no fences and no guards. It is noticeable that cars are left with keys in the ignition and the windows down. Local people, and their goats, wander freely in and out of the camp. Young men sit close to the canteen for two reasons. First, one exterior wall of the canteen is painted matt black and the cook keeps a piece of chalk with which he writes the names of the local men as they arrive. The next morning the foreman walks up to the canteen and calls out the names in order – first come, first serve; if the man is not there to start work the foreman skips to the next name. As the men take their place by the pickup trucks to go and do a day's work, their names are erased and the remaining names move up.

Second, people lingering near the canteen are invited to eat when the midday meal is cooked, regardless of whether they are working. Water is free; it comes from a clay jar cooler set by the canteen door that is equipped with a set of small stainless steel mugs. On days when things are not going well, the drillers arrange their crews in the pickups and drive 40 kilometres down to the coast to spend the afternoon and evening beach casting and having a barbecue. Many of the labourers had never seen the sea before the foremen took this initiative. The exploration programme, although small, has a full-time community liaison person who organises labour, holds meetings with local leaders and negotiates use of sites for drilling. He operates with a high degree of autonomy. The mine camp buys local goats for meat from shepherds up and down the entire valley on a broadly rotational basis. The economic impact of locally hired labour and water trucks, and locally bought goat meat means that cash income to the valley is 75% greater than before the project.

The oil exploration site is in a licence concession that has also been worked over, on and off, since the 1980s. In effect, the exploration phase on site may run for up to three years of a five-year licence. The drilling camp is a set of portable cabins laid to a grid and surrounded by two rings of chain-link fencing, the second fence set below a 2.5 metre-high security bund. At night the site is lit by arc lights, and the site and access route are guarded by small squads of armed military. Local people are not allowed into the site. There is minimal, if any, casual labour; the labour posts are typically negotiated through a gang master/subcontractor linked opaquely to a services company based in the capital city. The site provides water via a concrete tank some 50 metres beyond the perimeter fence. The services company, or others like it, supplies all the food, consumables and material for the drilling programme. Expatriate drillers arrive on a local dirt strip by aircraft, work their rotation and leave – wholly disconnected from the people among whom they operate. The oil camp has a GSM (global system for mobile communications) dial-up connection, no wireless internet, but a television set in every cabin. The oil company employs two community liaison staff to help negotiate land compensation, hiring of water trucks and limited local spend on civil engineering issues; these liaison staff operate under an expatriate.

The case for transparency is simple: companies know that they cannot reliably make good decisions on the basis of bad data; rural communities are no different. Unless they can see and understand the relevant information, they cannot make a rational, informed choice. Transparency does not come naturally to people in a competitive company environment where norms of information sharing lie firmly at the 'need to know' end of the spectrum – often for internal rather than external reasons. This contrasts strongly with the highly visible and accountable lives of rural communities. Traditional rural communities expect the same levels of transparency from an external entity as they consider normal within theirs. Reciprocal behaviour on all levels may be hard to achieve, but in terms of transparency of information, the onus is on the company to provide as much information as it can without losing commercial competitiveness or risking a conflict of interest.

In a context where good education is a rare privilege, educational grants can be used as a means to draw people toward a more progressive stance. In the example provided, a neo-colonial approach may suggest that the company can offer very attractive grants to the scions of the key business protagonists, thereby gaining considerable influence within the relevant families – a strategy not as unusual as it may first appear. At the other end of the spectrum, but equally challenging, is the notion that effective change can come only when the company's decision-makers have a clear and first-hand view of the perceptions of the people whose livelihoods are affected by the proposed developments. It is rare for top-rung decision-makers to spend time in and among rural communities; it is also rare for them both to listen directly to the communities' views and concerns, and to develop a first-hand understanding of the implications of the development upon those people's lives. The internal structures of a company usually act as a potent filter in translating the subtleties of peasant life into a handful of figures on a PowerPoint slide. The reality may be uncomfortable, but if the roles were reversed – as the golden rule says that they should be – the outcome will cut to the heart of the matter: in order to forge a positive working relationship, companies need to understand communities as much as communities need to understand companies, if only to minimise risk.

Trust is also based on access. Beyond the remit of professions and institutions, individuals in a society interact in ways that broadly equate to the lives of rural communities. Trust between individuals is also in part based on access – the ability to find and talk directly to the people who matter. It is relatively simple for a company to communicate with a community, but much harder for the reverse to happen. It is obvious that the only way to establish trust is diligently to establish a direct link between the company and the community by keeping an open door policy and by maintaining some continuity in the people involved. Companies come and go, staff rotate in posts, all of which is very different to stable and rather conservative rural communities. It may be a truism, but it is the personal relationships that will carry the weight of the negotiations and understandings, and these personal relationships take time to develop; they are only as solid as the stability and continuity that they are given. The implication is that the company must employ or elect an employee to be accessible within the community on a long-term basis, rather than a series of itinerant consultants or visitors from the office.

The higher ranking the person making the direct contact with the communities, the better. People understand authority and respond to the compliment that is implicit if faced with the interest of an important person. If the primary company contact in a community is no more than a messenger, the medium really is the message. The experience of community/company negotiations is littered with instances where the community has quite suddenly and significantly shifted its demands when finally faced with the ultimate authority, thus embarrassing and marginalising the individuals who worked on the lead-in, as well as destabilising the outcome. Where the primary contact from the company has held and demonstrated requisite authority, the predicted path is less likely to change: there is both credibility and continuity. To achieve this, the office-based engineers, geologists and financial analysts have to relinquish a modicum of control and authority to someone based in the field who may not have a natural science or technical profession. Trust starts within the company.

This is where it starts and ends: in the mindset of people at the helm of major investments. If, as is often the case in mining operations, they see the community as an integral part of their investment, the surrounding community and communication with it are integrated into every aspect of the operation. By contrast, in many oil operations the company simply does not need local people; its managers will see the community as irrelevant to their business interests, if not a threat, and the community is set on the periphery, beyond the security and asset protection mechanisms. It is not hard to alter the mindset, since being open cuts costs and risks. It merely requires a shift in beliefs, away from trust in fences and programmes, and toward a simpler trust in human nature.

2. Land issues

The environmental impact consultant clambered up a steep ravine at the top of which was a Chinese drilling rig. The slopes below the rig site were liberally scattered with cable drums, oil drums, old tyres, oil filters and thread protectors – the hard waste debris of any drill site. The waste pit that the drillers had created sat right on the edge of the escarpment and its earthen wall was failing – a mixture of cuttings and hydrocarbons was just beginning to ooze downslope, a breach that ultimately deposit hydrocarbons into the valley below. The consultant squatted on one end the wall and took a few photos, then rather gingerly moved to the other end of the pit and took some more. On the way he passed a group of rig workers, one of whom followed him to the edge of the pit and finally spoke up:

"Who are you? Are you a geologist or an engineer?"

"Neither, I'm looking at the environment."

The rig worker looked a bit perplexed, then said: "There's no environment here, no bear, no tree, just desert and mountain! No environment."

How people see land determines how they use and value that land, and vice versa. In the example above, and quite apart from the many flocks of goats in the valley below, the area contains over 120 species of plant – a baseline environmental descriptor not dissimilar to habitats in temperate Europe. The rig worker apparently could not see the value in the natural environment in the same way as the local people.

Land is literally the foundation of nations and who owns what, how and why is at the core of the development struggle in which industrial developments are merely a sideshow. Land tenure itself is the single most significant driver of equity and capital in any emerging market; the land owned by the poor in developing countries is often the largest single asset of all. To quote Hernando de Soto: "In Haiti, untitled rural and urban real estate holdings are together worth some $5.2 billion. To put that sum in context, it is four times the total of all the assets of all the legally operating companies in Haiti, nine times the value of all assets owned by the government, and 158 times the value of all foreign direct investment in Haiti's recorded history to 1995."[1]

---

1   De Soto, Hernando, *The mystery of Capital: why capitalism triumphs in the West and fails everywhere else*, Basic Books (2000).

2.1 Tenure, tribes and usufruct

Land falls broadly into two kinds of ownership: public and private. But there is a third kind of land that is both prevalent and complex in nature: communally owned land or, more rightly, communally used land. The levels of ownership essentially lie in parallel to the productive value of the land. Thus:
- low productivity = low value = low levels of cadastral survey and individual ownership; and
- high productivity = high value = high levels of cadastral survey and individual ownership.

There are whole sets of markers applying to both land tenure law and individual perspectives of ownership to describe the spectrum, from owned and demarcated land through to not owned and not demarcated. These markers include:
- cadastral survey data that describes boundaries and ownership;
- cadastral survey data that describes boundaries by design – that is, the area in question;
- cadastral survey data that describes boundaries by default – that is, areas surrounding the area in question;
- formal written records of ownership or records/accounts of primacy of rights of use ('formal' in the sense of being lodged with a governmental authority);
- informal written records of ownership or records/accounts primacy of rights of use (informal in the sense of being lodged with a community or tribal authority – this can come down to household-level records, deeds of inheritance or historical tracts);
- permanent visible markers – that is, fencing or walls;
- temporary visible markers – that is, signs on plants or planting/manipulation of specific plants;
- frequent use – that is, grazing or other use of natural resources;
- infrequent use – that is, very sporadic grazing and use of scarce natural resources (some areas of desert grazing are used only every five to 15 years); and
- simple community knowledge and acceptance – that is, common cognitive maps or indigenous geospatial awareness.

While land has the capacity to generate risk broadly in correlation to its value, the problems with land ownership tend to occur in the middle ground, in the areas that have value, that are subject to some form of ownership, and that are nevertheless not owned in the formal sense of having a known or recorded border whereon ownership changes.

Land tenure law is based in formal constitutional law. However, in many instances there is at least one more form of law in operation: religious law, or traditional or tribal law, with traditional laws taking precedence in rural areas. In general, the poorer the land and the poorer the people, the more likely they are to be organised around traditional or religious laws and, thus, the more likely they will fall outside the contractual frameworks created between companies and governments. This is not always the case, as people may own land of minimal inherent productivity but of strategic value, such as a pass from which they can extract a rent from people and goods in transit. The development of alternative routes – for example, a pipeline service track – may rob the community of an income.

Inheritance laws mean that single parcels of land can have multiple owners, such as a farm that has, by inheritance, been broken up into several fragments, but that may be 'owned' by one family, farmed by one person – essentially, a sharecropper – and yet owned by many whose ownership is expressed by retrieving, with equal or different percentage, shares of a harvest. A harvest can be divided since it is not derived from a specific piece of land within the farm. However, the impact of routing a pipeline through the farm raises the issue of ownership of a specific section of the farm and, therefore, the issue of whether loss of revenue/compensation is due to an individual or all owners.

One frequent problem in developing countries is the absence of an effective national land registry that operates a service at a cost that is accessible to ordinary people. The only formal systems for registering land exist for urban areas (a result of both cost and access). Beyond the formal state structures, tribal systems can be very robust and reliable, but the parallel systems of tribal land tenure and the state's insistence on an open market can come into conflict.

Where tribes and tribal organisations are very strong, they are highly capable of looking after their interests and of coordinating with companies. Tribes frequently have formal arrangements for dealing with conflict and mediators are paid for their services in settling disputes between families and communities. While there is no standardisation and the quality of outcomes is varied, there is a broadly understood set of precedents and strategies to manage conflict.

Usufruct (or rights of use) is very often applied to areas of grazing land, marginal cropland or forest land where the intensity of resource management is low. In a jungle, each and every important tree can be 'owned' by an individual or a family; the management of non-timber forest products, and the management and rights of individual families to hunt in specific areas is clearly defined and frequently demarcated. Where, for instance, the key resource is sparse, low-density grazing under highly variable rainfall, there is still primacy of rights of use: "The land 'belongs' to tribe X". However, any other person is formally allowed free passage and grazing on a transient basis, on a reciprocal basis or on terms otherwise agreed. The harsher the environment, the more mutual assistance and resource-sharing is prevalent.

The relative stability of peoples living on the least economically productive land should not be underestimated. A recent line survey for a proposed pipeline ran across a small valley on the floor of which a few graziers camped with goats and sheep. The graziers live with the land, as opposed to living off it; this dislocation is central to the outside perception that they are poor or primitive. The men were asked for the name of the valley and they replied in their strongly accented dialect. Although they are illiterate and neither they nor their forebears have ever attended school, the name that they gave was a direct transliteration of an inscription on a rock at the other end of the valley, some seven kilometres away. The inscription is possibly 2,700 years old and in a language unspoken for over two millennia. Their toponymy suggests an unbroken verbal tradition for a lineage of occupation by people who have only minimally altered the landscape.

These elements can be mapped out in the chart below.

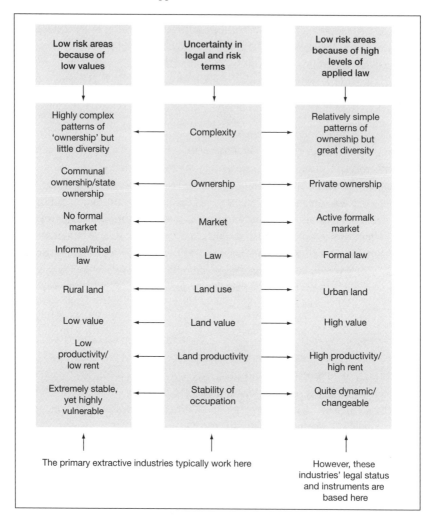

It is not a hopeful scenario because there is often a gap between the company's perception of land, land ownership and land access, and the perception of the relevant communities.

2.2 Risk and conflict

The risk of conflict over land grows with each situation described below:
- Introducing legal dislocations and disparities;
- Introducing increases in land values or rents in a formal legal vacuum; and
- Entering environments with existing land conflicts.

The paradox for the primary extractive industries is that they move onto land with a legal licence to operate – a licence couched in constitutional rights, but also primarily framed by perceived resource values and monetary values. However, the legal framework (the practical application of formal land law) stopped somewhere in the vicinity of the capital city, to be gradually and increasingly replaced by a complex resource use-based indigenous legal system or systems as one moves into the hinterland.

Risks can arise as a result of transposing the expectations held within formal rights, where individuals own or buy access to land on the basis of formal law, into an environment populated by people with very different concepts of ownership. Where people live off the land on a communal basis, they belong to their land, rather than the other way round.

Another form of risk comes in the manner by which companies increase the value of land, merely by introducing a high level of economic activity into an environment where there was previously very little economic activity. Thus, a pipeline running through a desert increases the value of that desert out of proportion with the land tenure systems that otherwise govern that environment. When this happens, conflict can occur due to a breakdown in the traditional land tenure and usufruct systems to deal with the added stresses and demands of competing elements in the surrounding communities.

State systems of registering land can be corrupted with new land titles generated and sold in a way that simply adds to the confusion. The interests of commercial ventures, many of them working with state sanction, try to reduce costs by achieving ownership or control of land through the most advantageous means. The interests of tribes typically force commercial interests to comply with their local tribal concepts of ownership and raise land values to a mutually acceptable level. Problems can arise in the valuation of land where there is no monetary market for land. The Guidance Notes for the International Finance Corporation's Performance Standards points out that:

> *Compensation for land and other assets should be calculated at the market value plus the transaction costs related to restoring the assets. In practice, those who suffer negative social and economic impacts as a result of the acquisition of land or land use rights for a project may range from those having legally recognized rights or claims to the land, to those with customary claims to land, and those with no legally recognized claims, to seasonal resource users such as herders or fishing families, hunters and gatherers who may have interdependent economic relations with communities located within the project area. The potential variety of land or land use claimants renders the calculation of full replacement cost difficult and complex.*[2]

Risks also arise from operating in environments with existing conflicts over land. The pattern of land tenure conflict falls into two main areas: rural and urban. Within this division there are two types of conflict: those involving internal or local actors,

---

[2] International Finance Corporation, Guidance Note 5, Land Acquisition and Involuntary Resettlement, "International Finance Corporation's Guidance Notes: Performance Standards on Social and Environmental Sustainability" (2007).

and those involving external actors. Each type of conflict carries specific characteristics.

	Rural areas	Urban areas
Internal actors	• Indigenous conflicts over land or critical resources – long running? • Indigenous conflict management structures	• Political or legal conflicts over land holdings or developments • Rights-based conflicts • Institutional and formal legal conflict management structures
External actors	• Resource-based conflicts, 'land grabs' • Resettlement schemes • Rights-based conflicts • The meeting of formal institutional and indigenous conflict management structures?	• Political and legal conflict over significant developments • Rights-based conflicts • Institutional and formal legal conflict management structures

Almost by default, the primary extractive industries predominantly fall into the 'rural and external' category, which is characterised by that interface between the institutionally based formal legal instruments and the apparent chaos of indigenous community-based conflict management mechanisms.

In rural areas, conflict over land can arise as a result of contested ownership subsequent to heredity – that is, disputed claims to ownership of land within families – or disputes over more marginal pieces of land in tribal borderlands. Marginal land may have minimal value apart from providing grazing following rainfall. However, rural land conflicts also occur over the rights to work. Similar to the desert pipeline example above, the introduction of an access track and a drilling should not be underestimated. Likewise, the ability to supply water to a drill site significantly changes the potential value of otherwise rather insignificant pieces of land.

Tribespeople with otherwise limited livelihoods will fight bitterly over the right to work on the basis of land ownership. If they can assert ownership of the land, however marginal or useless, as 'owners' or custodians of the land, the people may feel that they have the right to provide goods and services to the company operating on that piece of land. And because companies under tight schedules typically pay high prices for their goods and services, it is not uncommon for tribes to fight to assert their rights.

In practical terms, the vulnerability of land to conflict over ownership is likely to correlate with land use. A field is a discrete area, developed in the sense that it is worked by the owner or share-cropper. Thus, ownership of a field is both known and can be defended on the basis of visible physical boundaries: stone markers, a bund, trees or simple sticks. Usufruct (rights of use) for grazing land is also known, but tends to have fewer well-defined physical boundaries. For instance, claiming "this valley is ours" is fairly straightforward, but up on the watershed the demarcation becomes diffuse. Likewise, if a river or river bed forms a boundary, that boundary may not be exact as questions may remain as to, for example, which side of the river bed, which can be 200 metres wide, is the boundary. The level of detail in land ownership is prescribed by the intensity of land use or the distribution of critical resources such as a water well or a specific set of trees. However, grazing land is typically owned or controlled by communities within which there is a hierarchy that can be exploited.

Land use is not homogenous; among farms and villages there will be landless labourers who vie for access to woodland resources or areas of land that are not farmed in the sense of having owners of demarcated fields. What one sees are farms. The first contact with people in the area will be with villagers and farmers. Yet the community includes a sizeable portion of farm workers whose livelihoods in part depend on access to commons. This unseen group of people lives below the level of legal formality and, while they are extremely vulnerable, they are also hard to draw into concepts of compensation for assets lost.

There is also an inherent tension between farmers and hunter-gatherers. As Hugh Brody puts it: "The profound dichotomy that has shaped the agricultural era may indeed therefore lie in an opposition between nomads and settlers, between people for whom home is place of timeless constancy, a centre in which humanity itself arose, and those who are on the move and, if at rest, rest only while preparing for further movement. The paradox, of course, is that this is the divide between the settled hunters and the nomadic farmers."[3]

Beyond internal policies and the requirements of law, companies do not really have a choice about whether to consider land rights in their dealings with rural communities. Land is nearly always a highly contentious and charged issue, and populations all over the world have the capacity significantly to increase costs for companies if they are not happy with the outcome.

Companies cannot address issues of land ownership or become active participants in resolving land conflicts without becoming, by default, stakeholders in the outcome – good or bad. However, companies can link protagonists to independent legal support mechanisms or human rights organisations. The law is important because, as de Soto points out:

What the poor majority in the developing world do not have, is easy access to the legal system, which, in the advanced nations of the world and for the elite in their own countries, is the gateway to economic success, for it is in the legal system where property documents are created and standardized according to law. That documentation builds a public memory that permits society to engage in such crucial economic activities as identifying and gaining access to information about individuals, their assets, their titles,

---

[3] Brody, Hugh, *The other side of Eden: hunter-gatherers, farmers and the shaping of the world*, North Point Press (2001).

rights, charges and obligations; establishing the limits of liability for businesses; knowing an asset's previous economic situation; assuring protection of third parties; and quantifying and valuing assets and rights. These public memory mechanisms in turn facilitate such opportunities as access to credit, the establishment of systems of identification, the creation of systems for credit and insurance information, the provision for housing and infrastructure, the issue of shares, the mortgage of property, and a host of other economic activities that drive a modern market economy.[4]

2.3 Valuing land

The precautionary principle of 'doing no harm' is not simple to analyse or address in the context of developments on rural land. For instance, if land that was previously used for agriculture was sold for well below market value to a commercial concern, the outcome can be viewed in two ways:

- The commercial concern is of national importance in the development of a national market and framework. Therefore, it is wholly expedient to see the entity establish a base and create a modern state; the greater common (commercial) good should prevail.
- The farmers have been duped into selling their land below market value and lost a part of their livelihood to an external corporate interest. As a result, there is increasing disparity and social inequity, and this is dysfunctional in the development of the state.

There is of course a middle path to be followed between the two extremes outlined above, and there is plenty of scope for reconciling interests and achieving mutual benefit. This is exactly the concern of Guidance Note 5 on the International Finance Corporation performance standard on land acquisition and involuntary resettlement quoted above. However, the Guidance Notes confirm that:

If the affected people reject the client's offer of compensation that meets the requirements of Performance Standard 5 and, as a result, expropriation or other legal procedures are initiated, the affected people may be offered compensation based on the assessed value of the land, which may be less than the compensation required under Performance Standard 5. The matter may proceed to litigation and may take a number of years to be resolved. The court's final determination may confirm compensation based assessed value. Because there is a risk of impoverishment from loss to the income base or livelihood of the affected people or communities from a protracted process and depressed compensation, IFC will ascertain whether such expropriation is consistent with Performance Standard 5 by requesting information on the level of compensation offered by the government and the procedures used under such expropriation. In addition, the client should explore opportunities during these expropriation processes to play an active role in collaboration with the responsible government agency to achieve outcomes that are consistent with the Objectives of Performance Standard 5. Whether the client will be permitted to play an active role will depend in part on the applicable national law

---

4   See footnote 1 above.

and the judicial and administrative processes and practices of the responsible government agency.[5]

The problem lies in the paradox that people who live closest to the land are often the poorest, the most vulnerable and the least asset-rich. The people who have houses, incomes and education have assets and abilities beyond the land – that is, things that can be replaced and replicated in another environment. By contrast, a hunter-gatherer has minimal assets and very few of his/her skills are transferable to another environment. His or her identity is tied to his/her environment. To put it another way, it is impossible to compensate an Eskimo for loss of ice since one cannot reasonably replicate that cultural identity.

The onus is on the company to address the issues in an inclusive and appropriate manner since the public acceptability of development projects – and free prior and informed consent to projects by indigenous peoples – is rooted in international human rights law and international agreements on sustainable development, the principles of which lie at the core of corporate social responsibility.

2.4 What to do

The core issue is to consider that it is not so much the cost of trying to address land tenure issues that should be examined, as the cost of not addressing it.

(a) *Slowing down/studying the field*
Rural people cannot rapidly adapt to the introduction of commercial concerns. Commercial concerns cannot rapidly gain an adequate understanding of land tenure systems. The development of solid and qualitative understandings takes effort and time, and is often an uphill task. It can only pay off, but the time and effort required should not be underestimated.

(b) *Strengthening land registration facilities*
It should be simple to fund geographical information system and cartographic laboratories, and develop reasonably robust systems. Land titles are meaningless unless there is effective judicial process to arbitrate disputes and this judicial process is backed by the supreme authority. In most developing countries, this is highly unlikely.

(c) *Strengthening organisations working on land rights issues*
It is possible to fund the activities of non-governmental organisations dealing with human rights, but these groups typically occupy very restricted political space. It is hard to develop effective civil society actors and care must be taken to preserve impartiality, thus funding them via a third party. The company must be pragmatic: there are opportunities to develop skills at grassroots level and to strengthen legal access, but the levels of traction achieved are minimal when faced with powerful interests.

---

5   See footnote 2 above.

(d) **Strengthening legal access mechanisms**
It has been argued that recourse to the law can be effective only if the law is built from the ground up, not the other way round. The links with land law are obvious since the social contracts enshrined in traditional or informal land law should be the basis for formal law. Thus, the utility of the law will be tied to its function in the societal context. However, this approach is slow. It requires a given level of literacy and, above all, broad buy-in from national governments. In most developing countries, the bond between government and the population is weak and in many areas simply non-existent.

(e) **Strengthening the national asset**
In the end, by default if not design, strengthening the national asset to work with confidence and inclusion is the most cost-effective option for corporate actors. National offices and staff will have the capacity to work effectively in situations where there are complex land issues involved. Companies have been successful where they have consistently focused on a very simple humanitarian policy. Around this, and where appropriate, companies can address land issues through careful negotiation at village level. However, a structured debate around the potential to include land conflict resolution issues in the company's mandate, as agreed by corporate interests and by governmental partners, could be useful.

About the authors

Katie Baehl
Associate, Baker Botts LLP
katie.baehl@bakerbotts.com

Katie Baehl is a second-year associate at Baker Botts LLP. Her practice focuses on international and domestic oil and gas transactions. She earned her BA (in plan II honours and history) and her JD degree from the University of Texas at Austin.

David Blatchford
Partner, Infrastructure Development Partnership LLP
dblatchford@infradev.co.uk

David Blatchford is an environmental scientist with 30 years' business management and environmental and social consulting experience. He has worked mainly with multinational companies in the energy and mining sectors.

Mr Blatchford has developed and managed consulting practices in Australia, the United States, the United Kingdom (all as a vice president of Dames & Moore/URS) and Africa. From 2001 to 2008 he was closely involved in the Baku-Tbilisi-Ceyhan pipeline project in a wide variety of technical capacities. During this time he also consulted for ChevronTexaco, BP, the International Finance Corporation, Peru LNG and the European Bank for Reconstruction and Development. He retains a consultancy role with the Baku-Tbilisi-Ceyhan pipeline.

For the past two and a half years Mr Blatchford has worked on ExxonMobil's Papua New Guinea liquefied natural gas project in a number of environmental and social roles. He is currently the environmental and social lender interface lead for the project.

William E Browning
Partner, Infrastructure Development Partnership LLP
wbrowning@infradev.co.uk

William E Browning has 28 years of experience in the oil and gas industry and is a partner with Infrastructure Development Partnership LLP in London. IDP provides an array of project development support activities, including non-technical risk assessment and mitigation strategy, commercial, structuring, financing and environmental advice and strategy.

Mr Browning directed the legal work for the largest upstream development in the South Caspian Sea from inception to full production and for all export development from the South Caspian between 1995 and 2002. From 2002, he served as the Baku-Tbilisi-Ceyhan pipeline's negotiations manager, concluding with its successful $2.6 billion project financing. More recently, he has provided project development advice for a number of developments, including a greenfield urea plant development in Peru, a cross-border gas pipeline from Trinidad to other East Caribbean islands and the TransAdriatic pipeline.

Mr Browning is a member of the Texas Bar and holds a BA from Washington & Lee University and a JD from the University of Texas.

About the authors

R Coleson Bruce
Associate, Baker Botts LLP
coleson.bruce@bakerbotts.com

R Coleson Bruce grew up in northwest Arkansas and attended Cabrillo Community College before transferring to and obtaining his BA degree from Yale University. Mr Bruce received his JD degree from the University of Texas at Austin and is now a second-year associate in the oil and gas section of the global projects department of Baker Botts LLP, counselling clients in a broad range of energy-related projects and transactions. Before joining Baker Botts LLP, Mr Bruce served as a law clerk to the Honourable William R Wilson of the US District Court for the Eastern District of Arkansas.

Alexandre Chavarot
Head of clean energy finance, Clinton Climate Initiative, William J Clinton Foundation
achavarot@gmail.com

Alexandre Chavarot is head of clean energy finance at the Clinton Climate Initiative (a programme of the William J Clinton Foundation), where he also coordinates solar activities.

Mr Chavarot is responsible for the structuring and financing of large-scale pilot projects promoted by the Clinton Climate Initiative in the fields of solar power and carbon capture and storage. The initiative acts as a facilitator between governments, industry and finance to address the issues that would typically prevent these projects from being implemented.

Prior to joining the initiative in 2009, Mr Chavarot spent 15 years in investment banking, mostly with Lazard Ltd in London, where he was responsible for project investment advisory activities within the firm's energy and infrastructure group.

Mr Chavarot is a graduate of the *Institut d'Etudes Politiques de Paris*, and holds a master of public policy degree from the Harvard Kennedy School of Government and an MBA from INSEAD.

Tom Cummins
Associate, Ashurst LLP
tom.cummins@ashurst.com

Tom Cummins is an associate in Ashurst LLP's litigation department in London and a member of the firm's international arbitration group. He has advised clients on a range of disputes, with an emphasis on the energy and mining sectors. Mr Cummins has experience of a number of different forums, including English court litigation, expert determination and international arbitration.

Thomas J Dimitroff
Partner, Infrastructure Development Partnership LLP
tdimitroff@infradev.co.uk

Thomas J Dimitroff specialises in structuring and negotiating complex cross-border oil and gas transactions. He worked for BP plc for 10 years as a member of the core negotiating teams that delivered the Baku-Tbilisi-Ceyhan and South Caucasus pipelines. He was also the regional adviser to BP plc for Africa, the Middle East, Russia, the Caspian and Turkey, advising on non-technical risks posed to new ventures and existing operations across the region.

Since leaving BP, Mr Dimitroff has worked on a variety of large-scale cross-border oil and gas pipeline developments and upstream acquisitions on behalf of project development companies and private equity funds. He has advised governments in complex oil, gas and related infrastructure negotiations, and has provided global extractive companies with comprehensive assessments on social impact and governance risks.

Mr Dimitroff holds a BA in philosophy from Loyola University of Chicago, an MA in law from St Edmunds College, Cambridge, a BCL from Jesus College, Oxford and a JD from IIT Chicago-Kent College of Law.

About the authors

Kevin Gardiner
Managing director and head of global investment strategy, Barclays Wealth
Kevin.Gardiner@barclayswealth.com

Kevin Gardiner joined Barclays Wealth in September 2009. He is a member of the investment committee that shapes asset allocation views for client portfolios, and leads a team of analysts based in London, New York, Singapore and Mumbai.

Prior to joining Barclays Wealth, Mr Gardiner was global head of equity strategy at HSBC's investment banking unit in London, which he joined in 2003. He has worked at several other investment banks and the Bank of England in a financial market career spanning over 20 years. In 1994, while working as an economist at Morgan Stanley, he wrote the Celtic Tiger Report on the Irish economy.

Mr Gardiner was educated at the United World College of the Atlantic, the London School of Economics and Cambridge University. He is a governor at Atlantic College and a retained speaker on the programme of the CFA Institute (Global Association of Investment Professionals).

George F Goolsby
Partner, Baker Botts LLP
george.goolsby@bakerbotts.com

George F Goolsby has more than 36 years of experience in the international hydrocarbon transportation industry. He specialises in cross-border projects. He currently advises on the Eastern Caribbean gas pipeline project. Previously, he worked on the North Border pipeline project crossing Canada and the United States, the so-called 'peace pipeline' involving Egypt and Israel, and the 'early oil projects' connecting offshore Azerbaijan in the Caspian Sea to ports on the Black Sea. Later he coordinated the legal team for the 1,200-mile-long Baku-Tbilisi-Ceyhan oil pipeline and advised with respect to the South Caucasus gas pipeline, both of which traverse Azerbaijan, Georgia and Turkey.

Mr Goolsby formerly headed Baker Botts LLP's Moscow office and is resident in the firm's Houston office. He is a member of numerous legal and energy industry associations. He earned both his BA and JD degrees from the University of Texas at Austin.

Deborah L Grubbe
Owner and president, Operations and Safety Solutions LLC
deb.grubbe@comcast.net

Deborah L Grubbe is the owner of Operations and Safety Solutions LLC, a consultancy firm that specialises in safety and operations troubleshooting. Ms Grubbe is the former vice president of group safety for BP. She was trained in the characteristics of safe operations during her 27-year career at DuPont, where she held corporate director positions in engineering, operations and safety. Ms Grubbe is a member of the National Aeronautics and Space Administration Aerospace Safety Advisory Panel, and served as a consultant on safety culture to the Columbia Shuttle Accident Investigation Board. She sits on the Purdue University College of Engineering Dean's Advisory Council, is a trustee of the National Safety Council and is chair of the Institute for Sustainability.

Ms Grubbe obtained a bachelor of science in chemical engineering with highest distinction from Purdue University, and received a Winston Churchill Fellowship to study chemical engineering at Cambridge University in England.

Leigh Hancher
Of counsel, Allen & Overy LLP
Leigh.Hancher@Amsterdam.AllenOvery.com

Leigh Hancher is professor of European law at the University of Tilburg and of counsel at Allen & Overy LLP, Amsterdam. She holds an LLB (Hons), an MA in socio-legal studies and a PhD in law. She specialises in European energy law and has authored numerous academic and practical

About the authors

publications in this area of the law. She has advised a number of clients on the application of EU law to large cross-border energy infrastructure projects. She acted for the power generators planning the first links between the Nordic and the North European markets in the late 1980s and she remains a leader in this area of the law.

Judith H Kim
Partner, Ashurst LLP
judith.kim@ashurst.com

Judith Kim is a partner in the energy, transport and infrastructure group of the Dubai and Abu Dhabi offices of international law firm Ashurst LLP. Ms Kim specialises in energy and petrochemical project development and mergers and acquisitions involving energy assets. She has advised on a number of energy and petrochemical transactions in the Middle East and North Africa, Asia, Australia and South America.

Ronnie King
Partner, Ashurst LLP
ronnie.king@ashurst.com

Ronnie King is a partner in Ashurst LLP's litigation department in London and heads the firm's international arbitration group. He has considerable experience of multi-jurisdictional litigation, including forum shopping and anti-suit injunctions. He works for clients in a variety of industries, especially in the power and energy, and insurance sectors.

Mr King regularly appears as an advocate in international arbitrations and also sits as an arbitrator.

Yashar Latifov
Partner, Infrastructure Development Partnership LLP
ylatifov@infradev.co.uk

Yashar Latifov is a partner in Infrastructure Development Parnership LLP, a consultancy working in the energy and mining sectors to help clients acquire and implement complex infrastructure and upstream projects. He holds degrees in petroleum production, reservoir engineering and project management, and has taken active part in major regional cross-border pipeline infrastructure projects for over 15 years while based in Azerbaijan, the United Kingdom, Turkey and Russia.

In particular, Mr Latifov has been a leading management member of the Baku-Tbilisi-Ceyhan crude oil pipeline project from its concept inception to construction completion, as well as of a number of leading hydrocarbon pipeline infrastructure transactions from the Caspian region and in Eastern Europe.

In addition, he has been heavily involved in a number of upstream and midstream development projects in the former Soviet Union, based on innovative technology application and advanced management practices.

Martin Lednor
Partner, Infrastructure Development Partnership LLP
mlednor@infradev.co.uk

Martin Lednor has over 27 years' experience in the oil, gas and mining business in both industry and consultancy positions.

From 1995 to 2000 Mr Lednor managed environmental and social issues on behalf of a multinational oil consortium associated with the development of pipeline export options in Azerbaijan and Georgia. From 2000 to 2008 he fulfilled various environmental and social management roles for the Baku-Tbilisi-Ceyhan project, initially with responsibility for the Turkish section of the project and latterly as the environmental and social lender interface coordinator. This was a position he then undertook for 18 months for the Papua New Guinea liquefied natural gas project.

About the authors

Charles Lindsay
Partner, Allen & Overy LLP
charles.lindsay@allenovery.com

Charles Lindsay is an English solicitor and a partner in Allen & Overy LLP's energy and infrastructure projects department. He has practised in project financing for over 15 years in London, New York and Frankfurt. He advises lenders, project sponsors, export credit agencies and governments in relation to a variety of energy and infrastructure projects. In particular, he has advised on numerous oil and gas pipeline projects, including the Azeri early oil export routes through Azerbaijan, Georgia and Russia, Sasol's Mozambique-South Africa gas pipeline project, the proposed Bosphorus bypass project in Greece and Bulgaria, the Bicentenario oil export pipeline project in Colombia and the proposed trans-Adriatic gas pipeline project in Greece, Albania and Italy.

Anthony FS Ling
Director, LPD Strategic Risk Ltd
tonyling@lpdrisk.com

Anthony FS Ling (CBE) is a director of LPD Strategic Risk Ltd. Previously, he worked for BP in high-risk countries worldwide. His recent work has included managing security for the Azerbaijan, Georgia and Turkey pipeline projects and leading the US Iraq Energy Infrastructure Security Review.

Mr Ling's security concept and that of LPD Strategic Risk is to provide assessment and mitigation services to help clients work in a peaceful environment by engaging all stakeholders – particularly local communities, responsible non-governmental organisations, host government security forces and interested foreign governments.

He was involved in the discussions that resulted in the Voluntary Principles on Security and Human Rights, which he uses as rules of engagement. He co-authored the oil industry guide to *Pipe Line Security for Oil and Gas Producers* and is the author of *Fire Arms and the Use of Force*.

Geoffrey Picton-Turbervill
Partner, Ashurst LLP
geoffrey.picton-turbervill@ashurst.com

Geoffrey Picton-Turbervill is a partner in Ashurst LLP's energy, transport and infrastructure department in London and heads the global energy team. He has over 20 years' experience of working in the international oil and gas industry, advising on mergers and acquisitions, greenfield projects and commercial agreements. Mr Picton-Tubervill regularly acts for both governments and international energy companies on cross-border energy projects and has extensive experience both in the upstream sector and of pipeline and refining/petrochemical projects. He established Ashurst's office in New Delhi, India, in 1994. In addition to the United Kingdom, he has advised clients on projects throughout the Middle East, Africa and Asia.

Mr Picton-Tubervill is recognised in independent guides as a leading international energy lawyer and was recently described in one as "at the top of his trade"; he "has vast experience and handles the most complex transactions".

Glen Plant
Legal consultant
Door tenant 20 Essex Street (at present non-practising)
Glenplant160@btinternet.com

Glen Plant advises foreign (mainly energy) companies, intergovernmental organisations and governments on public international and EU law and policy, with an emphasis on law of the sea, environmental protection and energy production and transport. Pertinent examples include the external mid-stream legal work (since 1996) for the producing Caspian Basin international petroleum consortia and others interested in the maritime segments of the Caspian fossil-fuel export infrastructure. Dr Plant's extensive publications include several pieces on the

Bosporus and Dardanelles, most recently in the *Encyclopaedia of International Law* (forthcoming). He draws on a broad range of legal and policy experience, including with the UK Foreign and Commonwealth Office and the United Nations.

Dr Plant has taught at leading UK and US universities (most recently the London School of Economics, where he was founding director of the Centre for Environmental Law and Policy from 1991 to 1996). He holds the Hague Academy diploma of international law.

Stuart F Schaffer
Partner, Baker Botts LLP
stuart.schaffer@bakerbotts.com

Stuart F Schaffer is a tax partner in the Houston office of Baker Botts LLP and is the chair of Baker Botts' global projects department.

Mr Schaffer specialises in international and corporate taxation, with an emphasis on planning and structuring projects, joint ventures, mergers and acquisitions, investment funds and financings in the energy industry. He has particular expertise in cross-border transactions, including both outbound investments by US companies and funds in foreign assets, projects and businesses, and inbound investments by foreign companies and funds in US assets, projects and businesses. Mr Schaffer has a great deal of experience in structuring international energy projects, such as oil and gas exploration and production, electric power, pipeline and liquefied natural gas projects.

Henry Thompson
Community engagement consultant, Oxania Ltd
oxania@gmail.com

Henry Thompson is a development consultant with over 25 years' experience. His core areas of expertise include rural livelihoods in arid and semi-arid lands, pastoral systems and large-scale water resource planning and management. For the past decade he has primarily worked in the extractive industries sector in Yemen, covering environmental and community issues, environmental legislation and oil revenue management. He is increasingly involved in governance issues and analysis geared towards foreign policy development.

Arve Thorvik
Managing partner, Thorvik International Consulting AS
arve@thorvik.eu

Arve Thorvik is the managing partner of Thorvik International Consulting AS. From its offices in Brussels and Oslo the company works with large international clients in the areas of policy advice and government relations in the fields of energy, environment and climate change. The company's largest clients are involved in the development of major energy infrastructure.

Before establishing his own consultancy in 2008, Mr Thorvik worked for 11 years for Statoil in Norway, Switzerland and Belgium, holding positions as, among other things, senior vice president for health, safety and environment, and vice president for EU affairs.

Prior to joining Statoil, Mr Thorvik had worked in senior management positions in Norwegian industry associations.

After studying political science, Mr Thorvik served 16 years in the Norwegian foreign service, with postings in Lagos, Geneva and Washington DC.